Technological Change and Network Effects in Growth Regimes

In this new volume it is argued that network effects are much more common than usually assumed, and that they have a profound impact on many aspects of economic systems, especially technological change and economic growth. The analysis and modelling of this interrelationship is the central focus of this book.

While there exists a vast body of literature on economic growth, the theories put forward so far have had limited success in explaining observed patterns of economic growth. 'Growth cycles' in particular continue to elude standard economic models, although evolutionary economics has made some progress. Seeking to fill the gap, Torsten Heinrich's innovative approach uses microeconomics to explain heterogeneous sectoral dynamics on the meso level, and then aggregating these to observed macroeconomic growth rates. In this way, it is shown that an evolutionary model of technological change with network effects can explain not only commonly observed asymmetric industry structures, monopolies and oligopolies, but also 'growth cycles'.

The book includes a comprehensive account of the most influential economic growth theories, a discussion of the research on network effects as well as an introduction to the methodology, the model, and a case study on the recent emergence of information and communication technology.

This important new volume will be relevant to all those interested in theoretical economics, growth theory, innovation economics, agent based modelling and industry dynamics.

Torsten Heinrich is Postdoctoral Lecturer and Research Assistant in the Faculty of Business Studies and Economics at the University of Bremen, Germany.

Routledge advances in heterodox economics

Edited by Wolfram Elsner *University of Bremen*
and Peter Kriesler *University of New South Wales*

Over the past two decades, the intellectual agendas of heterodox economists have taken a decidedly pluralist turn. Leading thinkers have begun to move beyond the established paradigms of Austrian, feminist, institutional–evolutionary, Marxian, post Keynesian, radical, social and Sraffian economics – opening up new lines of analysis, criticism, and dialogue among dissenting schools of thought. This cross-fertilization of ideas is creating a new generation of scholarship in which novel combinations of heterodox ideas are being brought to bear on important contemporary and historical problems.

Routledge advances in heterodox economics aims to promote this new scholarship by publishing innovative books in heterodox economic theory, policy, philosophy, intellectual history, institutional history and pedagogy. Syntheses or critical engagement of two or more heterodox traditions are especially encouraged.

1. **Ontology and Economics**
   Tony Lawson and his critics
   Edited by Edward Fullbrook

2. **Currencies, Capital Flows and Crises**
   A post Keynesian analysis of exchange rate determination
   John T. Harvey

3. **Radical Economics and Labor**
   Frederic Lee and Jon Bekken

4. **A History of Heterodox Economics**
   Challenging the mainstream in the twentieth century
   Frederic Lee

5. **Heterodox Macroeconomics**
   Edited by Jonathan P. Goldstein and Michael G. Hillard

6. **The Marginal Productivity Theory of Distribution**
   A critical history
   John Pullen

7. **Informal Work in Developed Nations**
   Edited by Enrico A. Marcelli, Colin C. Williams and Pascale Jossart

8. **The Foundations of Non-Equilibrium Economics**
   The principle of circular and cumulative causation
   Edited by Sebastian Berger

9. **The Handbook of Pluralist Economics Education**
   Edited by Jack Reardon

10. **The Coming of Age of Information Technologies and the Path of Transformational Growth**
    A long run perspective on the 2000s recession
    Davide Gualerzi

11. **Cultural Economics and Theory**
    The evolutionary economics of David Hamilton
    William M. Dugger, William Waller, David Hamilton and Glen Atkinson

12. **The Cultural and Political Economy of Recovery**
    Social learning in a post-disaster environment
    Emily Chamlee-Wright

13. **The Foundations of Institutional Economics**
    K. William Kapp
    Edited by Sebastian Berger and Rolf Steppacher

14. **Alternative Theories of Competition**
    Edited by Jamee K. Moudud, Cyrus Bina and Patrick L. Mason

15. **In Defense of Post-Keynesian and Heterodox Economics**
    Responses to their critics
    Edited by Frederic S. Lee and Marc Lavoie

16. **The US Economy and Neoliberalism**
    Alternative strategies and policies
    Edited by Nikolaos Karagiannis, Zagros Madjd-Sadjadi and Swapan Sen

17. **Technological Change and Network Effects in Growth Regimes**
    Exploring the microfoundations of economic growth
    Torsten Heinrich

This series was previously published by The University of Michigan Press and the following books are available (please contact UMP for more information):

Economics in Real Time
A theoretical reconstruction
John McDermott

Liberating Economics
Feminist perspectives on families, work, and globalization
Drucilla K. Barker and Susan F. Feiner

Socialism After Hayek
Theodore A. Burczak

Future Directions for Heterodox Economics
Edited by John T. Harvey and Robert F. Garnett, Jr.

Are Worker Rights Human Rights?
Richard P. McIntyre

Technological Change and Network Effects in Growth Regimes

Exploring the microfoundations of economic growth

Torsten Heinrich

Routledge
Taylor & Francis Group
LONDON AND NEW YORK

First published 2013
by Routledge
2 Park Square, Milton Park, Abingdon, Oxon OX14 4RN

Simultaneously published in the USA and Canada
by Routledge
711 Third Avenue, New York, NY 10017

Routledge is an imprint of the Taylor & Francis Group, an informa business

British Library Cataloguing in Publication Data
A catalogue record for this book is available from the British Library

Library of Congress Cataloging-in-Publication Data
Heinrich, Torsten.
Technological change and network effects in growth regimes : exploring the microfoundations of economic growth / Torsten Heinrich.
p. cm.
Includes bibliographical references and index.
1. Technological innovations–Economic aspects. 2. Diffusion of innovations–Economic aspects. 3. Economic development. I. Title.
HC79.T4H453 2013
338'.064–dc23 2012030957

ISBN: 978-0-415-63110-5 (hbk)
ISBN: 978-0-203-09712-0 (ebk)

Typeset in Times New Roman
by Cenveo Publisher Services

Printed and bound in the United States of America by Publishers Graphics, LLC on sustainably sourced paper.

Contents

Figures

Tables

Symbols

The following two lists of mathematical symbols and operators present a comprehensive listing and explanation of the symbols used in Chapters 2, 4, and 5. The variables used in these chapters are unique and do not change their meaning except for index variables (like i, j) and general symbols (like x for endogenous variables of systems of equations, n for finite numbers of entities and η for abstract parameters). In Chapter 3, however, which discusses a number of growth theories in the economic literature, the notation will mostly follow the original notation of the respective authors (except for cases in which that would be confusing). Therefore, the variables and their meanings will change from section to section in Chapter 3 and will be explained there in the context of the respective model. Nevertheless, some conventions are more or less universally applied, e.g. to denote output by Y, capital by C, investment by I, the technology level by A or a, labour by L, utility by U, and depreciation by δ. (Chapters 1, 6 and 7 do not contain mathematical considerations).

Operators	*Explanation*
∂	the symbol for partial derivatives (not to be confused with δ (delta))
∇	the vector differential operator (or gradient of the vector) $\nabla F(x_1, ..., x_n) = \left(\frac{\partial F(x_1,...,x_n)}{\partial x_1}, ..., \frac{\partial F(x_1,...,x_n)}{\partial x_n} \right)$
$\int$	the Riemann integral operator (for indefinite or definite integrals)
$\prec, \succ$	choice preference operators ('preferred to', etc.)
$\in, \notin, \setminus$	set operators ('element in set', 'not an element in set', 'set without ...')
$det(\mathcal{A})$, $tr(\mathcal{A})$	matrix operators (determinant of matrix $\mathcal{A}$, trace of matrix $\mathcal{A}$)
$Re(x)$	the real part of a complex number
$Im(x)$	the imaginary part of a complex number
$\sharp X$	number of elements of a set X (or absolute frequency of an empirical value X)

Operators	*Explanation*
$E(x)$, $Var(x)$, $Mean(x)$	stochastic operations: expected value, variance and mean of a stochastic variable x
$lim_{x \to y}$	the limit operator (the limit of a function of x with x approaching y)
$max\ x$	the maximization operator (maximize a function x)
$\sum, \prod$	the sumation and product operators
$\dot{x}$	growth rate of a time-dependent variable x_t, i.e. if discrete, $\dot{x} = \frac{x_{t+1} - x_t}{x_t}$, if continuous $\dot{x} = \frac{dx(t)}{dt}$
$\tilde{x}\, \bar{x}\, x'$	modified functions of variables of the same type as the unmodified variable x

Symbols for chapters 2, 4 and 5:

Symbols	*Explanation*
a	technology or technology level (in Chapter 3 also A)
A, B, B_1, etc.	distinct entities (bundles of goods, products, standards, technologies, industries)
$\mathcal{A}$	transformation matrix of a linear dynamic system (e.g. the game matrix of a game in replicator dynamic evolutionary game theory)
$AC(LAG)$	autocorrelation coefficient of an autocorrelation over LAG periods
b	size of the subtree of a technology tree in a model in Chapter 5
B	subspace of the phase space of a dynamic system (e.g. fixed of infinitesimal environment of a particular point in phase space)
c	costs
ch_t	number of users already committed at time t (i.e. adopted one standard or technology)
C	consumption
$\mathcal{C}_1, \ldots, \mathcal{C}_n$	constants
d	dimensionality of a system
$D(\Xi)$	distribution of adopters of a technology between two periods $D(\Xi) = \{\Xi, N - \Xi\}$
e	Euler's number
E_T	kinetic energy
E_V	potential energy
$f(i,t)$	evolutionary fitness of agent type i at time t
$F(x)$, $G(x)$	abstract functions of variables x

Symbols	*Explanation*
$g(\kappa)$	user κ's (expected, believed) bandwagon effect (how many users κ can convince by adopting a technology to decide for this technology as well)
h, H	parameters (used to discuss homogeneity of functions)
HI	normalized Herfindahl index, $HI = \frac{\sum_i s_i^2 - \frac{1}{n}}{1-\frac{1}{n}}$ (s_i denoting the shares of the n firms or technologies, etc.)
$\mathcal{H}$	Hamiltonian function
i	index variable
I	investment
$\mathcal{I}$	identity matrix
j	index variable
$\mathcal{J}$	Jacobian matrix
K, k	capital stock, upper case (K) denotes macro-level
L, l	labour force, upper case (L) denotes macro-level
$\mathcal{L}$	Lagrangian function
$m_{b,t}$	number of technologies with a subtree of size b at time t in a model in Chapter 5
$MRS(x_j, x_n)$	marginal rate of substitution between the jth and the nth good with quantities x_j and x_n
n, N	number of elements in finite sets (of, e.g. agents in a model, etc.)
$N_{ch(\kappa)}$), $N_{ch(\kappa),A}$)	number of users who have chosen a standard (or in case of $N_{ch(\kappa),A}$) specifically standard A) before user κ chooses
p	abstract probability or relative frequency (also a share of an agent type in a population in evolutionary game theory, i.e. the probability of meeting such an agent)
$p_{i,t}$	population shares of agent type i at time t (in evolutionary game theory)
p_{rep}	probability for an indefinitely repeated game to be repeated for at least one more turn
q_j	population shares of a sector in multi-sector replicator dynamics models (the shares in the first sector being denoted by p_i)
r	stand-alone utility of a network technology (or standard or product), i.e. the utility this technology yields with a network size 0
$\Re$	real numbers
S_{CN}	number of possible combinations of choices of a number of agents

Symbols	*Explanation*
s_i, s_{TEC}	shares of a type of agent in the population, usage shares of a technology or standard, industry shares of a firm, etc. (in Chapter 3, s denotes the saving parameter)
t	time parameter
T	time differences and time values
$\mathcal{T}$	infinitesimal transformation
TEC	index variable for technology types
U, $u(r, n)$	utility, utility function
$v(n)$	network utility of a network technology (or standard or product) as a function of the network size
$V(x_{t+T})$	Lyapunov function
x_1, x_2, X_t	abstract endogenous variables, e.g. independent variables of dynamic systems, also quantities of goods in standard optimization problems
Y, y	output/production/income level, upper case (Y) denotes macro-level
$z_{i,\kappa}$	price of a good with network effects (or a good granting access to a standard or technology with network effects), may be negative (sponsored)
α	power law exponent
β_t	random vector sequence (with elements $\beta_{i,t}$) in Eggenberger–Pólya urn schemes
γ	parameter in a path dependent growth model in Chapters 4 and 5
Γ	a game
δ	depreciation
Δ_κ, Δ_t	in Chapter 4 net intrinsic preference of a specific user κ or (with otherwise homogeneous users) a user at time t, e.g. $\Delta = r_A - r_B$. In Chapter 3 where the notation tries to stay close to the original growth models discussed there, Δ usually denotes differences.
δ_t	time preference, discounting of future consumption
ε	arbitrarily small parameter
ζ_t	random vector sequence (with elements $\zeta_{i,t}$) in Eggenberger–Pólya urn schemes, composed of the purely random vectors β and a probability term
η, η_1, η_2	parameters
θ	specific technology
$\Theta_{i,t}$	set of technologies available to firm i at time t
ι	parameter for Eggenberger–Pólya urn schemes with discrete rules, number of balls to be drawn at a specific time

Symbols	*Explanation*
κ	specific user
λ	eigenvalue
Λ	Lyapunov exponent
μ_h	Hamiltonian costate variable (multiplier)
μ	Lagrangian multiplier (shadow price)
ν	correlation coefficient
ξ	empirical frequency of lock-in/break-out events
Ξ	number of first period adopters as a parameter in the distribution function of adopters of a technology in time, $D(\Xi, N - \Xi)$
$\Pi_{i,t}$	payoff of player i at time t
ρ_{inno}	probability of a successful innovation in the growth model in Chapter 5
ϱ	inverse demand function of the jth good (with quantities x_i for any good i, hence x_j for good j), thus $\frac{\frac{\partial F(x)}{\partial x_j}}{\frac{\partial F(x)}{\partial x_n}} = \varrho_j(x)$
σ_i	particular strategy of player i
σ_i^*, σ_i^P	particular strategies with equilibrium properties, e.g. leading to a Nash equilibrium (σ_i^*) or a Pareto optimum (σ_i^P)
σ_{-i}	particular combination of strategies of all players except i
Σ_i	set of strategies of player i in a game
τ	time differences and time values
Υ	normalization coefficient for power laws
ϕ_t	average fitness in a population at time t (in evolutionary game theory)
φ	risk parameter
χ	compatibility parameter in multi-sector models
ω	parameter for Watts–Strogatz small world networks (probability of a regular edge being rewired in the construction of a small world network from a regular grid or line network)
Ω	number of states a system may take

Series editor's foreword

This monograph has an extremely ambitious objective. It attempts to explain, and eventually synthetically reproduce, empirical GDP time series, which apparently are most irregular, non-stationary and result from complex structures, and have tended to become even more turbulent in recent times. Although modern post-World-War-II economics has aimed, first and foremost, to explain and support growth and development, mainstream growth theories have not succeeded in even coming close to reproducing the growth behaviour of real-world economies. And although some economists and models have come close to developing important ideas and formal model elements that are useful today in cutting-edge complexity models, growth models have basically been too simple over the decades. Only what nowadays is considered 'complexity economics' appears to be capable of generating the theoretical and formal tools of proper modelling, computer calculation and computer-based reproduction of ever more turbulent empirical aggregate behaviour. This has required new approaches to disaggregation and microfoundation that could not be provided in the world of simplistic 'mainstream' economics with its perspective on unique and stable equilibria of more or less perfect 'markets'.

The changes required took place stepwise in diverse areas such as growth theory itself, e.g. through evolutionary growth models, through disaggregation of the economy into inter-industry interactions or through complex dynamic modelling. Furthermore, they took place through directly interactive modelling approaches, such as game theory, with an evolutionary and institutional perspective, through agent-based modelling and computer simulation of such models, which no longer are calculable in deterministic ways. Finally, they also took place through the ideas of emergent structure or institutionalization and the recognition of new complexity structures in modern economies, such as functional and spatial fragmentation of value-added chains, increasing speed and turbulence of innovation processes, and the new quality of the collective character of the productive forces, as mirrored by the net-character of the new digital and microelectronic technologies.

Taken together, all these changes have made up the strands of modern complexity economics, and they have triggered an understanding of the decision problems, processes and outcomes of modern decentralized spontaneous economies (i.e. capitalist 'market' economies), different from the received simplistic 'harmony of the perfect market'.

In particular, individual decision makers are far less than perfectly informed, and are in fact strongly uncertain, searching in local areas, directly interdependent in 'wicked' (anti-)coordination and social dilemma problem structures, forced to coordinate and cooperate, while individualistic incentives often prevent them from doing exactly that. Moreover, successful coordination of technological and behavioural standards often turns out to be overtly sub-optimal, while better coordination may not be feasible because of lock-ins. This goes along with monopolistic and oligopolistic industry structures, irregular and complicated sequences of first-mover monopolizations and sometimes lock-outs by oligopolistic rivalry, in sum leading to complex patterns of aggregated behaviour with irregular sequences of innovation- and coordination-led growth, lock-in or turbulence-induced stagnation.

In addition, the degree of openness of knowledge (information) (including open-source phenomena in information and communications technology (ICT) and, particularly, software technologies) turns out to be critical here, as it influences the speed of diffusion and the ease of potential break-out.

While all these diverse aspects have been developed in distinct but overlapping fields of modern complexity-economic research – growth theory, dynamic systems research, industry population research, innovation economics, information economics, network-externalities research, models of technological and/or institutional 'lock-ins', evolutionary and institutional economics, evolutionary macro modelling, post-Keynesian microfoundations, game theory, simulation studies, and others – there have been only a few attempts at integrating the most important issues here: growth modelling together with complex dynamic and agent-based modelling, technological change together with net-technologies, and emergent structure, reflecting coordination/standardization success together with lock-in and break-out – and in this way reproducing most-complex aggregate phenomena.

This is the very approach of this monograph. It is obvious that an attempt at merging these diverse and distinct branches of modern complexity economics requires a most intimate and settled understanding of cutting-edge theories, models and methods. The present study navigates most fruitfully and enlighteningly among these fields, and most importantly distils the important net-value knowledge from these strands and combines them into its own cutting-edge theoretical approach, model and simulation, with fundamental, and at the same time practical, implications. The approach requires dealing with complexity at a highly advanced level. There is no doubt that such modelling has to deal with complexity issues such as path dependence, phase transitions, bifurcations and (deterministic) 'chaos' in general, and thus also with idiosyncrasies, multiple equilibria, complex system orbits and emergence/autopoiesis.

The basic concern of this study is thus, in short, the aggregate effect of technological change under network-externalities and multi-agent interactions on the growth path of the proper (emergent) economic aggregates.

The study starts off from the empirical phenomena of aggregate behaviour to be explained and synthetically reproduced, but so far not analysed at a proper level of theoretical integration of growth, micro-interaction, net-technologies, irregular sequences of lock-in and break-out, and emergent structure: and furthermore, super-linear growth on average, occasional and sudden decline, and persistent and possibly irregular fluctuations. The literature that has comprehensively addressed these issues so far is small and quite recent, but can be counted on to define the current boundaries of economic complexity research.

The chapters of this book demonstrate deep reflections on and settled dissections of the foundations and history, aberrations and also viable elements of those fields to be looked at later on. These sections not only could easily be restructured as advanced textbooks about important areas of complexity economics, they might also serve as potential standalone research monographs on their own.

The book deals with the (mistaken) physical analogies of nineteenth century neoclassical maximization and unique-equilibrium economics, discussing field theory and the Lagrange algorithm of classical mechanics, energy physics and thermodynamics – its restrictive assumptions rooted in the energy conservation assumption, convexity requirements, the auctioneer implication, etc. This 'benchmark' study leaves virtually nothing to be retained in complexity economics. A sharp paradigm change is required. Therefore, game theory and evolutionary modelling are introduced as reflections of the real-worlds of direct interdependence and cumulative systems. Usual evolutionary game theory (EGT), replicator mechanisms and the simulation approach are introduced as well. The dynamics of different decision structures, analysed in a population framework, are explained with the eigenvalues of their dynamic equation systems; and their potential properties, like (multiple) equilibria, fixed points/attractors, stability, and bifurcations/phase transitions, are introduced; phase portraits in simplexes are illustrated and the simulation approach presented. Finally, emergence at some aggregate ('meso') level is discussed.

There is a comprehensive, thorough, and enlightening representation of the development of economic growth theories, from conventional Harrod–Domar and AK models to endogenous growth models, and of their insufficient treatment of technological change. The critique of the 'neoclassical mathematics' involved is cutting-edge, following Mirowski, Hands, Chen, Keen and Albin/Foley. 'Final frontiers' are provided, in contrast, drawing on forerunners like Verdoorn and Thirlwall, but also Marx, Georgescu-Roegen, Arrow's learning-by-doing and Schmookler, by evolutionary models in the succession of Nelson and Winter's approach to technology evolution and diffusion, by the coevolution of the industry structure (Conlisk, Silverberg), and, finally, by Kauffman's biological NK/technology-landscapes approach. While Nelson–Winterian neo-Schumpeterian approaches have the disadvantage of tending towards strict (though perhaps terminated) monopolizations, Kauffman's approach may explain fast occurrences of innovations that may quickly lock-in at local optima – and may also be clustered in time.

The theory of growth cycles refers back to Slutsky, Samuelson, Goodwin, Kaldor, Lorenz, Minsky, Keen, Freeman and Lucas. In the end, growth cycles can be modelled to be irregular, recurring perturbations, with different wavelengths and amplitudes, and clustered in time.

The state of the art of complex evolutionary growth modelling still appears unconvincing to the author of this study in the light of what may be seen in the macroeconomic data series and what is known about the meso- and micro-levels of the economy. Specifically, the industry structure, the network effects, and the economic-organizational bounds to growth and technological and structural change – most importantly lock-ins – are still mostly not accounted for. In this sense, this study strives to integrate network externalities in a complex model of system dynamics.

Another cutting-edge, complex field of modern economics, i.e. net-technologies and net-externalities, is reviewed, systematized, reconstructed and its elements distilled for an integrated model to be developed. As in W. B. Arthur's famous technology choice model, what can be conceptualized as net-externalities, is termed and formalized as more conventional 'economies of scale' – effective in this case, though, in a cross-firm, industry- or economy-wide setting.

The shortcomings of conventional treatments of net-externalities – both in neoclassical and neo-Schumpeterian frameworks – are discussed. They are seen in the 'pure' monopolistic positions that too-simple modelling of cumulative processes resulting from net-technologies typically entails. The Katz–Shapiro, Farrell–Saloner and Arthur approaches to net-externalities are reconstructed, a comparison with the older issue of 'bandwagon' effects is investigated, and the collective-good dimension (in a prisoners' dilemma frame) is analysed. Furthermore, individual firm strategies (coming up with net-technologies and implied cumulative standardization and lock-in/lock-out issues, such as first-mover strategies, firm-based open-source strategies and related private sponsoring of technologies, planned obsolescence, standard tying, piracy, deliberate 'gate-keeping' of technology clusters, and others) are discussed and integrated into a more comprehensive picture of the economy shaped by net-technologies. The discussion proceeds here to the core issues of economic development under technological change based on net-technologies. These are the opportunities of lock-out, or, more exactly, the issue of optimal speed of technological change between too much turbulence (too early lock-out that prevents reaping the efficiency effects of coordination) versus sticking too long to a standard that might already have become outmoded (locked-in); and furthermore, the issue of concurrent technologies and segmented populations of firms/adopters of those different technologies, which in turn relates to segmented populations, or, in a game-theoretic perspective, basically is equivalent to persistent mixed strategies (mixedstrategy equilibria). Lock-in can also assume the form of 'adverse industry selection', with a typical asymmetric industry structure with one major monopolistic firm surrounded by narrow oligopolistic structures.

In addition, there is a focus is on issues of managing those complex processes, either by 'enlightened' and coordinated action of the private agents involved, or by 'enlightened' policy action. However, this is neither a management nor a policy analysis.

The study contains a formal analysis of system properties such as non-ergodicity and path dependence, underpinning those properties with Pólya-urn processes and generalized urn schemes, in this way modelling net-technology diffusion with urn schemes. And again, integrating different sources of knowledge and models from the literature leads to a complex system with 'chaotic' (i.e. critically non-predictable) properties.

Similar to the outstanding position of Kauffman landscapes within complexity and evolutionary theory, here, finally, Duncan Watt's 'small worlds' modelling of 'optimal'-diffusion structures contributes an essential foundation for explaining non-global lock-ins, local (technology) clustering, and resulting asymmetric industry structuring. Case study material illustrates the complex and irregular process of technological change, lock-in and break-out.

Standardization and lock-in are also connected to the conception of institutionalization, referring to the Original Institutional Economics tradition, which has the advantage of elaborating an interactive dichotomy between 'instrumental solution' and 'ceremonial encapsulation' – which in turn appears to be a fruitful conceptualization of the material developed so far.

A resulting process indeed displays the critical properties of path dependence and complex patterns in the sense of 'deterministic chaos', generating stable but irregular aggregate behavioural patterns, while supporting some degree of persistent diversity, i.e. different paths for different industries and spatial areas (i.e. clusters in topologies). Technological paradigm change may then occur on a regular (or irregular but recurring) basis. Altogether, this then generates differentiated growth paths that sufficiently approximate reality.

The particular contribution of the present study is a comprehensive, integrated model of technological change under net-technologies and resulting growth paths. It analyses network efficiency, network effects on the consumption side, and in research and development (R&D).

The microfoundation/macro-behaviour discussion also focuses on the possibilities of stable aggregate patterns resulting from net-externalities. A most interesting theoretical issue involved here is the knowledge provided by biochemistry (Prigogine) on properties of dissipative structures, which emerge from the complex micro-level and make emergent structure and persistent patterns at the system level feasible. Implied here is the issue of openness of the economy as a subsystem of a major system (which also includes the social and natural subsystems). Again, a major field of theoretical discussion (which in fact has taken place in the history of economic thought – see, e.g. Georgescu-Roegen and Kapp, and, more recently, Chen) opens up here.

Simulations that are run clarify phenomena like patterns of alternating slow and rapid growth, related wavelengths, and 'asymmetric' impacts of technologies in terms of firm size, agglomeration size and cluster/network size. A full developed model is simulated. The discussion of the simulation results recapitulates core theoretical issues elaborated in the thesis so far, such as Nelson–Winter-type informational openness and technology diffusion, and addresses firm structure, technologies and technology-applying plants, labour efficiency, research impact, production, output, capital stock, resulting technological change and industry structure.

Network externalities and lock-in do occur in the simulation results, and so do break-up and technology switch, but also significantly long periods of monopolization and asymmetric industry structure (oligopoly). Path dependence is stronger under net-effects compared to an economy without net-effects; and open-source technologies – as predicted – tend to prevent monopolization and to support more frequent break-out. Obviously, there is no finite equilibrium but continuing development far from equilibrium states. There are recognizable aggregate development patterns, and periods of slow versus fast growth can be traced back to the process of implementation of net-technologies with phases of innovation, standardization, exhaustion and break-out.

Net-externalities, rather than being disadvantageous to growth, are, in fact, advantageous – they are the very process of technological change. Standardization and even lock-in appear not only to be unavoidable but to be the very reason and basis for generating and exploiting efficiency and growth.

Finally, it should be noted that the aggregate level emerging is coordination below the level of the whole population or the 'whole nation' – it is 'meso'-sized.

A case study on the ICT sector and its impacts on growth provides an enlightening history and reconstruction of the ICT and internet economy, standard wars, lock-ins and oligopolistic structures, degrees of openness, standard tying, private first-mover strategies, private sponsoring, opensource firm strategies, and piracy. The areas of application cover operating systems, application software, the browser and server industries – and the open source movement in particular. In all, a high degree of socialization of productive forces becomes obvious, and, therefore, individualistic firm strategies often appear to be hampering progress and growth. In particular, many instances of individualistic private oligopolistic regulation of the internet illustrate how global societal progress for people is in fact highly vulnerable and in danger. Arbitrary interventions by providers against critical sites, infringed net neutrality, or deliberate artificial non-interoperability (to push their own technology to become a standard) appear to be ubiquitous features of private regulation of ICT hard- and software and the internet in the interest of individualistic mega-powers. Existing governments do not appear to be inclined to protect internet freedom and democracy against these oligopolistic private powers. The study particularly discusses strategies of open source communities – while public intervention in favour of broader societal goals, such as net neutrality, guaranteed service, etc., cannot be expected from governments that strive to raise IPR protection for established vested ICT interests and which have their own secret services and surveillance interests.

Thus, in the end, broader perspectives of policy intervention and regulation, even broader societal problems of (tacit) collective decisions on the speed of technological change, and the degree of turbulence wanted (and bearable by society) appear on the horizon.

In all, this study is an excellent cutting-edge reconstruction of core parts of modern complexity economics.

Wolfram Elsner
Bremen, August 2012

Acknowledgements

The current work has been accepted as a PhD thesis by the department of business and economics of the University of Bremen.

I am especially grateful to my advisor Prof. Dr Wolfram Elsner for his guidance and support over years and to the other members of my thesis committee, Prof. Dr Christian Cordes, Prof. Dr Dr Ulrich Krause, Prof. Dr Martin Missong, Dr Henning Schwardt and Yasar Damar. I would also like to express my gratitude to the participants at the Economics Brownbag Seminar at the University of Bremen, to my colleagues at the Institute for Institutional and Innovation Economics (University of Bremen) as well as to the discussants and participants at various conferences for many inspiring discussions.

I am indebted to Dr Henning Schwardt for valuable suggestions, to Dr Henning Schwardt and Jenny Augustin for proofreading the manuscript. I am grateful to Prof. Dr Andreas Pyka, Prof. Dr Ping Chen, Dr Matthias Greiff, Matthes Hackbusch as well as Miriam de Blasi for many helpful comments.

Further, I would like to thank Natalie Tomlinson, Emily Kindleysides and other staff from Routledge for guiding the publication of this work.

I wish to extend special thanks to my friends, my parents and my sister for their encouragement, patience and support.

Last, but not least, I would like to acknowledge the financial support I received from the University of Bremen.

Acknowledgements

1 Introduction

The poor craftsman and the wealthy inventor

1.1 Opening story

Our history books are filled with success stories about great innovations, such as agriculture, writing, seafaring, metal working, the printing press, the steam engine, electricity, the combustion engine, computers and genetic engineering, to name a few. All of these, and many other less- well-known or forgotten ones, may have shaped the history of humanity in ways that go well beyond even what the history book will admit. I would like to recount one of the fabulous tales (largely following the account in He, 1994), the tale of a financially troubled craftsman in the late European middle ages. His name was Johann Gutenberg, the story of his early life remains largely unknown. Historical records of lawsuits, however, establish that he had been working on the printing press with movable letters since at least 1439. It took him another 15 years – always on the edge of bankruptcy and often sued by various people for debts he was unable to repay – to make his invention workable and economically efficient. In the subsequent two decades, the use of this technology spread throughout the continent, the numbers of volumes printed in European countries have risen exponentially ever since (Baten and van Zanden, 2008). It has been suggested that this technology was the true foundation of modern Europe, that it facilitated both the consolidation of European nation states and the modernization of the economic organization during the early modern era (Anderson, 1991 [1983]). Concurring popular theories hold that modern capitalism was driven by either the protestant ethic that came into being because of the religious reformation in Europe, or by the vast amounts of gold and silver acquired by Europeans in America after Columbus' (re-)discovery of that continent in 1492. The religious reformation relied crucially on the works of reformation leaders being printed and distributed among the literate upper class of the time. Furthermore, Columbus was reportedly inspired to undertake his voyage by reading a printed copy of Marco Polo's travel report. It is beyond doubt that the invention and subsequent widespread use of printing with movable letters in Europe sparked profound social, economic and technological changes, and a variety of further innovations.

Roughly 400 years earlier and 5000 miles away, another commoner (though probably a wealthy man), Bi Shen in Sung dynasty China, had the very same idea. Technical details of his set of movable letters for printing are well-documented and were well enough known in China to inspire other inventors to work on improving that technology during the following centuries. The technology, however, failed to be adopted more widely for economic purposes in spite of a reportedly flourishing printing business in China at the time (He, 1994). During the life of Bi Shen, China was a well-organized empire with a complex economy, standardized system of administration and academics, and advanced technology, while Europe was little more than a collection of what today would be called failed states, arranged around the persisting religious authority (the papacy) of a fallen empire (the Western Roman Empire). Yet following the invention of the printing press with movable letters – and at least partly as a consequence of it – Europe proved to be much more successful politically, economically and technologically in almost every respect.

What is usually given as an explanation for this phenomenon – the utter failure of a promising innovation in an economically more developed region and its great success in a less developed one – is of course nothing less than the essence of these regions' cultures, religious beliefs and economic organization (in neo-Schumpeterian terms, the innovation system) (He, 1994). These all-encompassing, deterministic explanations may, however, not exactly match up to the historic reality. As He (1994) details, the printing business without movable letters (i.e. block printing) was quite successful in China at the time, much more successful than in Europe at the time of Johann Gutenberg's innovation. It is also conceivable that printing with movable letters would have led to social and economic processes not unlike those sparked by the advent of the printing press in Europe. He offers yet another explanation: Bi Shen's invention may have faced fierce competition from a developed industry with a well-established standard (block printing) such that no one would have wanted to invest in movable letter printing due to uncertain economic prospects for this technology, while block printing was not widespread in Europe in Johann Gutenberg's times. (Furthermore, the very small number of different characters required for the Latin alphabet compared to Chinese script may have played a part.) In this way, it might well have been the mere fortunes of history that shaped the future of Europe, China and the world rather than religion, institutional setting or the cultural attitudes of these societies. David and other scholars (David, 1985, 1992; David and Bunn, 1988; David and Steinmueller, 1994) discuss a number of similar cases including the QWERTY keyboard, direct current and alternating current electricity, and hardware standards and operating systems in modern computer technology. There are many more examples, from the technologies often reviewed in history books, to the many standards that have evolved in a very short time during the rise of information and communication technology that will more closely be examined in Chapter 6.

1.2 Shedding light on the trajectories of technological change and economic growth

1439 is the year in which historical records confirm that Johann Gutenberg had started to work on his innovation – the printing press with movable letters – that would shape the future of the world and put the future of Europe on a trajectory of development very different from that of China and every other region of the world. Exactly 500 years later – in 1939 – the first influential theories of economic growth were published (Harrod, 1939; Samuelson, 1939a,b), theories that would eventually help to put economic growth theory on a trajectory of development ignorant of the effects of path dependence – even though the models of 1939 themselves were in no way equilibrium models like the successor theories developed by Solow (1956, 1957) or the proponents of the endogenous growth theory (Uzawa, 1965; Cass, 1966; Lucas, 1988), and even though the models of 1939 would likewise nourish the tradition of Kaldorian growth models that explicitly embraced increasing returns and thus different possible paths of development (Kaldor, 1940, 1972; León-Ledesma, 2000).

The data patterns to be explained in economic growth are basically the following three:

- *On average* superlinear growth (see the development of the GDP of Germany and the USA over the past decades in Figure 1.1 as an example).
- Occasional and sudden decline as a result of crises (as in 2009 in Figure 1.1). That may, however, be explained by financial sector instabilities and occasionally wars and political decisions, although some theories of economic cycles try to internalize this, e.g. Minsky (1980) and Keen (1995). This phenomenon will not be a central focus of the current study.
- Persisting and possibly regular fluctuations in the growth rate (as seen in Figure 1.2), though the regularity is doubtful. These fluctuations lead to a pattern of alternating slow and rapid growth (which is not obvious from Figure 1.1 due to the scale of that figure).

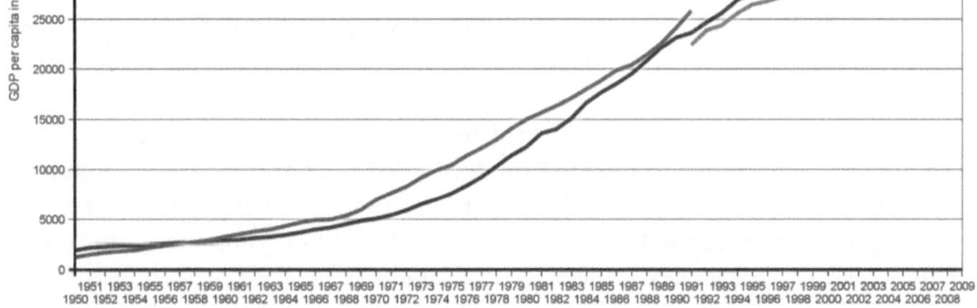

Figure 1.1 GDP per capita in Germany and the USA. (Data from the Bureau of Economic Analysis (2010) and the German Federal Statistical Office (2010))

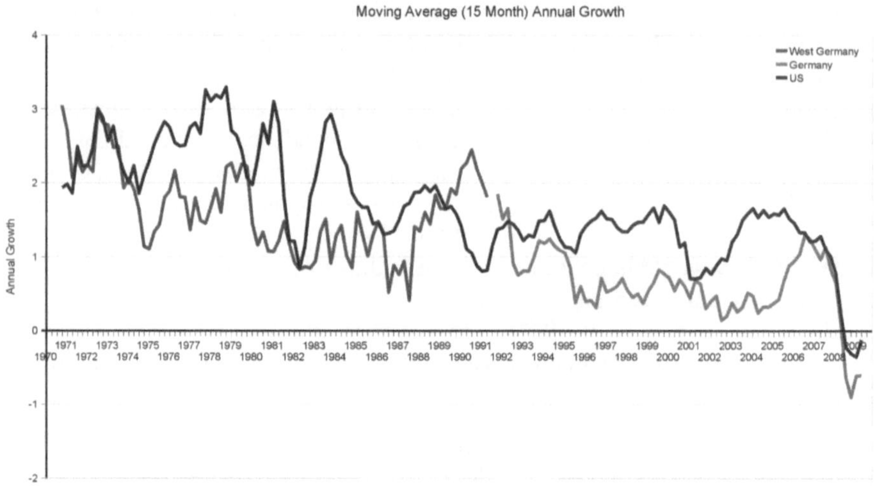

Figure 1.2 Annual GDP per capita growth rates of Germany and the USA (three-month data), with a 15-month moving average filter. (Data from the Bureau of Economic Analysis (2010) and the German Federal Statistical Office (2010))

Many efforts have been made to approach both economic growth and so-called 'growth cycles',[1] proposing a number of different mathematical mechanisms for both and offering different economic interpretations of their respective mechanism. The models currently favoured by the economic mainstream are mechanisms of endogenous growth; exponential growth with an optimal growth rate that is optimal in the sense of (approximately) maximizing utility in an infinite horizon optimization problem (Uzawa, 1965; Cass, 1966; Lucas, 1988). Random shocks are believed to cause a coordinated response by rational agents, thus forming constant cycle patterns (Lucas, 1972, 1975, 1981). This theory, the neoclassical approach, has however been heavily criticized (Albin, 1987; Mirowski, 1989; Albin and Foley, 1992; Foley, 1998, 2010; Keen, 2001; Chen, 2002; Wheat, 2009) and a number of other research traditions have offered alternative models, including evolutionary economists (mostly Schumpeterians, for instance Nelson and Winter (1974, 1982a) or Silverberg and Lehnert (1993, 1994)), Marxian radicals (e.g. Keen (1995)), and Kaldorians (Lorenz, 1987; León-Ledesma, 2000).

Among the few things that are more or less uncontested is the assessment that economic growth results mostly from technological improvements (including knowledge about technologies and thus human capital) rather than strict capital growth as has been modelled in early growth models. For evolutionary (and other Schumpeterian) approaches (Nelson and Winter, 1974, 1982a; Conlisk, 1989; Silverberg and Lehnert, 1993, 1994), technological innovation and diffusion[2] plays the crucial part in economic growth, Solow – whose model treated the technology level as an exogenous factor – admits in his empirical application of his model (Solow, 1957) that a substantial part of actual growth must be due to technological improvement, and the models of endogenous growth (Uzawa, 1965; Cass, 1966; Romer, 1986; Lucas, 1988) usually optimize over investments in 'human capital' that might just as easily be interpreted as an influence that works primarily through either innovation or the efficient application of technology. Economic cycles on the other hand remain a contested subject. Evolutionary theories generate 'circular' patterns (repeated alterations in the growth rate) through the pattern of innovation and the subsequent diffusion of innovative technology, while other explanations mostly seem to be dynamic systems that lead to clockwork-like precision in the model's cycles which may seem odd compared to the reality of 'cycles'.

None of these models, however, has tried to examine the patterns of technological change that obviously play such a crucial part in economic growth in more detail. As the above story of Johann Gutenberg and Bi Shen shows, the success of a technology may depend not only on the technology itself but rather also on specific – seemingly random – details of the environmental conditions. In particular, the presence or absence of similar established technologies has an impact on the chances of an innovative technology eventually establishing itself as an industry standard which in turn governs the adoption decisions of potential users. Mathematically, this relation, known as a *network effect* – and, as the network of a technology does also influence the chances of other competing technologies, a *network externality* (Katz and Shapiro, 1985) – is rather simple. It merely states that the probability of a technology being adopted by an agent increases with the number of previous adopters and the relative number of adopters compared to alternative technologies. This process is obviously path dependent: for several available technologies a process of (sequential) adoption decisions converges to the universal adoption of one of those technologies (David, 1985; Arthur, 1989). A two-dimensional variant of that system (with shares x and probabilities $p(x)$) is shown in Figure 1.3. Such a technology will, even with constant progress in technological research, be able to defend its position as the industry standard for a long time, a phenomenon known as *technological lock-in* (Arthur *et al.*, 1987; Arthur, 1989). Given the probably widespread occurrence of such lock-ins, an industry will only in rare historical events break-out from such a lock-in, which in turn leads to more efficient production (or consumption) and in turn an increased amount of available resources. If this amount is large enough, the effects of their – incidental – re-allocation to other industries may spark a similar break-out event there and thus trigger a chain reaction that may be adequately described as a 'technological paradigm change'. As a consequence, there may in addition to 'normal' slow processes of economic growth be a regime of rapid growth and technological change (see Figure 1.4), basically what may have happened in Europe as a result of Johann Gutenberg's innovation, and possibly again with the invention of the steam engine and more recently on a global scale (with the benefits not being limited to European economies) with electricity, information and communication technology and a number of other great innovations. This may of course also be the case at a lower level of

to models with multiple equilibria. Suitable tools to study both the micro-level of network externalities and direct interdependent situations in general are provided by game theory. To proceed to macro-level effects, it will also be necessary to include models from evolutionary growth theory, the theory of dynamic systems, and, in particular, complex systems.

While this is a powerful set of analytical methods, the inevitably resulting multi-agent-systems are to be expected to exhibit a level of complexity that is usually not conveniently solvable with analytical tools. Computer simulation of agent-based models provides a – relatively – quick and reliable solution for this case. Carefully executed simulations are able to give a comprehensive view of the entire phase space of the model and provide a good impression of the model's dynamics – though they are not an analytically exact method.

Simulations will be used extensively to study more complex models under consideration at various points work including, at some points, the reconstruction of the models and findings of previous authors. The source code is available from the author.

After a brief mathematical introduction in Chapter 2 to the methods that will be employed over the course of the present volume, Chapter 3 will proceed to discuss the different major traditions of economic growth theories. This account includes the Harrod–Domar tradition, the Solow–Swan model, models of endogenous growth, a discussion of empirical 'growth laws' in the tradition of Kaldor, Verdoorn and Thirlwall, similar considerations in the traditions of Marx, Georgescu-Roegen, Arrow and Schmookler, as well as a number of models of evolutionary growth and growth cycle models. Chapter 4 will discuss network effects, network externalities and lock-ins in the context of their respective literature traditions. Chapter 5 will develop an integrated model of network externalities, technological change and economic growth. This chapter will also discuss possible ways of integrating micro- and macro-levels. It will further construct several simulation models, evaluate the results and compare them to growth models from the literature tradition discussed earlier, especially evolutionary growth models which also work with simulation (in particular the models by Nelson and Winter). Chapter 6 will include a case study and proceed to review the findings of the models in Chapter 5 in the light of qualitative and (to a lesser extent) quantitative evidence from the recent development of the information and communication technology sector.

2 Methods for an analysis of economic growth and network effects

The current chapter establishes the analytic methods that will be used in subsequent considerations. This is not intended to be a comprehensive account of algorithms used in economics, but rather a brief introduction to the mathematical methods that will repeatedly be employed in both the discussion of growth theory in Chapter 3 and the analysis of network effects in Chapter 4, as well as in the resulting models in Chapter 5.

The following sections will proceed by giving an overview of the central method of neoclassical economics, i.e. the optimization in a conservative vector field in section 2.1, a short summary of the basic concepts of game theory in section 2.2, and evolutionary game theory in section 2.3. Subsequently, dynamic theory will be discussed in a little more detail in section 2.4 as it will be used extensively in the later chapters, and as it is also best suited to illustrate the method of simulation (section 2.4.2) that will also be essential for later considerations, especially the model in Chapter 5. The final section of (section 2.5) gives an introduction to the concept of structural emergence that follows from more complex dynamic systems (complex systems).

2.1 Field theory: a convenient and unique equilibrium

Writing from a heterodox perspective the current author can still not resist admiring the coherence of the neoclassical theory and the beauty of neoclassical mathematics. It truly is a pity that most students of neoclassical economics can never come to recognize this beauty – the usual neoclassic textbooks do not discuss it (Varian, 1992; Mas-Colell *et al.*, 1995). Those textbooks teach methods – first and foremost Lagrange's method of nonlinear optimization. The question why these methods are suitable to solve a particular problem (and which alternative methods might be used) is not addressed.[1] Nevertheless, it must be considered a valid and sometimes a good approach to certain kinds of economic questions (naturally, those where optimization with perfect information is indeed part of the phenomenon that is to be investigated) and it is able to offer explanations for a broad variety of economic issues, though its explanatory power is most likely overestimated by contemporary mainstream economics. Taking into account when economists started to develop this approach at the end of the nineteenth century with Marshall (1990 [1890]) and others,[2] and what was known and understood about the economic systems at the time, it must be recognized that the 'marginalist revolution', the development of the theory of a unique equilibrium in a conservative vector field (more or less adapted from mechanical physics (Mirowski, 1989)), was a major step forword.

As this approach and its methods are almost universally used in mainstream economics today, it is necessary and appropriate to discuss them in here. Other chapters will refer to this approach when discussing specific mainstream models of economic growth and network externalities as well as, occasionally, as a benchmark for models with very different approaches.

Neoclassical theory relies strongly on the assumption of universal rationality. The marginalist equilibrium analysis furthermore involves the theorem that everything is quantifiable in monetary terms – i.e. one-dimensional; everything has a price that can also be measured by assessing the respective agent's willingness to pay. The identity of that money is arbitrary, it is just one out of a number of goods (indexed i with quantities denoted x_i). While the *market* is in equilibrium, all goods are convertible into any other good and back without gains or losses. Given this, one of the goods may be set as *numéraire* to express all prices in quantities of that good. The market that establishes the equilibrium does, of course, in that form only work without indivisibilities, without imperfect information, without any other kinds of value losses in transactions,[3] with well-behaved preference and utility functions, and, most importantly, without direct interdependence. That latter requirement means there is a (hypothetical) central auctioneer who collects all bids and then determines the market price before the actual transactions begin.[4] In effect, the auctioneer provides the agents with the correct market prices for free. As a result, there is a decreasing aggregated demand function (aggregated over the agents' individual supplies) and an increasing aggregated supply function with a unique equilibrium as a result.

A well-behaved utility function in turn especially requires[5] concavity, that is, for any two bundles of goods A and B, the utility of an arbitrary mixture $pA+(1-p)B$ (with $0<p<1$) must never be exceeded by the utilities of *both* bundles,[6] which provides that the isoclines of the utility functions are concave while the budget function (i.e. the endowments of the agents with money or simply goods) is linear. As a result, there is a well-behaved nonlinear optimization problem of the form as depicted (for the simple two-commodity case) in Figures 2.1 and 2.2, with utility as the objective function to be maximized and the budget as a restriction. Of course this problem may be of higher dimensionality, and such problems may be defined for all kinds of settings[7] as long as the objective and restriction functions are of the required form. The problems are conveniently solved with Lagrange's method.

Lagrange's method of nonlinear optimization, as frequently used in economics, including in models of neoclassical endogenous growth, was initially developed and used by physicists of the nineteenth century, at the time of the development of thermodynamics. The state of a mechanic system is described using kinetic and potential energy, the sum of which is conserved in closed systems (Hamilton's principle). The system is in an equilibrium at space coordinates where effort and

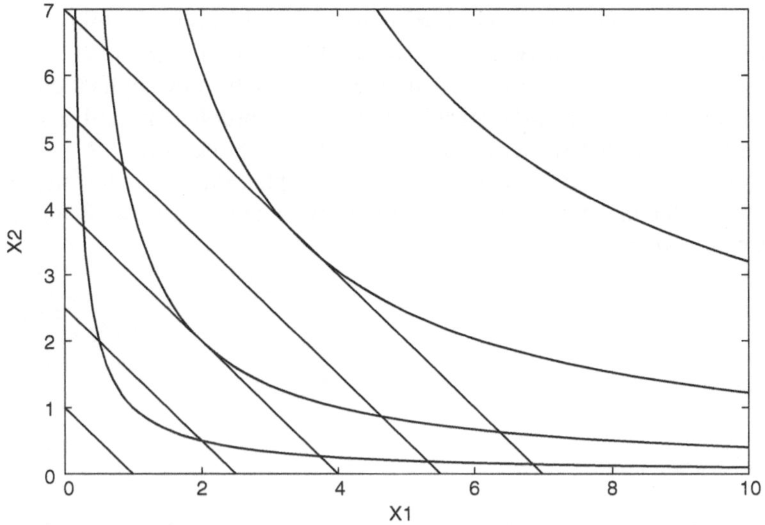

Figure 2.1 Neoclassical maximization problem with budget function isoclines (linear) and indifference curves. Tangential points are utility maxima for the respective budget line (budget function isocline).

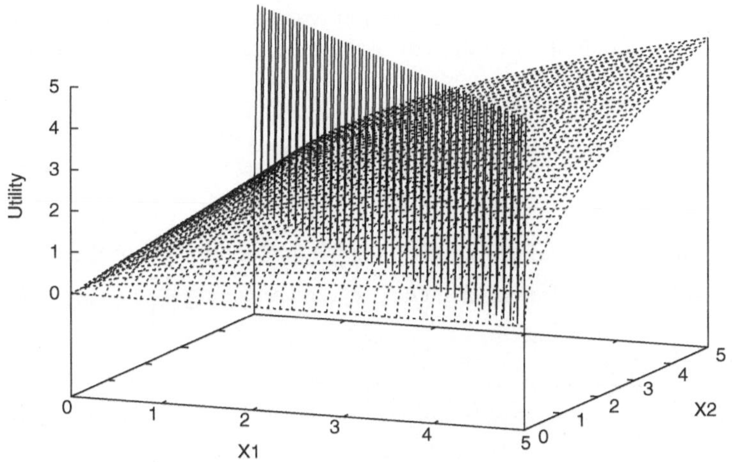

Figure 2.2 Three-dimensional representation of the maximization problem shown in Figure 2.1 with the utility function and one budget function isocline.

resisting forces are equal. Mirowski (1989) details both the mathematics and the history of the adoption of the physical methodology into equilibrium economics, while others discuss the conceptual and mathematical analogies in greater depth (Hands, 1993; Smith and Foley, 2008). For mathematical details of the following analysis see Sato (1981), Mirowski (1989), Hands (1993, 2006), Gandolfo (1997) and Matsuyama (1991). In economics, the problem to be solved by application of this method comes down to the maximization (or as a dual problem minimization) of a function in a restricted space. For neoclassical growth theories, this means the maximization of utility by choosing an investment plan subject to the budget restriction[8] where the income results from earlier investments. Specifically there is an objective function

$$F(x_1, x_2) \rightarrow \max \tag{2.1}$$

with a restriction[9]

$$G(x_1, x_2) \leq 0 \tag{2.2}$$

which is, if binding, specifically supposed to be $G(x_1, x_2) = 0$. Differentiation with respect to x_1 yields

$$\frac{dG(x_1, x_2)}{dx_1} = 0 = \frac{\partial G(x_1, x_2)}{\partial x_1} + \frac{\partial G(x_1, x_2)}{\partial x_2}\frac{dx_2}{dx_1}. \tag{2.3}$$

Furthermore, in equilibrium the objective function is supposed to be constant for infinitesimal changes in one of its variables (taking the interdependence with the other one into account)

$$\frac{dF(x_1, x_2)}{dx_1} = 0 = \frac{\partial F(x_1, x_2)}{\partial x_1} + \frac{\partial F(x_1, x_2)}{\partial x_2}\frac{dx_2}{dx_1}. \tag{2.4}$$

Resolving one of these two total differentials for $\frac{dx_2}{dx_1}$ and applying it to the other yields (e.g. for resolving the differential of $G(x_1, x_2)$ and replacing $\frac{dx_2}{dx_1}$ in $F(x_1, x_2)$)

$$\frac{dx_2}{dx_1} = -\frac{\frac{\partial G(x_1, x_2)}{\partial x_1}}{\frac{\partial G(x_1, x_2)}{\partial x_2}} \tag{2.5}$$

$$\frac{\partial F(x_1, x_2)}{\partial x_1} = \frac{\partial F(x_1, x_2)}{\partial x_2}\frac{\frac{\partial G(x_1, x_2)}{\partial x_1}}{\frac{\partial G(x_1, x_2)}{\partial x_2}}. \tag{2.6}$$

Now, defining what is to become the Lagrange multiplier

$$\mu = \frac{\frac{\partial F(x_1,x_2)}{\partial x_2}}{\frac{\partial G(x_1,x_2)}{\partial x_2}} \tag{2.7}$$

and substituting μ into (2.6) leads to

$$\frac{\partial F(x_1,x_2)}{\partial x_1} = \mu \frac{\partial G(x_1,x_2)}{\partial x_1} \tag{2.8}$$

while, from the definition of μ, it follows that

$$\frac{\partial F(x_1,x_2)}{\partial x_2} = \mu \frac{\partial G(x_1,x_2)}{\partial x_2}. \tag{2.9}$$

Equations (2.8) and (2.9) are equivalent to

$$\nabla F(x_1,x_2) = \mu \nabla G(x_1,x_2), \tag{2.10}$$

more commonly written as

$$0 = \nabla F(x_1,x_2) - \mu \nabla G(x_1,x_2) \tag{2.11}$$

That is, in the vicinity of equilibria, the gradients of the objective function and the restriction have to be parallel, a result that holds also for higher dimensional systems and systems with more than one restriction. In order to proceed with this optimization and obtain valid results the function must be integrable. If this is the case, integration leads to the usual Lagrangian function

$$\int 0 = \int \nabla F(x_1,x_2) - \mu \int \nabla G(x_1,x_2) \tag{2.12}$$

$$\mathcal{L} = \mathcal{L}(x_1,x_2,\mu) = F(x_1,x_2) - \mu G(x_1,x_2) \tag{2.13}$$

where $\mathcal{L}$ is a scalar quantity which is constant as long as μ is well-defined.[10] However, the condition holds also for a more general multiplier, as used in the Hamiltonian

$$\mathcal{H}(x_i,\dot{x_i},\mu_h,t) = F(x_i,\dot{x_i},t) + \mu_h(G(x_i,\dot{x_i},t)) \tag{2.14}$$

which describes a dynamic system with t being the time, $\dot{x_i}$ the change of the x_i in time, and μ_h the Hamilton multiplier. Of course, this approach is especially useful for time-dependent systems like neoclassical models of endogenous growth, as discussed in the context of growth theory in section 3.2.

The quantity $\mathcal{H}$ is preserved for all transformations, a property called economic conservation laws (Sato, 1981). The critical point about this analysis is that this is borrowed from mechanical physics, where the constant Hamiltonian equals the sum of the kinetic energy (E_T) and potential energy (E_V)

$$\mathcal{H} = E_T + E_V \tag{2.15}$$

For economics, this usually (as long as the restrictions are budget constraints, which is the case both in the endogenous growth models under consideration and the standard neoclassical microeconomics) means that net utility (E_V) minus net expenditure ($-E_T$) equals a constant term, a conservation law that is not generally justified. This issue is discussed in more detail in Mirowski (1989) and Hands (1993).

As Sato (1981) details, the necessary and sufficient condition for integrability, and equivalently the economic conservation laws, is that the utility function of the optimization problem[11] be invariant under $n-1$ (for an n-dimensional system) unconnected infinitesimal transformations. With infinitesimal transformations

$$\mathcal{T}_j = \frac{\partial}{\partial x_j} - r_j(x)\frac{\partial}{\partial x_n} \qquad \text{(Sato, 1981, Ch.4, Eq.80)}$$

interpreted by Sato as taste change (Sato, 1981: 149–152), this yields[12]

$$\mathcal{T}_j F = \frac{\partial F(x)}{\partial x_j} - \varrho_j(x)\frac{\partial F(x)}{\partial x_n} = 0 \qquad \forall j \neq n \qquad \text{(Sato, 1981, Ch.4, Eq.81)}$$

or, equivalently the usual solution of Lagrangian optimization[13]

$$-\frac{\frac{\partial G(x)}{\partial x_j}}{\frac{\partial G(x)}{\partial x_n}} = -\frac{\frac{\partial F(x)}{\partial x_j}}{\frac{\partial F(x)}{\partial x_n}} = -\varrho_j(x) = MRS(x_j, x_n) \qquad \forall j \neq n. \tag{2.16}$$

Hence, invariances must hold simultaneously, given by Sato (1981) as a Lie bracket

$$[\mathcal{T}_i, \mathcal{T}_j] \equiv 0 \qquad \text{(Sato, 1981, Ch.4, Eq.84)}$$

$$[\mathcal{T}_i, \mathcal{T}_j] = \left(\frac{\partial}{\partial x_i} - \varrho_i(x)\frac{\partial}{\partial x_n}\right)\frac{\partial}{\partial x_i} + \left(\frac{\partial}{\partial x_j} - \varrho_j(x)\frac{\partial}{\partial x_n}\right)\frac{\partial}{\partial x_j}$$
$$+\left\{\left(\frac{\partial(-\varrho_j(x))}{\partial x_i} - \varrho_i(x)\frac{\partial(-\varrho_j(x))}{\partial x_n}\right) - \left(\frac{\partial(-\varrho_i(x))}{\partial x_j} - \varrho_j(x)\frac{\partial(-\varrho_i(x))}{\partial x_n}\right)\right\}$$
$$\frac{\partial}{\partial x_n} \equiv 0 \qquad \text{(Sato, 1981, Ch.4, Eq.84)}$$

As the first two terms are zero (the simple infinitesimal transformations), this yields the Antonelli or Samuelson integrability conditions

$$\frac{\partial \varrho_i(x)}{\partial x_j} - \varrho_j(x)\frac{\partial \varrho_i(x)}{\partial x_n} = \frac{\partial \varrho_j(x)}{\partial x_i} - \varrho_i(x)\frac{\partial \varrho_j(x)}{\partial x_n}. \qquad \text{(Sato, 1981, Ch.4, Eq.85)}$$

that are fulfilled if the Hessian matrix – the matrix of the second derivatives – of the objective function

$$Hessian = \begin{pmatrix} \frac{\partial^2 F(x)}{\partial x_1 \partial x_1} & \cdots & \frac{\partial^2 F(x)}{\partial x_1 \partial x_n} \\ \cdots & & \\ \frac{\partial^2 F(x)}{\partial x_n \partial x_1} & \cdots & \frac{\partial^2 F(x)}{\partial x_n \partial x_n} \end{pmatrix} \qquad (2.17)$$

is exactly symmetric

$$\frac{\partial^2 F(x)}{\partial x_i \partial x_j} = \frac{\partial^2 F(x)}{\partial x_j \partial x_i} \qquad \forall i \neq j. \qquad \text{(Sato, 1981, Ch.4, Eq.85}')$$

The model is beautiful, and the result is also convenient in order to keep this (micro) level simple and to move on to considerations at arbitrary levels of aggregation (macro). The assumptions, however, are heroic. Real-world utility 'functions' have been found to be very different, much less well-behaved, not necessarily continuous and integrable, and indeed heuristic rather than perfectly rationally optimizing (Simon, 1955; Tversky and Kahneman, 1972; Kahneman and Tversky, 1973). The economic conservation law will also be effected by this, and with direct interdependence and the central auctioneer gone, the whole setting changes. The theory may be viewed as one out of many theoretical constructs, with advantages and disadvantages, that may be suited to conducting investigations, but is unsuited for description of the big picture.

2.2 Game theory: considering direct interdependence

For problems with direct interdependence and possibly also discrete choices (the assumption of divisibility does not generally hold), game theory is a much better-suited approach. There is still the assumption of rationality, though this does not necessarily translate into utility in that sense. It must be possible for outcomes for agents to be ordered on an ordinal scale while it is still convenient to use cardinal values in practice.[14]

Game theory requires the definition of a game Γ with a number of agents (indexed i), well-defined rules (including the order of decision making among the agents and the information sets available to each agent), a set of strategies Σ_i (particular strategies denoted $\sigma_i \in \Sigma_i$, combinations of the strategies of different agents denoted σ without index) for every agent as well as the payoffs (outcomes for the agents) for all agents for any possible combinations of strategies $\Pi_i(\Gamma, \sigma_j)\ \forall j \forall i$. The task is to derive a set of equilibria in the game. An equilibrium is a result – hence a combination of strategies of all agents – with a certain stability for instance, such that the agents do not change or (if that is impossible anyway) regret their decisions. In turn, it may be concluded that such an outcome is more likely than others. There are different concepts of equilibria for different types of games. For the current thesis it shall be sufficient to give a basic and most common example, the Nash–equilibrium[15] and a very brief account of some extensions. For more detailed discussions of game theory, see Gintis (2000), Hargreaves Heap and Varoufakis (1995) or Easley and Kleinberg (2010).

A Nash equilibrium is defined as an outcome $\sigma^* \; \forall i$ where none of the involved agents has reason to unilaterally change her strategy. The sufficient condition for this is that for all agents the strategy σ_i^* leading to this equilibrium conveys a better outcome (higher payoff) than any other possible strategy $\sigma_i \in \Sigma_i \setminus \sigma_i^*$, hence[16]

$$\Pi_i(\Gamma, \sigma_i^*, \sigma_{-i}) \succ \Pi_i(\Gamma, \sigma_i, \sigma_{-i}^*) \quad \forall \sigma_i \in \Sigma_i \setminus \sigma_i \quad \forall i. \tag{2.18}$$

Nash equilibria are neither necessarily unique nor optimal outcomes, even in terms of the (not very restrictive) Pareto optimality.[17] Consider for instance the following two-person normal-form game[18] Γ_{PD} conveniently written in matrix form – see Table 2.1.

This game, a *prisoners' dilemma*, has one unique Nash equilibrium (($\sigma_{1,2}, \sigma_{2,2}$), lower right); the only Pareto optimal solution, however, is the strategy combination ($\sigma_{1,1}, \sigma_{2,1}$), upper left).

With *mixed strategies*, players may choose to play any of their strategies with a certain probability (thus mixed). There is at least one Nash equilibrium for any normal form game (Nash, 1950).

Two notable and straightforward refinements of the Nash equilibrium are the *trembling hand perfect Nash equilibrium* (Selten, 1975) and the *proper equilibrium* (Myerson, 1978). Both add a small perturbation to the original game, thus deriving a perturbed game $\widetilde{\Gamma}$ where agents still choose a strategy σ_i^*, and will most likely succeed in playing this strategy, but with a small probability ε (the 'trembling hand') end up playing any other strategy in their set of strategies Σ_i. The player cannot control her own 'trembling hand' but should take into account the 'trembling hands' of her opponents. For trembling hand perfectness, ε is an abstract arbitrarily small probability, for properness, the ε for any strategy also depends

Table 2.1 Game in matrix notation

		Player 2	
		$\sigma_{2,1}$	$\sigma_{2,2}$
Player 1	$\sigma_{1,1}$	$\Pi_2(\sigma_{1,1}, \sigma_{2,1}) = 2$ $\Pi_1(\sigma_{1,1}, \sigma_{2,1}) = 2$	$\Pi_2(\sigma_{1,1}, \sigma_{2,2}) = 3$ $\Pi_1(\sigma_{1,1}, \sigma_{2,2}) = 0$
	$\sigma_{1,2}$	$\Pi_2(\sigma_{1,2}, \sigma_{2,1}) = 0$ $\Pi_1(\sigma_{1,2}, \sigma_{2,1}) = 3$	$\Pi_2(\sigma_{1,2}, \sigma_{2,2}) = 1$ $\Pi_1(\sigma_{1,2}, \sigma_{2,2}) = 1$

upon the payoffs resulting from this strategy. Note that the Nash equilibrium of the above game Γ_{PD} is both trembling hand and perfect and proper.[19]

The same concepts still apply for repeated games, as long as there are finitely many repetitions. The last period is identical to a non-repeated game and may thus (for certain games) be predicted by all involved agents with perfect certainty.[20] Hence, as all actions taken in the last period are anticipated, the second-to-last period also appears as identical to a non-repeated game and may be predicted. This is repeated until the very first period of the game is reached; the method is known as *backward induction*.

The situation changes, however, for infinitely or indefinitely repeated games. Now, a player desiring to attain the Pareto superior result can use her future strategy choices as reward and punishment. A strategy configuration σ^* is now feasible if any agent has either no incentive to deviate from this configuration (if it maximizes her one-shot payoffs) or if another agent who would loose from this deviation is able to exercise a *credible thread* (the so-called Folk theorem) (Rubinstein, 1979). In the above game, the desired strategy combination is, of course, the Pareto optimal one (mutual cooperation); to respond to the other strategy (defection) by also playing this second strategy (defection) is credible as a thread, as it increases the payoff for the player herself while inflicting losses on the opponent. This led to Axelrod's (1984) analysis that cooperation may evolve on its own in indefinitely repeated games, even with even rather simple strategies that involve exercising the credible thread (such as tit-for-tat, starting with cooperation and then always mirroring the opponent's previous action in the next repetition).

Another refinement of the Nash equilibrium, discussed in the context of evolutionary game theory below, is the concept of an evolutionary stable strategy developed by Maynard Smith and Price (1973).

2.3 Evolutionary modelling: alternatives to neoclassical equilibrium models

Though some kind of rationality is not entirely unrealistic when considering intelligent decision makers, this remains a rather daring concept with complex explicit or implicit further assumptions about motivations, preferences, information sets, etc. Rationality, as required for both neoclassical economic equilibrium models and non-evolutionary game theory, always involves the foretelling of possible outcomes as consequences of certain actions (and then choosing the actions appropriately). An interesting and powerful alternative concept is to eliminate the rationality assumption entirely and reduce the assumptions concerning behaviour of agents to simple idiosyncratic rules, analogous to genes shaping simple characteristics of biological organisms. Applying an evolutionary selection mechanism can then be used to drive out agent types with less favourable outcomes (however outcome may then have to be defined) and create both a stable population and an endogenous notion of rationality (that will then depend only on the definition of outcome and the way selection works). Deep analogies may be drawn between such models for economic and social behaviour and evolutionary biology. These include an agent type as an analogy for a species (more generally a group of organisms with a common genetic trait), social and economic success as an analogy to biological evolutionary selection, and the creation of new types of agents as an analogy to mutation in biological systems. Note that agent types in evolutionary economic or social models do not necessarily represent humans or even organizations, but rather concepts of behaviour, methods, ways of administration, organization, etc. Agents may consequently also not represent 'instances' of these types (like biological organizations are phenotypical manifestations of a genotype), but rather an environment through which behaviours, ideas, concepts, etc., float in a more or less successful way. This is also analogous to the fact that biological evolution does not so much select between individuals or species but rather between what is actually able to retain a stable state through a turbulent evolutionary process – the genes and genomes. This analysis is due to Dawkins (1976) who saw the analogy to social (and thus implicitly economic) systems.

The setting of evolutionary economic models is generally a population of agents, with a changing or constant number of types distinguishable by their strategy choices or, more generally, their actions when faced with certain kinds of situations. Furthermore, there are interactions between those agents, and they are awarded a payoff (or assigned a value of evolutionary fitness) as a result of their performance in these interactions. A replicator mechanism then shifts the shares of the agent types in the population according to their payoffs or fitnesses relative to the average – this may be accomplished by adding agents to the system or by letting agents switch their type (more illustratively, their strategy). The central questions then are how the population evolves, if an equilibrium or stable pattern is approached, and under which conditions an agent type (or strategy) is evolutionary successful. The resulting system is usually rather complex mathematically, potentially involving as many independent variables as agent types or even agents in the system, or multiples of that number if more than one variable is necessary for an agent type or agent. This complexity in turn raises another issue, which was considerably less important in simple game theory and also neoclassical equilibrium models: that of how to analyze the system when faced with limited computation power and time.

There are three major approaches:

- The analysis of evolutionary stable strategies (Maynard Smith and Price, 1973). Another refinement of the Nash equilibrium in game theory discussed above in section 2.2, it considers a population of repeatedly and randomly matched agents that play a symmetric game. According to their type, the agents choose a (pure or mixed) strategy. Selection follows according to the attained payoff. However, the course of the game is not what matters; consideration focuses on which strategies are successful in this setting, or rather how a strategy ought to be in order to perform well. In particular, a strategy is said to be evolutionary stable if it cannot be driven out of a population it dominates by a small group of invaders.[21] If the agents are assumed to choose their strategies rationally and to imitate successful strategies of other agents (learning), stability properties at the strategy-level may influence the dynamics and the stability or instability of the system to a large extent.[22] Evolutionary stable strategies that will dominate the system once they have achieved control of it (defending it against small groups of 'intruders') and neutrally stable strategies that cannot be driven out of the system are two examples of this approach (Maynard Smith and Price, 1973). Any evolutionary stable strategy is a Nash equilibrium of the underlying game[23] and an attractor (stable fixed point) of a (hypothetical) corresponding dynamic system.[24]

- Replicator dynamics. Unlike than evolutionary stable strategies, replicator dynamics is not about *n* two-person games but rather one *n*-person game. It also addresses the agent types directly; in fact, in this case the shares of the population are sufficient to describe the model, the exact number of agents is irrelevant. There is a global fitness function depending on the population shares $p_{i,t}$ (for agent type *i* at time *t*), the setting and the agent types' characteristics. Thus for every agent type, a fitness value $f_{i,t}$ is computed, the average fitness in the system being consequently

$$\phi_t = \sum_i p_{i,t} f_{i,t} \tag{2.19}$$

The replicator equation may be defined appropriately to increase the shares of agents with higher relative fitness, and decrease the shares of those with lower relative fitness. It can be either a difference or differential equation and vary considerably in exact form. The two most basic and convenient examples are (as a difference equation)

$$p_{i,t+1} = \frac{f_{i,t}}{\phi_t} p_{i,t} \tag{2.20}$$

and (as a differential equation)

$$\frac{\partial p_i(t)}{\partial t} = (f_i(t) - \phi(t))p_i(t). \tag{2.21}$$

For a more detailed discussion of both the method in general and particular replicator equations, including a number of variants with different properties, see Nowak (2006). The global fitness equation and equation (2.19) could of course simply be substituted into the respective replicator equation (2.20 or 2.21) thus eliminating the fitness terms; however, the fitness variable also provides a convenient measure of the evolutionary performance of the respective agent type that changes in time with the changing dynamic properties of the system. In effect, replicator models are simply systems of difference or differential equations, to which the methods of dynamic systems analysis (see below, section 2.4) are applicable. Fixed points (equilibria) are defined accordingly (for difference equations)

$$p_{i,t+1} = p_{i,t} \quad \forall i \tag{2.22}$$

and (for differential equations)

$$\frac{\partial p_i(t)}{\partial t} = 0 \quad \forall i. \tag{2.23}$$

The stability properties of those fixed points are also assessed as in other dynamic systems; for details, see section 2.4.

- Simulation – though this is not specific for evolutionary modelling. See section 2.4.2 for details.

Evolutionary modelling as an approach to economic and social systems is able to make predictions for the macro level with micro level models – in fact with rather simple (i.e. simplified) micro-level structures. The macro-level dynamics are in turn different from the micro-level structure but do still interact with it. So far, the approach may not seem very impressive compared to neoclassical microeconomics. However, the model is able to generate much more complex stable results than neoclassical models with much less restrictive assumptions. To illustrate this, Figure 2.3 shows the equilibrium properties of three games transformed into replicator dynamic settings (the prisoners' dilemma, a coordination game and an anti-coordination game).[25] For instance, anti-coordination games (a type of game to which the Hawk–Dove games also belong) generate structured populations as a stable equilibrium, i.e. a part of the population plays strategy σ_1, the other part σ_2.[26] For coordination games, there are two equilibria, the pure strategy Nash equilibria; which of the two the game converges to is not ex ante predictable. The prisoners' dilemma on the other hand has only one unique equilibrium that becomes a global attractor in a dynamic system context. The result increases in complexity with the complexity of the underlying game. A two-person normal form game with three strategies (the rock–paper–scissors game shown in the simplex in Figure 2.6c) supports a stable limit cycle. With more players or a more complex rule setting, the result may again be very different.

Of course, evolutionary modelling is not limited to an evolutionary game theoretic context. Though it is probably possible to describe the models in game theory terms, for most models, including the replicator models of evolutionary growth theory discussed in section 3.4, this is not exactly what the authors had in mind. Nevertheless, for the current section on the methodology of evolutionary modelling, the approach via game theory is very easy and illustrative – later on, in section 4.3, game theory based replicator dynamic settings approaches to network externalities will be considered.

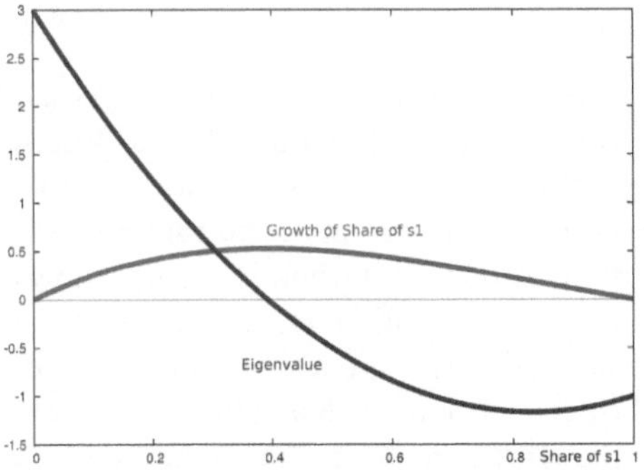

(a) Prisoners' dilemma

Player 1 \ Player 2	σ_1	σ_2
σ_1	1, 1	5, 0
σ_2	0, 5	2, 2

(b) Prisoners' dilemma matrix

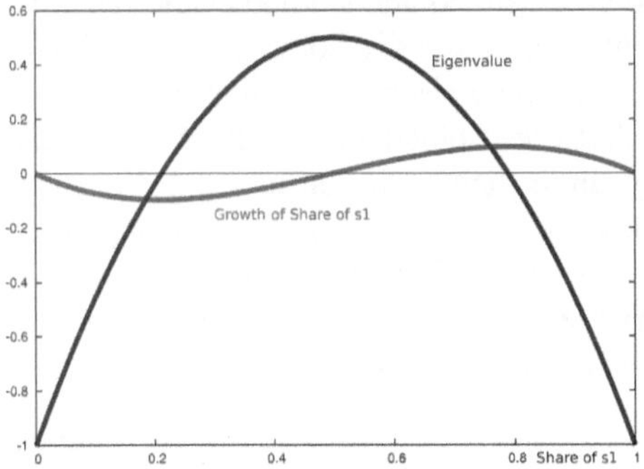

(c) Coordination game

Player 1 \ Player 2	σ_1	σ_2
σ_1	1, 1	0, 0
σ_2	0, 0	1, 1

(d) Coordination game matrix

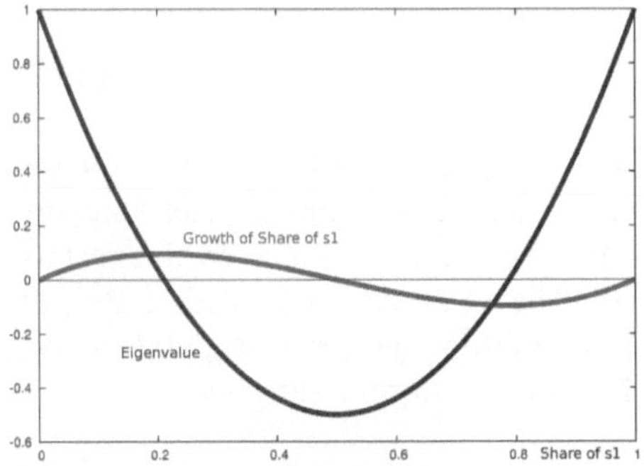

(e) Anti-coordination game

Player 1 \ Player 2	σ_1	σ_2
σ_1	0, 0	1, 1
σ_2	1, 1	0, 0

(f) Anti-coordination game matrix

Figure 2.3 Game incentive structures (right), dynamics and eigenvalues (left) of the corresponding replicator equation (as a differential equation) as a function of the population share of strategy 1.

2.4 Chaos and complexity theory: more than the sum of its parts

2.4.1 Dynamic systems, fixed points, stability

The advent of modern dynamic theory inspired a number of economic theories and models (basically with the writings of Lyapunov (1907) and the slow diffusion of the method from mathematics through natural sciences into economics) that were very different to the earlier static equilibrium approaches, and pathbreaking for a variety of later models and research traditions.[27] Dynamic theory allows for modelling the development of systems in time, and hence also enables the description and prediction of disequilibrium models (i.e. models that do not, or not necessarily, converge towards an equilibrium). Systems with a unique and global equilibrium (a global attractor) are just the most simple case in a continuum of possible dynamic systems. In a way, neoclassical equilibrium theory thus seems to cover only a fraction of what can possibly be defined and studied using dynamic theory – though dynamic systems can also become very complex and difficult to solve for many cases. However, dynamic theory is a still rapidly developing field of research and there may be much more sophisticated methods available in the future. For a brief history of dynamic theory and an account of recent developments, see Mira (1997). The following considerations are given in greater detail by Markus and Yamabe (1960), Gandolfo (1997) and Foley (1998).

Common dynamic systems (with variables x_t, t denoting time) are defined as either systems of difference equations (discrete in time) of the form

$$x_{t+1} = F(x_t) \tag{2.24}$$

or differential equations (continuous in time) of the form

$$\frac{\partial x(t)}{\partial t} = F(x(t)). \tag{2.25}$$

In both cases, suppose the system is n-dimensional, hence contains n variables, x is consequently an n-dimensional vector and F is an n-dimensional vector function, usually with real dimensions $F : \Re^n \rightarrow \Re^n$. The space generated by the vectors x is called the *phase space* of the system. The computation of static states, or equilibria, or fixed points x^* in the phase space of these systems is straightforward. For difference equations (i.e. system (2.24)), an equilibrium satisfies

$$x_{t+1} = F(x_t) = x_t, \tag{2.26}$$

for differential equations (i.e. system (2.25)), that condition is

$$\frac{\partial x(t)}{\partial t} = F(x(t)) = 0. \tag{2.27}$$

However, equilibria may be very different in nature. A stable equilibrium, also called a stable fixed point or an attractor, is an equilibrium x^* with an environment

$\mathcal{B} : x \pm \varepsilon$ (where ε is an n-dimensional vector of arbitrary small length $|\varepsilon|$) such that the dynamic movement starting from every point in $\mathcal{B}$ strictly converges to x^*

$$lim_{T\to\infty} x_{t+T}(x_t) = x^* \qquad \forall x_t \in \mathcal{B}. \tag{2.28}$$

The paths of dynamic movement a system takes, starting from a particular point in phase space, are called *trajectories*. Thus, while an attractor is an equilibrium that is approached by *all* trajectories within (at least) an arbitrary small environment $\mathcal{B}$, this is not the case for other types of equilibria. An equilibrium from which the trajectories move away in an immediate environment $\mathcal{B}$ is called a *repeller*.[28] An equilibrium with an immediate environment $\mathcal{B}$ within which the trajectories neither approach the equilibrium nor move away from the it (i.e. they do not leave $\mathcal{B}$) but form closed paths eventually reaching the starting point again (*limit cycles*) is called the centre of a limit cycle. An equilibrium with approaching and diverging trajectories is called a *saddle point*. The crucial question is usually, of course, whether the equilibrium is stable, that is, an attractor.

To assess stability, the eigenvalues λ of the system's Jacobian matrix $\mathcal{J}$, the matrix of the first derivatives

$$\mathcal{J} = \begin{pmatrix} \frac{\partial F(x_{1,t})}{\partial x_1} & \cdots & \frac{\partial F(x_{1,t})}{\partial x_n} \\ \cdots & & \cdots \\ \frac{\partial F(x_{n,t})}{\partial x_1} & \cdots & \frac{\partial F(x_{n,t})}{\partial x_n} \end{pmatrix} \tag{2.29}$$

are computed with

$$0 = det(\mathcal{J} - \lambda\mathcal{I}) \tag{2.30}$$

where $\mathcal{I}$ denotes the n-dimensional unit matrix. Every eigenvalue characterizes a dynamic movement of the system. For systems of difference equations, this movement is guaranteed to be contracting if the modulus of the eigenvalue is smaller than 1 ($|\lambda| < 1$) and expanding if it is larger than 1 ($|\lambda| > 1$). In turn, if the modulus of the *dominant eigenvalue* (in this case the eigenvalue with the largest modulus) is smaller than 1 in the vicinity of the equilibrium, the equilibrium is an attractor.[29] For systems of differential equations, this movement is guaranteed to be contracting if the modulus of the real part of the eigenvalue is negative ($Re(\lambda) < 0$), and expanding if it is positive ($Re(\lambda) > 0$). In turn, if the dominant eigenvalue (in this case the eigenvalue with the largest real part) is negative in the vicinity of the equilibrium, the equilibrium is an attractor.[30] Furthermore, non-zero imaginary parts of the eigenvalues translate into circular motion of the system, and in turn the system potentially shows cycles[31] if $\exists\lambda_i : Im(\lambda_i) \neq 0$.

Linear systems of difference equations of differential equations (that is, $F(x_t)$ is a vector of linear functions of the vector x_t, i.e. the n variables $x_{i,t}$ $i = 1, \ldots n$) lead to globally constant eigenvalues, since, for linear systems, the Jacobian matrix is a matrix of simple scalar numbers (real or complex). The stability properties of the system are the same over the entire phase space. As the systems are always of the same dimensionality as the number of dependent variables, there is only one solution for linear systems. *Path dependence* may nevertheless occur in divergent systems with an unstable equilibrium (a repeller), with the trajectories diverging to all possible directions depending on the exact location in phase space (relative to the repeller) where the system was initialized. Non-linear systems of difference equations or differential equations may be much more complex. As the Jacobian is no longer a matrix of scalars, but a matrix of functions of the variables $x_{i,t}$, the eigenvalues vary depending on the location in phase space. In turn, there may be a much larger numbers of equilibria and very localized behavioural patterns of the trajectories. Consider, as an easy example the discrete logistic map

$$x_{t+1} = \eta x_t(1 - x_t) \qquad 0 < \eta < 4, \tag{2.31}$$

a quadratic first order difference equation (one-dimensional system). With parameter η, the behaviour of the resulting system changes. There are two equilibria,

$$x_1^* = \frac{1-\eta}{\eta}$$
$$x_2^* = 0\eta = 0 \tag{2.32}$$

The single eigenvalue is

$$\lambda = \eta(1 - 2x) \tag{2.33}$$

which is monotonically falling in x, i.e. there are clear (upper and lower) boundaries to the stable area of phase space, therefore the stable area may be written as the interval

$$\left[\frac{\eta-1}{2\eta}, \frac{\eta+1}{2\eta}\right] \tag{2.34}$$

For $\eta < 1$, the first equilibrium $x_1^* = \frac{1-\eta}{\eta}$ is not valid (negative, i.e. not reached by the map), the second equilibrium $x_2^* = 0$ is stable. At $\eta = 1$, a *phase transition* occurs, x_2^* becomes unstable and x_1^* becomes a stable fixed point, which remains the case for $1 < \eta < 3$. At $\eta = 3$, the stable equilibrium x_1^* leaves the stable area of phase space, thus the system no longer has any stable equilibria, enters a limit cycle and begins oscillating between two points $x_{LC} = \frac{\eta+1}{2\eta} \pm \sqrt{\frac{\eta^2-2\eta-3}{4\eta^2}}$; the phase transition at $\eta = 3$ is the first of a cascade of period-doubling bifurcations, eventually, interrupted by small windows of shorter period lengths (most notably a period length of 3), thereafter the system reaches a chaotic state (May, 1976). The situation is illustrated in Figure 2.4.

This example also illustrates the practical choice of methods when investigating more complex dynamic systems. While it is simple to investigate the system's

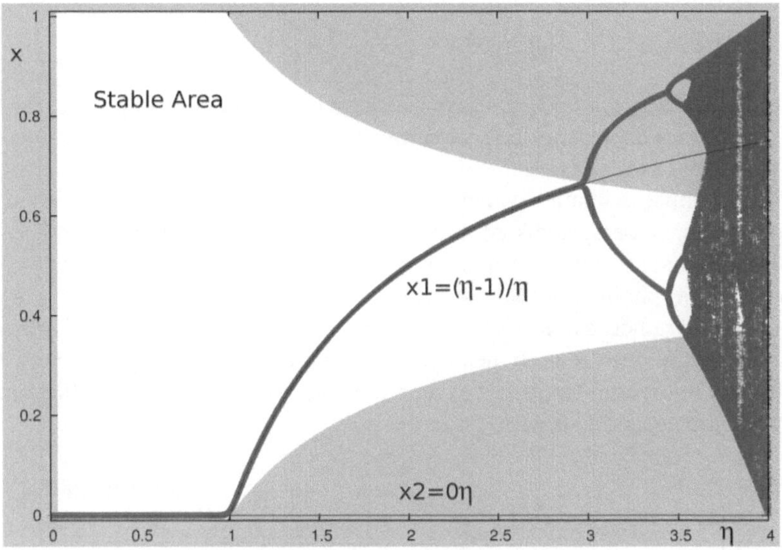

Figure 2.4 Bifurcation diagram of the discrete logistic map.

behaviour for values of η for which the system has stable equilibria (i.e. lower values), this is very difficult for values between about 3.5 and 4. For Figure 2.4, the system was, however, simply simulated. After a while, the system gets sufficiently close to its equilibrium or limit cycle – if that exists for the respective parameter value. Hence, the system may simply be run for a sufficiently dense field of parameter values η no matter which initial variable value x_0 it is initialized with; then, after giving the system an initial period, several periods are recorded. When plotted as in Figure 2.4, this yields the *phase portrait* for a dense spectrum of parameter values, and is therefore also able to indicate changes in the *phase portrait*, such as bifurcations, in great detail, hence providing a *bifurcation diagram*. The simulation code is available from the author on request.

For more complex systems, it may be impossible to first solve the system and then to assess its stability using the now known equilibria, Jacobian and eigenvalues. In this case, it is still possible with empirical or simulation data to use *Lyapunov's second method* (Lyapunov, 1907; Grosche *et al.*, 1995; Gandolfo, 1997), measuring whether trajectories converge or diverge. This is done by defining a *Lyapunov function* $V(x_t - \overline{x_t})$, where x_t and $\overline{x_t}$ are two different possible states the system may take at time t; the Lyapunov function applies some metric (for instance, simply the Euclidian norm) to quantify the distance. Furthermore, the states of the system following from x_t and $\overline{x_t}$ at a later time $t + T$ are computed and the development of the distance of the two trajectories is assessed. Supposing that this is approximately continuous in time, the distance develops approximately as

$$V(x_{t+T} - \overline{x_{t+T}}) \approx e^{\Lambda(T)} V(x_t - \overline{x_t}) \tag{2.35}$$

where Λ is called the *Lyapunov exponent*. The larger the Lyapunov exponent, the faster the trajectories diverge, if the Lyapunov exponent is positive $\Lambda > 0$; if it is negative, the trajectories converge.

While more simple dynamic systems are relatively easy to handle, the higher dimensional systems or systems of higher degree quickly become complex. *Complexity* in this sense refers to non-trivial patterns that result from dynamic systems; complexity should not be thought of as a binary property (something is or is not complex) but rather as a relative term. Empirically, it may most conveniently be measured in the information needed to describe a system exhaustively. Specifically, the number of states Ω a system may take relates to the most bits of information N are needed to describe it in the way

$$N = log(\Omega), \qquad \text{(Bar-Yam, 1997, 0.5.1)}$$

or, equivalently

$$\Omega = 2^N. \qquad \text{(Bar-Yam, 1997, 0.5.2)}$$

The system is further said to be *chaotic* or to exhibit *deterministic chaos* if it is not conveniently predictable. This, in turn, may be assessed by evaluating the Lyapunov exponents. Sufficiently large positive Lyapunov exponents, as detailed above, indicate that arbitrarily close trajectories will at a later state be at a considerable distance from one another. Deterministic chaos is also indicated by a scattering of autocorrelation coefficients (of course a discrete consideration) such that they do not show a consistent pattern over time (hence with the iteration of the system) and in different lags of time (hence in the autocorrelation spectrum).[32]

2.4.2 Simulation as a method for dynamic systems

As in the consideration concerning the discrete logistic map above, the investigation of Lyapunov exponents and many other approaches to complex dynamic systems involve simulation. Simulation is not an analytically exact method; it is rather an alternative for cases in which analytically exact methods are not feasible. As mentioned above, agent-based models may become fairly complex. Stable patterns may as a consequence also become more complex, not taking the form of simple fixed points at which the system is stable, but limit cycles and other more complex patterns, such as structural emergence (Prigogine, 1975; Prigogine and Stengers, 1984). In case of increased complexity, direct simulation of the system at hand will produce some trajectories and give an idea of the systems dynamics in general. Simulation may be repeated as often as needed for as many different initial conditions as desired. The more dense the starting conditions, located in phase space, for which the system was simulated, the better is the resulting understanding of the systems dynamics. For simple systems, a few repetitions of the simulation suffice to understand the system (e.g. to detect a global attractor and establish that it indeed is probably a global attractor), still it is impossible to tell only from simulation whether the system is indeed as suggested by the simulation results, or whether (against incredible statistical odds) the simulated cases are exceptions from the system's normal behaviour. Simulation does, however, have advantages, that offer good reasons for employing simulation not only to approximate final results – i.e. even if the global equilibrium is computable. Simulation allows us to track not only the end point but the entire trajectory, and hence the dynamic movement of the system. If enough different initial conditions of the system are simulated, a fairly comprehensive phase portrait of the system can be acquired.

In order to get optimal results, simulation also requires a carefully considered model, attentive programming and – most likely – incremental development of both model and program. Generally, simulation should go through six phases, (1) the proper identification and definition of the system under investigation,[33] (2) the formalization of that system to clearly defined entities and interrelations, (3) the programming of that system in a suitable programming language, (4) the decision on which data values or ranges of values to feed into the program, (5) running the simulation program for every input data set gathered in phase (4), and finally (6) the evaluation of the results. Based on how conclusive and certain the outcome is, and how satisfying the result is as an answer to the original problem, this sequence may be repeated several times (i.e. incremental development). Two things have to be paid attention to in particular:

- Validation. The model should really resemble what was to be modelled. This is basically to search for programming errors that may result from a number of sources, including simple mistakes, program-specific issues (such as variable overflows, rounding issues), erroneous specification of time (especially when trying to program models of continuous time to be run on necessarily discrete computers) or elements of the model, etc.
- Verification. The results should be feasibly robust. In order to establish this, the simulation has to be run with a sufficiently large set of different initial conditions, preferably covering large part of the phase space, the simulation runs should be long enough (to avoid mistaking the initial stage for the behavioural characteristic of the program) with sufficiently many agents (to avoid small group effects), and, if stochastic terms are involved, the simulation runs should be repeated sufficiently often to average out the (possibly with path dependence amplified) impact of randomness. All this is, however, also a question of computation time and power – a balance has to be found for cases in which the required computing capacity is found to exceed the available resources (including time).

One final important issue for computer simulation is the choice of the technical methods and tools required to perform the simulation (assuming it is already established that computer simulation is the proper course of analysis). This includes the question whether the hardware equipment at hand is sufficient, if access to better equipment is needed, or if either the simulation program or the data configurations for which the simulation is conducted, are to be restrained. The simulations discussed in the course of this study are mostly rather simple models requiring limited resource; for the programs in the model chapter (Chapter 5), however, the number of iterations had to be limited to a smaller number for particular settings. Generally, the simulations in this book do not claim to provide an exhaustive and solid account of the models studied, that would be capable of completely replacing analytical investigation, but rather they offer panoramic views both of the dynamics the systems under investigation are subject to, and of the phenomena that occur in such systems.

The programming language used for the simulations of the current study is Python (Version 2.7). Python is a scripting language, and thus quick to write compared to lower-level languages (such as C, C++ and Fortran), and flexible with the programs or scripts that usable in any platform with a Python implementation. Yet, compared to other scripting languages, Python is very powerful,[34] with generally good performance in memory requirements, processing efficiency (speed)[35] and stability.

2.4.3 Examples from evolutionary game theory

In section 2.3, it was mentioned that systems of agent-based approaches to evolutionary game theory settings may become very complex, which may leave no other option but to simulate the system. For this to be accomplished the system has to be transformed into a dynamic system (if not yet the case), preferably a discrete one, i.e. a system of difference equations. Simple systems of this type just use a symmetric game, the payoffs of which give the fitness values and apply a replicator in the form of a difference equation. The population shares of the agent types $p_{i,t}$ are the dynamic variables. Define the variables as a vector

$$p_t = \begin{pmatrix} p_{1,t} \\ \dots \\ p_{n.t} \end{pmatrix} \tag{2.36}$$

and let the game matrix be $\mathcal{A}$ as defined above, in section 2.3. The expected payoff, hence the fitness of type i, computed is with this matrix as $f_i = p_{i,t}\mathcal{A}p_t$, while the average fitness is given as $\phi_i = p_t\mathcal{A}p_t$. Substituting these in equation (2.20) yields the replicator equation.

$$p_{i,t+1} = \frac{p_{i,t}\mathcal{A}p_t}{p_t\mathcal{A}p_t} p_{i,t} \quad \forall i \tag{2.37}$$

Different game matrices $\mathcal{A}$ yield a wide variety of possible behaviours of dynamic systems. The simulation is available from the author on request.

Table 2.2 Rock–paper–scissors game in matrix notation

		Player 2		
		Rock	Paper	Scissors
Player 1	Rock	105 105	130 100	100 130
	Paper	100 130	105 105	130 100
	Scissors	130 100	100 130	105 105

Consider a simple form of the well-known rock–paper–scissors game as given in Table 2.2. Hence,

$$\mathcal{A} = \begin{pmatrix} 105 & 100 & 130 \\ 130 & 105 & 100 \\ 100 & 130 & 105 \end{pmatrix} \tag{2.38}$$

The phase space of the three-dimensional replicator system is conveniently depicted in a simplex, a triangular space where each side stands for one of the agent types, and the distance from this side indicates the population share of the respective agent type (see Figure 2.5). The starting points of the simulation runs are denoted by a square, the lines departing from these squares show the trajectories of the system. The system with the game matrix given above for

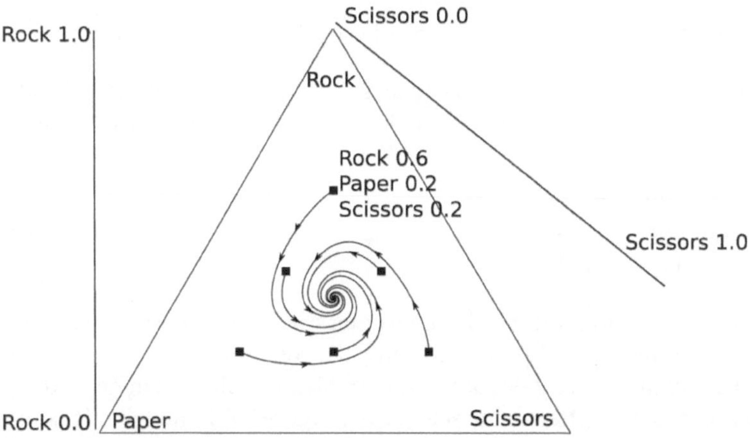

Figure 2.5 Phase portrait in a three-dimensional simplex (rock–paper–scissors game). Every edge serves as the 0% baseline of one of the agent types (rock at the bottom, paper at the right, scissors at the left); the opposite vertex is the 100% state for this agent type (rock at the top, paper left, scissors right). Starting points of the simulations are given as squares, trajectories are given as lines (here with additional arrows).

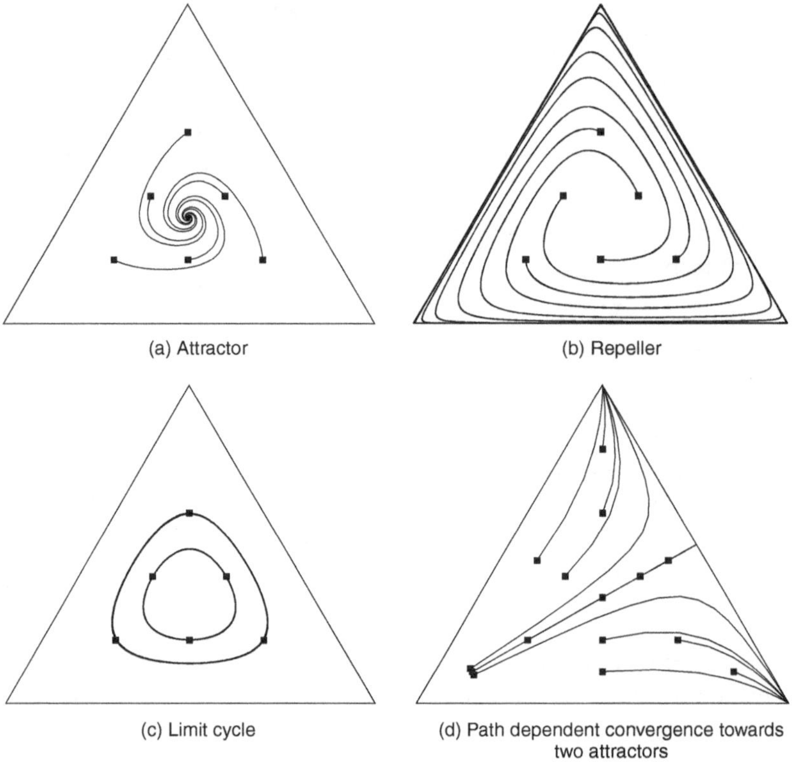

Figure 2.6 Example phase portraits – three-dimensional (three-strategy) simplexes.

instance, obviously has an attractor, with equal shares for all agent types, with all trajectories converging (see Figure 2.6a). Consider another setting with a game matrix

$$\mathcal{A} = \begin{pmatrix} 105 & 100 & 112 \\ 112 & 105 & 100 \\ 100 & 112 & 105 \end{pmatrix} \tag{2.39}$$

For this setting, shown in Figure 2.6b, the fixed point still exists. However, it is now a repeller with the trajectories diverging towards the edges of the simplex. Stable limit cycles (as seen in Figure 2.6c) are also possible, as well as path dependent convergence towards two different attractors as in Figure 2.6d (though this latter one is not a rock–paper–scissors game).

2.5 Structural emergence, the meso-level, and the quest for proper microfoundations of macro-level phenomena

Where the micro and macro branches of economic theory come from may be a somewhat philosophical question. In fact, those were simply the two levels the classical scholars had the opportunity to gather data about. The micro-level data were straightforwardly available from witness reports, firm-level data and occasionally the scholars' own experience. The macro-level data (Adam Smith's 'nation' that is supposed to have wealth), on the other hand were just what officially collected data happened to be available. The question that remains to be solved is how to bring the two levels together, especially if the models are fundamentally different. The neoclassical approach is, of course, to say that the structures on those levels are not that different after all; evolutionary game theory on the other hand puts more effort into modelling the micro-level, and suggests that the higher-level orders itself to a stable state, in turn imposing that structure on the micro-level (for instance, forcing an agent to switch her strategy). This is supported by the concepts of dynamic theory, especially structural emergence.

The macro-level (more concretely several macro-levels, or, more accurately, administration-level perspectives on the economic system) is probably not just an arbitrary level of aggregation for which data happens to be available. Administrative entities are somewhat coherent economic systems at least at the country-level – that is, interactions within the entity are much more frequent than interactions with agents outside that entity. Partly, this is obviously a result of customs duties and geographical proximity. The same coherence may, however, also hold for other, and less-official, entities like economic sectors, social groups, etc. These are in turn self-organized structures, which means that models of social and economic systems should also address both the endogenous self-organization of such entities and the question of whether and how micro-level patterns translate into phenomena at aggregated levels.

There are indeed a number of models addressing structural emergence in economic systems that attempt to explain a wide variety of things, including social class, segregation and persistent social inequality (Schelling, 1971; Axtell *et al.*, 2000; Dessalles and Phan, 2005; Bowles and Sethi, 2006), money (Howitt and Clower, 2000), industry structure (Saviotti and Pyka, 2008), technology adoption patterns (Arthur *et al.*, 1987; Lee *et al.*, 2006; Pegoretti *et al.*, 2009), and geographic clusters (Krugman, 1997) including geographic cluster formation in ('digital economy') sectors that do not require personal interaction (Quah, 1999a). Given the research of historians and political scientists on the emergence of nation state in the early modern period (Anderson, 1991 [1983]), it has also to be concluded that even administrative entities, the borders of which roughly follow the usage patterns of certain languages in geographical space, are the result of a fairly recent self-organization process.[36]

Despite these various models, an integrated approach is still missing.[37] The models (and certainly also the obvious complexity of the system) do, however, suggest that structural emergence may be one of the characteristic social and economic systems. Stable structures may emerge at various levels, giving rise to a hierarchical system of organizational entities that all exhibit a certain coherence in themselves, and exist at different levels of aggregation (which might be termed macro- and meso-levels). The models mentioned above only provide some early insights into the question of how such a system might work, and especially how different levels can exist and what prevents the automatic self-organization effects (responsible for aggregation to a meso-level) from including larger structures, hence raising the entity to the highest possible macro-level. An interesting approach is the discussion on group size and cooperation: in a dilemma-prone situation, groups smaller than a certain size may succeed in inhibiting free-riding, while in larger groups, social cohesion (and thus the negative consequences of being known as a free-rider) is too low (Elsner and Heinrich, 2009, 2011). Dilemma-prone situations in turn may arguably be a central aspect of human societies (i.e. social and economic systems in general) in terms of the common management of resources,[38] which with scarcity allows free-riding and takes the form of an *n*-person prisoners' dilemma.[39] Another important aspect may be the interaction structure, i.e. the network of interactions between agents or entities, especially the number of interaction partners and the clustering of this network. For example, networks with a shorter average lengths of paths between two random nodes are known to allow faster diffusion of knowledge and behaviours (Pegoretti *et al.*, 2009).

The explanation of structural emergence in social and economic systems and, in particular, the microfoundation of macro-level phenomena, may be an important element in the analysis of the origins of economic growth, and especially the consequences of network effects. Both the discussion of the literature on economic growth in Chapter 3 and network effects in Chapter 4, and the development and analysis of an integrated model of technological change, economic growth, and network effects in Chapter 5, will consider this aspect. The agent-based simulation method that will be a central aspect of the model in Chapter 5 is well-suited to investigating the effects of structural emergence, and hence aspects of micro-to-macro transition.

3 A literature review

Models of economic growth

By the end of the nineteenth century, light had been shed on many different aspects of economic systems. The classic economists understood the advantages of functional organization of labour (Smith),[1] comparative advantages in production, and trade (Ricardo). Early neoclassic economists added the functioning of price as a result of bargaining power (that is, of supply, demand and market power), including some mathematical theories to explain it (Cournot, Marshall). And some less-orthodox scholars provided powerful insights into the nature of money and value (Marx), the dynamics of social classes and labour organizations (Marx, Veblen) and the nature of informal institutions (Veblen) and many others.[2] Growth theory, on the other hand, was still very much neglected. Empirical observation provided evidence for more or less continuous growth of capital and production, and regular cycles in output. In order to construct a more elaborate theory to explain these phenomena, an understanding of the macroeconomic dynamics was necessary. This insight was only achieved with the writings of Keynes in the 1930s (Keynes, 1936).

Economic knowledge about growth up to this time included Smith's and Marshall's surplus over subsistence level achieved through functional organization, and the cycle of profit and investment.[3] What was known extended to Malthus' idea of the linear growth of production compared to a superlinear growth of population (thus forecasting the deterioration of per capita wealth, and hence poverty) as well as – less surprisingly – the common knowledge that long-term growth needs good institutions and governance. Marx's political philosophy of the economy added the idea of constant accumulation of capital. If capital in itself is not capable of production – is unproductive – and the rate of accumulation surpasses that of productive 'living labour' (Marx, 1963 [1885]) and any other means of increasing productivity, the crucial flaw of the modern organization of production – of capitalism – is the mistaken assumption of the productivity of capital and the distribution of the producer's surplus – profits – to the capital owners. This in turn leads to an impoverished working class, an unstable (falling) profit rate for the rich class, growth of production with constant demand (overproduction or underconsumption), and thus unstable growth interrupted by a pattern of crises (Marx, 1964 [1867], 1963 [1885]).

In this system there is no effect of technological change other than the increase of productivity equivalent to the growth of the labour force. Whether the model as imagined by Marx retains unstable growth, leads to a final crisis, or the recurring crisis destroys the redundant capital only to let the system start over again and again, is unclear. Philosophical and technical details of the theory remain subject to much discussion based on the exegesis of Marx's texts, but are ultimately not crucial to the functioning and the dynamics of the model. Theories are, of course, generally difficult to develop, as even the facts and numbers are not entirely clear and undisputed.[4]

Keynes and his followers introduced the use of dynamic systems into economics. The earlier concept of a single optimal equilibrium sufficiently describing the single possible economic state was replaced by the understanding of a state space containing the equilibrium, which may or may not be stable and does not instantly constitute the state of the system at any given time, but may be reached by the system after a certain period of time and along a certain – predictable – path. With this it is possible to compute the influence and interdependence of variables and their impact on the development of the system (multiplier effects). In order to compute time paths a notion of time is necessary. For convenience, this is usually done by introducing either a finite and fixed time lag, yielding an ordinary difference equation (or system of difference equations),[5] or an infinitesimal time lag which allows the system to be expressed as a differential equation or system of differential equations.[6] As the methods had thus come to include powerful tools to approach progress in terms of the scope of production (output), capital (wealth) at a theoretical level, a diverse branch devoted to the study of economic growth developed from the 1930s. This chapter will discuss the major contributions to this field of study from that time onward (with a brief excursus to the earlier model by Marx (1963 [1885])), starting with Harrod–Domar models and early AK models, then continuing with neoclassical growth theory, more empirically focused approaches and evolutionary growth theory, and finally a section on the study of growth cycles.

The objective of economic growth theory is, of course, to explain the empirically observed patterns of economic growth (see Figures 1.1 and 1.2), especially the three phenomena mentioned in the Introduction: on average approximately stable (though with small variations) growth rates which yield exponential growth; occasional rather sudden crashes; and a circular pattern in the growth rates, a smooth change between slow and rapid growth (growth cycles). Particularly influential models of growth include the early Keynesian approaches of Harrod (1939) and Domar (1946) and their successors (Kalecki, 1962; Frankel, 1962), as well as, more focused on growth cycles, models by Samuelson (Samuelson, 1939a,b), Kaldor (1940, 1972), and Goodwin (1967). Furthermore, there is the neoclassical growth theory with the Solow–Swan model (Solow, 1956, 1957; Swan, 2002 [1956]) and the models of endogenous growth by Cass (1965, 1966), Uzawa (1965), Lucas (1972, 1975, 1981, 1988) and Romer (1986). Notable alternative accounts were offered by Thirlwall and collaborators (Dixon and

Thirlwall, 1975; Thirlwall, 1980; Thirlwall and Hussain, 1982), Georgescu-Roegen (Georgescu-Roegen, 1975b,a), Arrow (1962), and Schmookler (1966). Schumpeterian and evolutionary growth theory takes a different approach: while other models focus on the dynamics of capital and output and treat the technology level as a more or less exogenous factor, Schumpeter emphasized the importance of processes of innovation (led by the Schumpeterian *entrepreneur*) and *creative destruction* to get rid of outdated technology and allow room for new innovations.[7] The resulting models (Nelson and Winter, 1974; Nelson *et al.*, 1976; Nelson and Winter, 1982a; Conlisk, 1989; Silverberg and Lehnert, 1993, 1994; Freeman and Perez, 1988; Kauffman *et al.*, 2000) are mostly agent-based, and use methods more close to evolutionary game theory and simulation than lower-order differential equations. Finally, a more or less separate branch of theory deals with growth cycles both empirically and theoretically (Slutzky, 1937 [1927]; Samuelson, 1939a,b; Goodwin, 1967; Kaldor, 1972; Lucas, 1972, 1975, 1981; Minsky, 1980; Perez, 1983; Freeman and Perez, 1988; Lorenz, 1987; Keen, 1995); who use mostly systems of difference or differential equations of varying complexity and order. Empirical methods in traditional growth theory primarily use regressions to fit empirical data to various models; more recent approaches to growth cycles include also non-static methods, such as the investigation of the autocorrelation spectrum (de Groot and Franses, 2008, 2009)

This chapter will proceed by discussing the development of the Harrod–Domar model and its immediate successors including early AK models, that all produce exponential growth with a constant growth rate, in section 3.1. Then, the core of neoclassical growth theory, the Solow–Swan model and the models of endogenous growth, will be discussed in section 3.2. Empirical and qualitative observations of real-world growth following Verdoorn, Thirlwall, Marx, Georgescu-Roegen, Schmookler and Arrow will be reviewed in section 3.3, before going into detail on evolutionary growth theory in section 3.4 and on theories of growth cycles in section 3.5. Note that with respect to the notation of variables, this chapter will attempt to remain as close as possible to the original choice of symbols in the respective models (and will, of course, also give the source of any quoted equation); the list of symbols on page iv therefore mostly does not apply to this chapter.

3.1 A dynamic theory without dynamism: Harrod–Domar models and early AK models

The first attempt to formally model economic growth as a general theory were early Keynesian models, later commonly known as Harrod–Domar models or – together with several neoclassical endogenous growth models – as AK models. The term 'AK models' is derived from these later attempts to combine the Harrod–Domar class of models with the Solow–Swan class of growth theories.[8] Here, however, 'AK' refers to the early Keynesian Harrod–Domar approach, while the later neoclassical contributions are denominated (also in line with the denomination in current textbooks such as Aghion and Howitt (2009)) 'models of endogenous growth' or 'endogenous growth theory' (EGT).

The general idea is that investment follows output, usually but not necessarily linearly, since a part of total income (in turn equaling output in equilibrium) is assumed to be saved and invested. Harrod (1939), developing the first model of the class in 1939, was aware that this does not hold in general (i.e. that saving does not ex ante generally equal investment) and in fact did not intend to describe economic growth with his theory. Rather, his model was designed to be a general dynamic theory. However, he was the first to use this relation of output and investment in a lagged dynamic model[9]

$$K_t = sY_t \tag{3.1}$$

$$Y_{t+1} = F(K) \tag{3.2}$$

though in Harrod's theory investment also depended on recent (rising or decreasing) consumption, a feature that was dropped in later AK models.

An approach more explicitly focused on the analysis of economic growth was made by Domar (1946) seven years later. Basically, Domar used the same interrelation of macro-level variables, also recognizing the general solution of the dynamic system and its properties, first and foremost, infinite growth of both output and capital at constant growth rates. Following the development of the more popular and – with equal marginal utility of input factors – more neoclassical Solow–Swan model, attempts were made to combine the two classes of growth models, the most influential being the paper by Frankel (1962) These reformulations not only shaped the current textbook representations of AK models[10] but also gave rise to the later models of endogenous growth.

AK models predict exponential growth of the considered macro-level variables. The solution of the simple representation as a system of difference equations as stated above, leads to an exponential development of the variables for capital K and output Y in time with the constant growth rate s. (See Figure 3.1.) The common interpretation of AK dynamics is that only one of the input factors – capital K and labour L – is binding at one time. That is, a fixed combination of the factors is required for production, thus the amount available of one of these is the current bottleneck, while marginal changes of the other do not lead to any change in output, the production function being as depicted in Figure 3.2. It is of course a serious shortcoming in these models to be limited to persistent growth with a constant growth rate, since this is a major simplification. Cycles and regular fluctuations of growth are not accounted for. Furthermore, the model fails to include other factors. The microfoundation of the model is limited to recognizing a time lag until changes in certain parts of the economic system affect other variables. Otherwise, the model restricts itself to the macro-level, neglecting individual decisions, direct interdependence, innovations as micro-level events and institutional emergence. It is, however, true and obvious from empirical data that economic growth is indeed positive and greater than linear over long periods of time, matching the exponential pattern in the models at hand; see Figure 1.1.

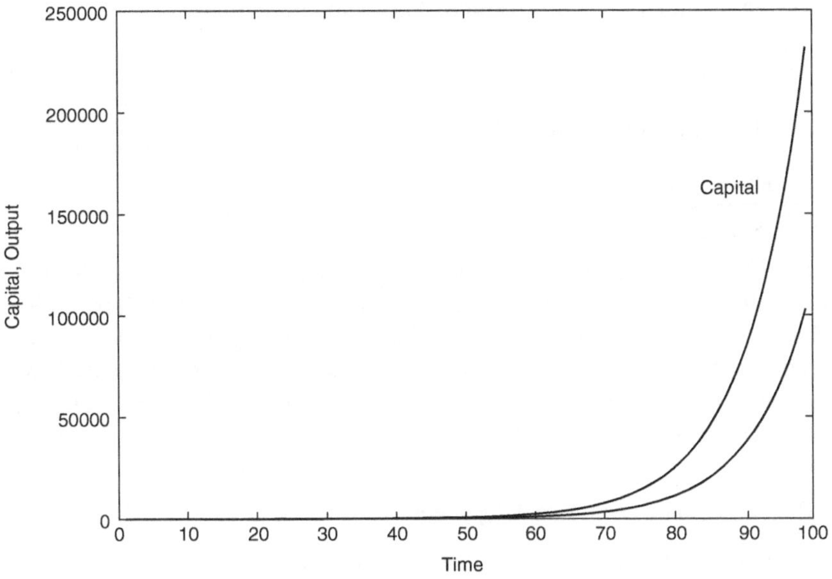

Figure 3.1 Growth of capital K and output Y in an AK model.

3.1.1 Harrod's model: a dynamic investment process

Harrod, writing in 1939, did not intend to create a theory of economic growth. His stated goal was instead to contribute to the development of a Keynesian dynamic macro-economic theory. This involved the circular dependence of overall production and capital investment, while the latter was assumed to change with the investors' expectations of future demand and revenue, Keynes' famous 'animal spirits' (Keynes, 1936). The stated interdependence yields a well-defined dynamic system with predictable behaviour, and possibly a computable stable and unique equilibrium.

Harrod (1939) originally used the formulation

$$sx = C(x_1 - x_0) + kx + K. \qquad \text{(Harrod, 1939, footnote to Eq. 2)}$$

x indicates the output, s is a parameter for normalizing the output in time (such that sx denotes the normalized output in one unit of time) x_t stands for the output at time t, C for the quantity of capital needed to produce one more unit of output, $kx+K$ for the 'long-range capital outlay' (divided into a fixed and a variable term), a concept roughly equivalent to capital stock and investment. Subtracting capital and dividing by Cx_0 yields

$$\frac{s - k - \dfrac{K}{x}}{C} = \frac{x_1 - x_0}{x_0} \qquad \text{(Harrod, 1939, 2)}$$

which is the growth rate of output. Harrod's formula may be transformed into a more convenient notation, which is more similar to later models:

$$Y = C(\Delta Y) + F(K), \tag{3.3}$$

or, resolved for the growth rate,

$$\frac{1}{C} - \frac{F(K)}{CY} = \frac{\Delta Y}{Y}. \tag{3.4}$$

Harrod explains at length that this is to be seen as a long-run dynamic equilibrium that the reality may vary from. Being a dedicated follower of Keynes, Harrod also adds that savings exceeding the needs of this equilibrium growth path are from this point of view futile, if not harmful to economic performance. Together with Domar's theory of the exponential growth of the capital stock, Harrod's theory forms the core of what became known as AK models.

3.1.2 Domar's model: exponential growth

Domar (1946) concentrated on the growth paths of the macro-economic variables assuming similar interrelations to Harrod. He defined the growth of capital as a differential equation

$$\frac{dK}{dt} = I - I\frac{s - \sigma}{s} = I\frac{\sigma}{s} \qquad \text{(Domar, 1946, 15)}$$

where σ is the potential increase in productivity, while the actual level is s, with $s > \sigma$. s is the nominal growth of capital; the productivity increase σ is realized while the remainder $s - \sigma$ is lost. Domar further assumed an investment function

$$\frac{dY}{dt} = \frac{dI}{dt}\frac{1}{\alpha} \qquad \text{(Domar, 1946, 3)}$$

where α states the Keynesian marginal propensity to invest. Next, the above consideration is formalized as $\frac{dY}{dt} = I\sigma$ yielding

$$I\sigma = \frac{dI}{dt}\frac{1}{\alpha} \qquad \text{(Domar, 1946, 6)}$$

and hence

$$\frac{dI}{dt} = \alpha\sigma I(t) \tag{3.5}$$

a differential equation with the general solution

$$I = I_0 e^{\alpha\sigma t} + \text{const.} \qquad \text{(Domar, 1946, 7)}$$

This in turn gives a capital growth path

$$K = \frac{\sigma}{s}\int Idt + \text{const} = K_0 + I_0\frac{\sigma}{s}\int_0^t e^{\alpha\sigma t}dt. \qquad \text{(Domar, 1946, 8)}$$

Y is obviously parallel to I, therefore all three significant macro-economic variables, production, capital and investment are subject to exponential growth. Domar's model thus predicts growth paths that can be slightly changed to use an appropriate production function to construct a model with the properties of what was later called the Harrod–Domar model or, more generally, the AK model (see e.g. Solow, 1956; Frankel, 1962; Aghion and Howitt, 2009; McGrattan, 1998; Mulder *et al.*, 2001), $Y = F(K) = AK$ and $\frac{dK}{dt} = sF(K) - \Delta K$. The first equation follows immediately from Domar's equations with $A = \alpha\sigma$, the second one depends on the production function. Of course the interpretation of A as technological change in terms of increasing capital productivity is not consistent with Domar's model setting, but was developed in Solow (1956) and other papers that finally gave rise to the typical textbook AK model.

3.1.3 Kalecki: discontinuous growth

Kalecki proposed a discontinuous growth function (termed 'utilization of the environment' by Kalecki, but readily identified with growth in the same paper) in 1962 (Kalecki, 1962). Advancing from Harrod's work (1939), Kalecki extended the model to cover catastrophic changes for a certain value of what he calls the 'utilization of the environment' or the 'intensity of innovations' ε. For 'normal' values of ε, there exists a unique and globally stable growth rate, and thus a very convenient equilibrium property. Reaching the critical value with increasing ε, however, the system loses its fixed points, and the dominant dynamic becomes one of ever accelerating growth. To match obvious restrictions of real economic systems,[11] Kalecki introduces an upper boundary to the 'intensity of innovations'.

Kalecki intended to show Harrod's original model to be either unstable or incapable of growth without a semi-exogenous 'utilization of the environment' factor. However, this rests on additional though convenient assumptions by Kalecki.[12] Kalecki derives his system from his reformulation of Harrod's equation[13]

$$I_{t+1} = (1-\lambda)I_t + \lambda C\Delta Y.^{14} \qquad \text{(Kalecki, 1962, 15)}$$

and his equation for the trend of output Y'

$$Y' = hI' + nY' + rK' \qquad \text{(Kalecki, 1962: p. 140 [without number])}$$

Resolving the output equation for Y' and re-substituting it into itself yields

$$Y_t = hI' + \frac{nhI' + rK'}{1-n} \qquad \text{(Kalecki, 1962, 21)}$$

which in turn is substituted into Kalecki's Harrodian investment function

$$I_{t+1} = (1-\lambda)I_t + \lambda Ch\Delta I' + \lambda C\frac{nh\Delta I' + r\Delta K'}{1-n}. \qquad \text{(Kalecki, 1962, 30)}$$

where ΔK – the increase in capital stock – is $\Delta K = I'$.

Kalecki repeats this for two alternative investment equations, one of which (the one he personally favors) depends on profits rather than output according to the above equation. All three, however, have in common that they resemble the structure

$$I_{t+1} = \alpha I_t + \beta \Delta I_t + \gamma \Delta I'_t + \delta I'_t \qquad \text{with } 1 > \alpha \geq 0, \beta, \gamma, \delta > 0$$
(Kalecki, 1962, 31)

thus

$$\frac{I_{t+1}}{K'_{t+1}} = \frac{\alpha I_t + \beta \Delta I_t + \gamma \Delta I'_t + \delta I'_t}{K'_t} \qquad \text{(Kalecki, 1962, 33)}$$

Assuming for simplicity that I moves along its trend $I = I'$ and that $\frac{I'_t}{K'_t} = \frac{\Delta I'_t}{I'_t}$ $\left(\text{and thus } \left(\frac{I'_t}{K'_t}\right)^2 = \frac{\Delta I'_t}{K'_t}\right)$ yields a quadratic equation for the growth rate $g_t = \frac{I'_{t+1}}{K'_{t+1}}$

$$g_{t+1} = \frac{I'_{t+1}}{K'_{t+1}} = (\alpha + \delta)\frac{I'_t}{K'_t} + (\beta + \gamma)\left(\frac{I'_t}{K'_t}\right)^2 \qquad \text{(Kalecki, 1962, 36)}$$

which has the fixed points $\left(\frac{I'_{t+1}}{K'_{t+1}} = \frac{I'_t}{K'_t}\right)$

$$g^*_{t+1} = \frac{I'_{t+1}}{K'_{t+1}}^* = 0 \quad \text{and } g^*_{t+1} = \frac{I'_{t+1}}{K'_{t+1}}^* = \frac{1-\alpha-\delta}{\beta+\gamma}$$
(Kalecki, 1962, 37, p.144)

In order to obtain a non-zero stable fixed point in a development equation $g_{t+1} = F(g_t)$, the function F must cross the 45°-line (the steady state line) from the left (from above). The derived function, however, passes the origin, is strictly increasing, and is strictly concave, meaning that none of the two derived steady state equilibria can be stable. In the first case, growth equals zero, the second crosses the 45°-line from below. The former is trivial, while the latter is unstable.

This, however, is changed by the introduction of a 'utilization of the environment' term ε as

$$\frac{I_{t+1}}{K'_{t+1}} = \frac{\alpha I_t + \beta \Delta I_t + \gamma \Delta I'_t + \delta I'_t + \varepsilon K'_t}{K'_t} \qquad \text{(Kalecki, 1962 37, p. 148)}$$

which results in a modified quadratic equation

$$g_{t+1} = \frac{I'_{t+1}}{K'_{t+1}} = (\alpha + \delta)\frac{I'_t}{K'_t} + (\beta + \gamma)\left(\frac{I'_t}{K'_t}\right)^2 + \varepsilon \qquad \text{(Kalecki, 1962, 40)}$$

The fixed points – not explicitly given by Kalecki – are

$$g*_{t+1} = \frac{I'_{t+1}}{K'_{t+1}}^{*} = \frac{1 - \alpha - \delta}{2\beta + 2\gamma} \pm \sqrt{\left(\frac{\alpha + \delta - 1}{2\beta + 2\gamma}\right)^2 - \frac{\varepsilon}{\beta + \gamma}} \tag{3.6}$$

These solutions exist as long as the function intersects the equilibrium line $\frac{I'_{t+1}}{K'_{t+1}} = \frac{I'_t}{K'_t} =$ (the 45°-line), that is as long as

$$\left(\frac{\alpha + \delta - 1}{2\beta + 2\gamma}\right)^2 \geq \frac{\varepsilon}{\beta + \gamma} \tag{3.7}$$

otherwise the radicand becomes negative and both solutions become complex numbers. That is, the function remains above the equilibrium line over its entire domain. In this area, the dynamics are strictly expanding, yielding ever accelerated growth. Letting ε, the intercept of the function $g_{t+1} = F(g_t)$, increase leads to a catastrophic change in the behaviour of the investment function for some critical value ε_{crit}, surpassing which the economy would grow beyond all limits. Kalecki artificially introduced a further obstacle to that growth, assuming that other factors than capital equipment (which the investment function depicts) might limit growth as well. He therefore assumes a vertical jump and a horizontal equilibrium growth path at a higher level.

Kalecki thus provided both a mathematically more elaborate and an economically more elegant variant of the Harrod–Domar model. Though the 'intensity of innovations' is not endogenous in Kalecki's model, the system describes two different dynamic regimes with a catastrophic transition. Kalecki's argument might be helpful wherever economic growth and technological change exhibit different growth regimes and there is more or less sudden changes between these regimes.

3.1.4 Recent textbook versions of the Harrod–Domar and AK models

Today, AK models seem overly simplified. They are, however, one of the few classes of growth models usually deemed worth discussing, even in rather neoclassical texts on growth theory (Barro and Sala-i-Martin, 2003; Aghion and Howitt, 2009). Also, the terms AK models and Harrod–Domar model are used more or less synonymously, in that the Harrod–Domar model is usually identified as historically the first instance of an AK model. These notations not only dominate the contemporary view on both Harrod–Domar models and AK models in general, they also provide relatively concise summaries of the functionality, results and predictions of the models of this class, although of course – given their own preferred newer growth theories – rather biased ones.

Taking the recent volume by Aghion and Howitt (2009) as an example, the Harrod–Domar model is discussed as a system that uses 'fixed technological coefficients' (Aghion and Howitt, 2009), namely

$$Y = F(L, K) = min\{BL, AK\}. \tag{3.8}$$

This model is adapted from one of the non-Cobb–Douglas cases studied by Solow in his seminal 1956 paper (Solow, 1956).[15] One of the two factors is limiting, no other mode of production but the fixed relation $\frac{K}{L} = \frac{B}{A}$ is possible. (See Figure 3.2.) This idea brings the model very close to Solow's assumptions. Obviously Aghion and Howitt intend to exemplify the working and the resulting dynamics of the Harrod–Domar approach without considering assumptions different from Solow's and Swan's models.

Departing from that setting, the exponential nature of the model presented is almost immediately obvious (first suppose $AK < BL$, i.e. $Y = AK$):

$$\dot{K} = sY - \delta K \tag{3.9}$$

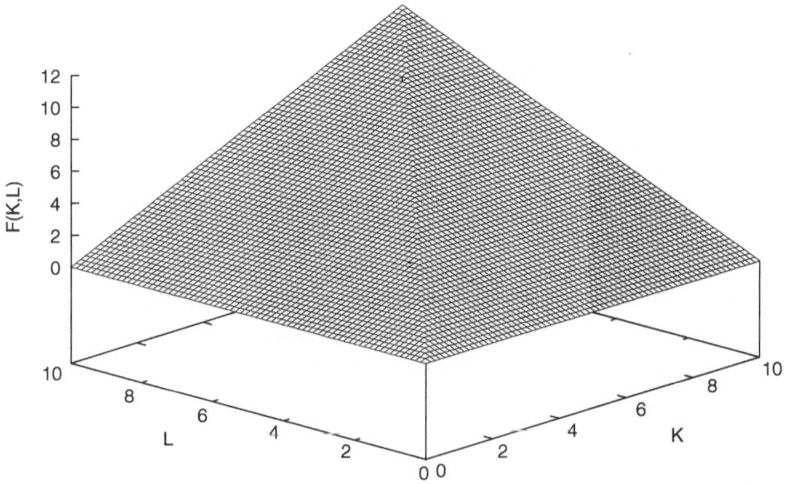

Figure 3.2 AK production function following Solow's Harrod–Domar-type example, $Y = F(L, K) = min\{BL, AK\}$.

$$\dot{K} = \frac{dK}{dt} = sAK - \delta K \tag{3.10}$$

Integration yields

$$\int \frac{1}{k}\frac{dK}{dt}dt = \int (sA - \delta)dt \tag{3.11}$$

$$K(t) = e^{(sA-\delta)t}. \tag{3.12}$$

Assuming capital growth sA to be greater than depreciation δ, this yields (for AK models) the characteristic exponential growth beyond all limits, with the constant growth rate

$$\frac{\dot{K}}{K} = sA - \delta = g. \tag{3.13}$$

If, however, AK exceeds BL, Aghion and Howitt's rendering of the model yields (whether explicitly intended by Aghion and Howitt or not) a capital growth governed by the size of the labour force which is almost Marxian in nature. With a small negative component of depreciation this growth effect converges to zero. For a constant labour force $L =$ const. this means:

$$\dot{K} = \frac{dK}{dt} = sBL - \delta K \tag{3.14}$$

$$K(t) = \frac{sBL}{\delta} + e^{-\delta t} \tag{3.15}$$

Consequently, capital converges to $\frac{sBL}{\delta}$.

If the labour force is assumed to be time dependent, for example constantly growing with rate n, hence $\dot{L} = \frac{dL}{dt} = nL$ and $L(t) = e^{nt}$ and this yields the differential equation

$$\frac{dK}{dt} = -\delta K + sBe^{nt}. \tag{3.16}$$

with the solution

$$K(t) = \frac{sBe^{nt}}{n+\delta} + e^{-\delta t}. \tag{3.17}$$

It follows that for $t \to \infty$, capital approaches a linear function of labour

$$K(t) = \frac{sB}{n+\delta}L(t). \tag{3.18}$$

3.1.5 AK models and technological change

Technology in the Harrod–Domar and subsequent AK models has largely been considered either to be constant or incorporated in the capital stock. This is how the first models, specifically those by Harrod and Domar dealt with it. Later additions add technology factors multiplying capital – and, if included explicitly, labour[16] – such that they enter the model in the same way as the propensity to invest s. Thus, in the case of technologically increasing capital productivity

$$Y_{t+1} = A_t K_t = A_t s Y_t \quad \text{with} \quad \frac{dA_t}{dt} > 0 \tag{3.19}$$

the growth rate sA_t increases over time. This yields a simple difference equation either for the capital stock K

$$\frac{dK}{dt} = sA(t)K \tag{3.20}$$

or identically for the overall production Y ($\frac{dY}{dt} = sA(t)Y$) with the solution

$$K(t) = e^{s\int A(t)dt} \tag{3.21}$$

in particular for $A(t)$ being a constant $A(t) = \text{const}$

$$K(t) = e^{sAt}. \tag{3.22}$$

Depending on the shape of $\int A(t)dt$, and assuming that for any likely case it grows faster than logarithmic, this predicts exponential growth, with the growth rate being $sA(t)$. The phase space of the growth rates of this model class therefore usually has two fixed points, one a repeller at the origin (not a stable equilibrium), the other one being the trivially stable constant growth rate, leading to divergent dynamics.

Solow, the champion of early neoclassical growth theory used to polemicize about growth models with quantities that 'exceed any stated bound before Christmas' (Solow, 1994, p. 50). He seemed to argue that increasing returns to capital for many growth models (of the AK class and the class of endogenous growth theories) led to stock variables (namely the capital stock) reaching infinity in finite time. This, however, is generally not the case – neither for earlier AK models nor for models of endogenous growth nor for other model classes – and the general property of exponential growth comes surprisingly close to a very good long-term estimation or real-world growth dynamics. (See Figures 1.1 and 1.2 for the constant-price GDP data series of Germany and the USA. The graphs appear to be – leaving out the short-term turbulence – exponential.) The model is also consistent with the long-term non-convergence that has been listed as an argument against the neoclassical Solow–Swan growth model.

Nevertheless, there are a number of issues that cannot be answered by this model or this dynamic or any common interpretation of this model: growth cycles and other turbulence in the empirical data, the complexity of the socio-economic system, as well as physical and energetic boundaries to growth. A growth regime as described by the usual early AK models is shown in Figure 3.1.[17]

To summarize, the Harrod–Domar and subsequent AK models

- predict exponential growth of the relevant macro-variables (capital and production, probably followed by others such as consumption)
- yield dynamics without attractors, just sustaining divergence governed by a stable and constant growth rate
- do not contain any mechanism to account for economic cycles or turbulence of any kind in socio-economic systems
- are unable to account for more complex structures unless they are reducible (by homogeneity assumptions) to trivial settings; there is virtually no microfoundation, no entrepreneurship, no individual incentives, etc., and no individual decision making – arguably no decision making at all before the advent of the endogenous growth theories.

Thus the representation of technological change in AK models is not sufficient; yet taking into account that at least the earlier models of this class predate the important advances in dynamic theory and game theory they are in line with other models of the time. Further, the idea of technology incorporated in stock variables such as capital opened the way for later more sophisticated descriptions of technological change, which add a better representation of the nature of innovations and the role of institutions at a micro-level, most notably Arrow's model of learning by doing (Arrow, 1962).

3.2 Neutral technical progress and endogenous growth: neoclassical and similar approaches to growth models

The earlier AK models (to the grief of some parts of the profession) did not fit neatly into the neoclassical theory of the time. Not only was it impossible to derive a stable equilibrium path, they could also not explain how an optimal combination of input factors is chosen to yield a time path of optimal modes of production. Defining all other factors of production as either not scarce or incorporated into capital, there was only one factor left, yielding a not dynamically stable time path.[18] It was also explicitly not the purpose of these models to develop another theory of equilibria and total rationality,[19] and the production function chosen was thus Keynesian, drawing on the circular relationship of income and production as well as (in Harrod's (1939) model) fluctuations of investment dynamics, while the Cobb–Douglas production function that Solow (1956) eventually used was not considered (although it was known at the time).

A true neoclassical dynamic of economic growth was finally developed by Solow (1956, 1957) in the late 1950s, and simultaneously but independently by Swan (2002 [1956]). The model Solow became famous for[20] was essentially a Cobb–Douglas function of two input factors, with one of the input factors – capital – depending (linearly) on capital and its stock being deprecated linearly as well. Since the Cobb–Douglas function is based on decreasing returns while depreciation remains linear, there is a general equilibrium if the other input factor – labour – is constant. In this case, growth (or possibly shrinking) would converge asymptotically towards this equilibrium, and thus be eventually halted. However, it is usually assumed that labour is not constant but growing at a certain rate – much like capital in AK models – thus yielding an optimal path of equilibrium modes of production as seen in Figure 3.3. The core of Solow's argument, as well as its weakest point, however, is technology. Technology is multiplied as an exogenous coefficient to one or both of the input factors. In his 1957 paper, Solow announced the idea of neutral technical change: one productivity factor increases, however leaving the marginal rate of substitution of the two factors intact, thus expanding the equilibrium in both factors.[21] Using empirical data the shortcoming of this point becomes clear. Solow (1957) estimates his model for macro-level data of the US economy in the first half of the twentieth century getting a more or less constant – but certainly not decreasing, as the model would predict with constant labour force – growth of capital, which accounts for about a half of the total output growth. The other half is introduced by exogenously raising the technology level. This result is more reminescent of the AK models – it includes technology as a (capital) productivity term and yields exponential growth. Second, the interpretation reduces the dynamics of the convergence to the Cobb–Douglas equilibrium to an unimportant feature, while the actual

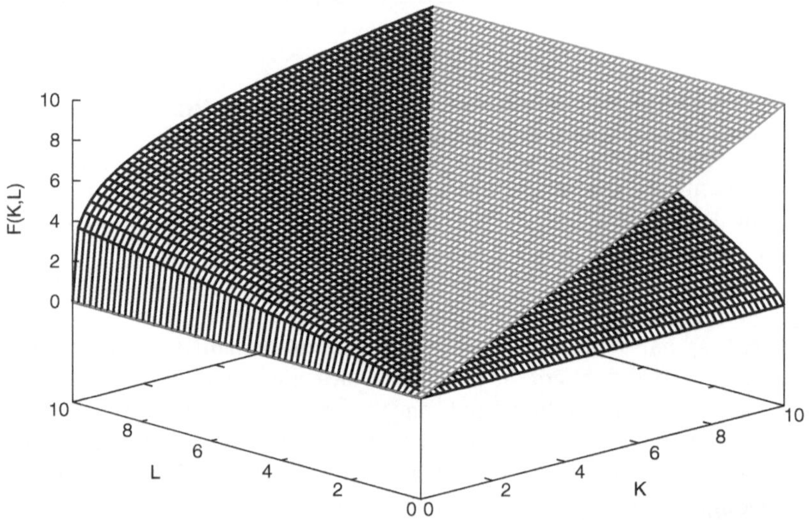

Figure 3.3 Equilibrium output and capital expansion path in the Solow–Swan model. The curved function is the Cobb–Douglas production function; the inclined plane function is the depreciation. Labour L is exogeneous (or constantly growing at an exogeneous rate). Hence, for every value of available labour a unique equilibrium exists. The intersection line of the two functions is the equilibrium expansion path.

development of the variables is governed by the technology coefficient and the subsequent reaction of other variables. And finally, the model becomes vulnerable to the criticism of just obtaining its implicit assumptions as a result and explaining very little.[22]

Starting in the 1960s, efforts have been made to cope with this problem. The solution is essentially the combination of Solow–Swan and AK approaches applying intertemporal infinite-horizon Ramsey optimization. These models are called endogenous growth theory, as they try to combine the equilibrium properties of the Solow–Swan class of models with endogenizing technological change inspired by the capital stock dynamics of the AK class of growth models, further retaining continuous growth from AK-type models, and adding a more specific microfoundation in terms of individual optimization by (more or less) homogeneous agents. Major advances in this direction include models by Cass (1965, 1966), Koopmans, Uzawa (1965), Romer (1986, 1990) and Lucas (1988) Endogenous growth theory continues to be presented as the state of the art of growth theory.[23]

3.2.1 Solow and Swan's model: neutral technical change

The idea of neutral technical change is due to Solow (1956, 1957). It is discussed as a special addition to his model of economic growth in his well-known 1956 paper (Solow, 1956); in 1957, Solow dedicated another paper to this particular issue (Solow, 1957). Neutral technical change in Solow's terms is defined as increasing the productivity without changing the shape and the properties of the isoquant system in the production possibility space, the quantities of all involved productive factors. Therefore, the isoquants are just multiplied, with a technology factor $A(t)$ being shifted upwards but retaining shape and properties. In his 1957 paper, Solow included an empirical study to underline the empirical relevance of its implications.[24]

Even more notable, however, is that Solow introduced the idea of a steady state equilibrium in economic growth. Using factors labour L and capital K (per capita $k = \frac{K}{L}$), it was derived from the interference of the degressive reproduction of the capital stock (functions F,f) per capita[25] $sF\left(\frac{K}{L}\right) = sf(k)$ and linear capital depreciation δK (and also linear population growth effects nK) (Solow, 1956). In contrast with Harrod's (1939) ultimately unstable growth pattern, such a model is inherently stable. For $sf(k) = nk + \delta K$ – the fixed point – the capital stock per capita is constant. For $sf(k) > nk + \delta K$, the capital per capita would expand, pushing the system toward the fixed point, and for $sf(k) < nk + \delta K$ the linear effect is stronger, and the system contracts, again towards the fixed point.[26]

The example for Solow's equilibrium system that was provided in Solow's original paper and has been invoked again and again is the Cobb–Douglas function

$$Y = F(K, L) = K^{\alpha} L^{1-\alpha} \qquad \text{(Solow, 1956, 1 and 7)}$$

with[27]

$$\Delta K = sF(K, L) - \delta K. \qquad \text{(Solow, 1956, 3)}$$

Up to this point, the growth model is purely macroeconomic. To compute the stable equilibrium – from the macroeconomic perspective the stable growth path – the macro-variables are divided by the size of the labour force[28] L, yielding a 'quasi-microeconomic' system of equations

$$\frac{Y}{L} = y = f(k) = \frac{K^{\alpha} L^{1-\alpha}}{L} = \left(\frac{K}{L}\right)^{\alpha} = k^{\alpha} \tag{3.23}$$

and

$$\Delta k = sf(k) - (n + \delta)k \tag{3.24}$$

As Solow notes, the microeconomic fixed point will be retained, once it is reached by the system. Furthermore, it will be approached within finite time if the system is not trapped in the unstable equilibrium point $(k, y) = (0, 0)$. Since

$$\dot{K} = \dot{Y} = \dot{y} + \dot{L} = \dot{y} + n = \dot{k} + n \tag{3.25}$$

and in the fixed points $\dot{k} = \dot{y} = 0$ the macroeconomic growth rate is

$$\dot{K} = \dot{Y} = \dot{y} + n = n \tag{3.26}$$

that is, the system follows an equilibrium growth path determined by the growth rate of the labour force n.

To include neutral technological progress, Solow applied a term representing the level of technology depending on the time t and otherwise exogenous to the system[29]

$$Y = A(t)F(K, L). \qquad \text{(Solow, 1956, 13)}$$

Solow included several possibilities for particular production functions, with technological progress exhibiting constant, diminishing or increasing returns to scale, his main example being $A(t) = e^{gt}$ (Solow, 1956, 1957). However, the representation of technological change in Solow's model remains exogenous. At the same time, Swan had the very same ideas, although published in a little-known Australian journal;[30] he also developed a growth model in per capita terms using a Cobb–Douglas function to get a stable growth path as a result. Swan even discussed technological progress in a neutral way[31] and recognized that technological progress may lead to increasing returns to scale.

The Solow–Swan model has been heavily criticised for both its prediction of convergence towards a stable equilibrium that does not fit the empirics (see

Figure 1.2), and the result that most of the actual growth is generated by raising technology levels, which in turn is exogenous to the model (Nelson and Winter, 1974). Nevertheless, the model managed to be almost universally recognized as the neoclassical benchmark for any theory of economic growth, and is presented as such in virtually every volume on growth theory (Antonelli and de Liso, 1997; Barro and Sala-i-Martin, 2003; Aghion and Howitt, 2009). Featuring 'well-behaved' (monotonically decreasing, concave upward, continuous in the first quadrant, differentiable, and integrable) neoclassical isoquants, with some supporting empirical evidence (presented by Solow himself (Solow, 1956, 1957)), and – within the neoclassical system – well-justified assumptions, the theory was at the time the best possible alternative for the neoclassical part of the profession (and probably others), later to be replaced by the endogenous growth theories discussed below. To put the results briefly, the key points of the Solow–Swan approach are:

- Convergence towards a stable equilibrium (in per capita terms) instead of growth. Assuming – as usual in these models – that the system is currently below the equilibrium, the convergence towards the fixed point may of course be equivalent to growth for the foreseeable future. Still, growth cannot be exponential as empirical data suggests – except for growth caused by the exogenous technology level.
- Lasting growth if the exogenous technological progress function develops favourably. In absolute (as opposed to per capita) terms, growth is also driven by population growth.
- The system remains within certain limits, and does not evolve beyond all boundaries – given the energy and physical boundaries this is one of the major advantages of the Solow–Swan model over AK and EGT models as addressed by Solow himself (Solow, 1994).
- Any kind of cycles, turbulence, etc., are not accounted for in the model – these must thus be interpreted as either random short term noise or as introduced by the exogenous technology term.

3.2.2 Cass, Uzawa and Lucas: endogeneous growth

Solow and Swan's model implicitly calls for the endogenization of the technological progress factor to derive an optimal intertemporal equilibrium of investments in technology. While Arrow (1962) provided a capital driven model that contained such a very simple intertemporal optimization but dismissed Solow's steady state equilibrium,[32] the task of integrating Solow's growth theory and Ramsey's intertemporal optimization was largely left to Cass (1965, 1966), Koopmans, Uzawa (1965), and later Lucas (1988) and Romer (1986).

Cass, Koopmans and Uzawa worked relatively closely together – therefore Cass' and Uzawa's formalisms shall for the purpose of the current thesis be sufficient while Koopmans' publications will be omitted. Lucas, publications much later, refined Uzawa's model.

The idea of applying a Ramsey type intertemporal optimization to an AK model in order to establish a growth theory that explains the empirically observed persistent exponential growth and persistent divergence between countries, but remains able to account for a neoclassical micro-level with homogeneous and optimizing agents, formed the core of Cass' (1965; 1966) model. Cass formulates the problem as the maximization of the general social welfare function discounted by rate δ

$$\max_{c_t} \int_0^T U(c_t)e^{-\delta t} \qquad \text{(Cass, 1966, 10)}$$

(U being a usual neoclassical monotonic utility function with decreasing returns, t being time and c_t being consumption at time t) under the constraints

$$c_t + z_t = f(k_{t-1}) \qquad \text{(Cass, 1966, 10)}$$

$$\dot{k} = z_t - \lambda k_{t-1} \qquad \text{(Cass, 1966, 10)}$$

λ being the constant decrease of per capita capital, that is depreciation and population growth combined, while k stands for capital and z for (re-)investment. The production function $f(k)$ is monotonic with decreasing returns. The dynamics of this model are equivalent to those of the textbook AK models – the simple exponential function as in the standard AK model if $\frac{z}{k}$ is constant,[33] or a more complex but still faster than linear growth path as long as $\frac{z_t}{k} - \lambda$ is not declining. The utility function in this case just serves the purpose of deriving an optimal non-trivial consumption and investment plan (such that both c_t and z_t are usually positive). Whether the social optimal investment and consumption plan (as derived in the model) is the same as the aggregate of the individual investment and consumption plans, does, however, depend on the utility and production functions $U(c)$ and $f(k)$. The utility function obviously maximizes individual utilities and aggregates them to a social level, while the production function is a macro-level function, leaving the question whether individuals really would decide accordingly, as they do not care about aggregated but rather only about their own utility while still being confronted with a macro-level production function beyond their control. That question was addressed in a slightly different model by Romer (1986) 20 years later while a number of economists discussed the problem of the aggregability of maximization problems, notably Föllmer (1974), Albin (1982), Albin and Foley (1992), Foley (1998) and (Kirman, 1997).

Furthermore, although not only affecting Cass' approach (Cass, 1965, 1966) at this point, neoclassical economics runs with unwanted side effects usually referred to as 'economic conservation laws' (Sato, 1981; Mirowski, 1989) following from the neoclassical optimization. These effects and their implausibility are discussed at length by Mirowski (1989) and Hands (1993); that problem will be taken up again at the end of this section.

In Cass' model of endogenous growth, the growth is specifically endogenous to capital, leaving open whether a higher capital, stock might be seen as an investment into better technology or whether it is just a quantitative change in the size of the machinery used for production. Following the reasoning of Arrow (1962) about technology and the effects of learning by doing (which will be considered in section 3.3.3) and recognizing the potential of Cass' methodology, Uzawa extended the model to shift the control variables of the optimization problem to incorporate the building of human capital equivalent to an increase in the technology level. Thus, in Uzawa's model technological progress is labour driven, in the sense that labour is divided into labour used directly for production L_P and a part is used for building human capital L_E (with $L = L_P + L_E$). This partition is chosen maximizing a discounted utility function

$$\int \frac{C(t)}{L(t)} e^{\delta t} dt \qquad \text{(Uzawa, 1965, 8)}$$

as is the savings rate s (the choice between consumption and investment that is also present in Cass' approach). This objective function is the same as Cass', given that Cass' per capita consumption is in this setting equal to Uzawa's ratio of total consumption and total labour force $c(t) = \frac{C(t)}{L(t)}$. The production function is similar to Cass' but contains human capital A as the driving force of technological change

$$y(t) = A(t)u(t)f\left(\frac{k(t)}{A(t)u(t)}\right) \text{ with: } u(t) = 1 - \frac{L_E(t)}{L(t)} = \frac{L_P(t)}{L(t)}. \qquad \text{(Uzawa, 1965, 16)}$$

Uzawa further uses the same AK-type development equation of capital as Cass

$$\dot{k(t)} = s(t)y(t) - \lambda k \qquad \text{(Uzawa, 1965, 14)}$$

yet complemented with an equation for A

$$\dot{A(t)} = A(t)\theta(1 - u(t)) \quad \text{with:} \quad u(t) = 1 - \frac{L_E(t)}{L(t)} = \frac{L_P(t)}{L(t)}. \qquad \text{(Uzawa, 1965, 15)}$$

$\theta(1 - u(t))$ is a degressively growing function defined for $0 < 1 - u(t) < 1$ leading to diminishing returns for the technology or human capital term A, with $u(t)$ being the share of labour in the production sector. The utility function comes down to

$$\int (1 - s(t))y(t)e^{\delta t} dt. \qquad \text{(Uzawa, 1965, 13)}$$

This setting yields two optimal equilibrium paths of growth computed by Uzawa, the exact properties of which are not important in this context; the picture of persistent growth inherited from the simpler AK models is, however, retained. Uzawa does succeed in properly endogenizing technological change into the model; however, it is notable that the human capital or technology addressed by Uzawa is fundamentally different from the term used by Arrow. Arrow points to a concept of technology as human capital, that is inherent in the capital goods, and growing in time as the labour force learns to employ the particular capital good efficiently, a process that cannot be slowed or accelerated (Arrow, 1962). Uzawa (1965), on the other hand, uses a more traditional conception that human capital may be bought for labour time; the conception is obviously meant to resemble a universal education system. This is perfectly apt for neoclassical optimization. Yet the concept fails to explain why and how the conditions for a stable, optimal equilibrium are met. Most importantly, it neglects the property observed by Arrow that at least some knowledge cannot be bought on a free market within neglectable time, but instead needs time to be developed.

Lucas (1988) continued to develop Uzawa's growth model. Using a different utility function[34]

$$\int e^{\delta t}\frac{1}{1-\sigma}(c(t)^{1-\sigma}-1)L(t)dt, \qquad \text{(Lucas, 1988, 1)}$$

a modified production function, and a slightly modified human capital or technology function

$$\dot{A(t)}=A(t)^{\zeta}\theta(1-u(t)). \qquad \text{(Lucas, 1988, 12)}$$

Uzawa's version is equivalent to this equation with $\zeta = 1$; ζ is a parameter for the linearity of returns of marginally more human capital with respect to the stock of human capital. Lucas assumes the real values to be $\zeta < 1$; that is, individuals start accumulating human capital at a relatively high level and continually reduce their efforts due to diminishing returns. However, in the long run and on an aggregated level (over the members of households), Lucas assumes the function to be linear, again $\zeta = 1$.

Lucas obtains an optimal equilibrium growth path depending on the growth of human capital $\zeta(1 - u(t))$ and the parameters of the production function β and γ. All other things being equal, growth is constant over time and between countries, thus sustaining different levels of wealth and income between countries. However, the model relies on the conditions of homogeneity and aggregability of abilities and utilities, the stability of the solution as a simplified benchmark case (perfect markets as attractor in a field of slight disturbances), and the existence of a continuous field of price-quantities and utilities. Furthermore, as mentioned before, only the most neoclassical possible representation of technology, knowledge or human capital is present in the model.

3.2.3 Romer: endogeneous growth and increasing returns

A similar model has been proposed by Romer (1986). Romer discusses evidence of increasing growth rates over very long times for both individual western countries and the leading economies in history. Consequently, increasing returns must play an important role in economic growth. He considers a firm-level production function[35]

$$F(\psi k_i, \psi K, \psi x) > F(\psi k_i, K, \psi x) = \psi F(k_i, K, x) \qquad \text{for } \psi < 1$$
(Romer, 1986: p. 1015 [without number])

where k_i is the 'knowledge stock' of firm i, K is the aggregate 'knowledge stock' and the vector x contains any other necessary input. The marginal productivity is therefore homogeneous in k_i and x, but increasing in K (and therefore also increasing in K at the macro-level).

Taking the usual Ramsey-type intertemporal optimization over an arbitrary utility function $U(c_1, c_2)$ (depending on the consumption in the two succeeding periods) that is maximized over $k \in [0, \overline{e}]$ and x subject to[36]

$$c_1 \leq \overline{e} - k$$ (Romer, 1986: p. 1016 [without number])

$$c_2 \leq F(k, K, x)$$ (Romer, 1986: p. 1016 [without number])

and

$$x \leq \overline{x}.$$ (Romer, 1986: p. 1016 [without number])

Thus, a Lagrangian[37]

$$\mathcal{L}(c_i, \mu_j, k, x) = U(c_1, c_2) + \mu_1(\overline{e} - k - c_1) + \mu_2(F(k, K, x) - c_2) + \mu_3(\overline{x} - x)$$
(Romer, 1986: p. 1017 [without number])

emerges.[38] The first-order conditions obtained from the derivatives with respect to the shadow prices $\mu_j \quad j = 1, 2, 3$ are the constraints, and derivation with respect to consumption yields

$$\frac{\partial \mathcal{L}}{\partial c_i} = \frac{\partial U}{\partial c_i} - \mu_i = 0. \tag{3.27}$$

More interestingly, maximizing with respect to x results in

$$\frac{\partial \mathcal{L}}{\partial x} = \mu_2 \frac{\partial F}{\partial x} - \mu_3 = 0 \tag{3.28}$$

and maximizing with respect to k obtains

$$\frac{\partial \mathcal{L}}{\partial k} = \mu_2 \frac{\partial F}{\partial k} - \mu_1 = 0 \tag{3.29}$$

At an aggregate level as seen by a neoclassical 'benevolent dictator' the last of these first-order conditions differs from the individual level[39]

$$\frac{\partial \mathcal{L}}{\partial k} = \mu_2 \frac{\partial F}{\partial k} + \mu_2 \frac{\partial F}{\partial K} \frac{\partial K}{\partial k} - \mu_1 = 0 \tag{3.30}$$

Thus, the marginal rate of substitution, from the first two first-order conditions derived as

$$\frac{\frac{\mathcal{L}}{\partial c_1}}{\frac{\partial \mathcal{L}}{\partial c_2}} = \frac{\mu_1}{\mu_2} \tag{3.31}$$

is different individually and also socially, since the relation $\frac{\mu_1}{\mu_2}$ is individually and socially obtained from the respective third (the different) first-order condition. Thus, the competitive investment and consumption path in Romer's model is not equivalent to the socially optimal one.

In the next step, the infinite horizon problem is constructed with a discounted (with discount rate $\delta > 0$) infinite series of utility of an infinite consumption and production plan

$$\int_0^\infty U(c(t))e^{-\delta t}dt \qquad \text{(Romer, 1986, p. 1018 [without number])}$$

for the first two constraints transformed as follows

$$c_t \leq F(k_t, K_t, x_t) - k_{t+1} \quad \forall t.$$

Romer shows that the discrepancy of the collectively and individually optimal condition still holds for the infinite horizon analysis, and proceeds to discuss the welfare implications of the derived difference between the competitive (individual) and social solution.

3.2.4 Aghion and Howitt: endogenous growth in a Schumpeterian setting

In the light of Schumpeterian thought and of attempts to apply Schumpeterian arguments to the theory of economic growth and technological change[40] (Schmookler, 1966; Nelson and Winter, 1974; Conlisk, 1989), Aghion and Howitt integrated a Schumpeterian model setting into the endogenous growth theory (Aghion and Howitt, 1992, 2009). Schumpeterian in this respect means that

1. economic growth is at least partly driven by growth of new efficient industries, and the related and simultaneous destruction of less efficient ones by creative destruction
2. research necessary for the creation of new technologies (new industries) is motivated by temporary monopoly rents available to the first firm to develop and introduce a new technology.

This Schumpeterian consideration might be classified to be itself evolutionary in nature, by evoking mechanisms similar to mutation and selection, the driving forces of genetic evolution. While this has been the preferred interpretation of the Schumpeterian argument for most of the other scholars who followed Schumpeter, Aghion and Howitt (1992) chose a more deterministic model without any evolutionary elements. The agents – in this model the firms – are homogeneous, producing a consumption good, an intermediate good and technology research. Furthermore there are three different kinds of labour, skilled, unskilled and specialized, all of which are exogenously given (no unemployment, no population growth). Unskilled labour is used to produce the consumption good. Skilled labour is divided between research and the production of the intermediate good according to the willingness of the firms to engage in research. Specialized labour is exclusively used for research, which in turn effects the efficiency of the production of the consumption good, modifying a technology parameter A. Thus if the production function for the intermediate good is a trivial one[41] ($Q_{inter} = L_{skill,inter}$), for the consumption good a production function arises as (exogenous unskilled labour normalized to unity)

$$Y = AF(L_{skill,inter}) \qquad \text{(Aghion and Howitt, 1992, 2.1)}$$

The technology parameter A in turn is an exponential function of the number of successful scientific breakthroughs b (γ being a constant)

$$A_b = A_0\gamma^b \text{with b=0,1,..,}\infty \qquad \text{(Aghion and Howitt, 1992, 2.3)}$$

where the flow of the scientific breakthroughs is distributed according to a Poisson distribution (λ being a constant and θ being a constant-returns production function)

$$b \sim Pois(\lambda\theta(L_{skill,resear}, L_{spec})). \qquad (3.32)$$

For any technology level A_b, there is a different intermediate product produced by the firm that was eventually successful in researching that technology. This firm obtains monopoly rents but will, however, lose this favourable market position as soon as a new technology is acquired. Aghion and Howitt detail many different aspects of this model, including predictions about which firm will invest in research (not the current monopolists since they currently – holding the monopoly – have not much to gain), wage effects, the effect of future expected research regarding to wage effects, etc. The fundamental working of the model, however, remains that of endogenous growth theory: there is persistent growth attuned to an intertemporal rational decision by a population of homogeneous agents, as long as the system does not fall into a stagnation trap with zero growth caused by special features or anomalies of the model.[42]

Aghion and Howitt succeeded in developing an, arguably in some aspects, neoclassical notion of Schumpeterian growth. By doing so, they underlined the flexibility of the modern theory of endogenous technological change that is capable of integrating and being combined with very different approaches. However, the fundamental flaws of the endogenous growth theory, both with respect to the lack of empirical evidence closely resembling these models, and the issues with neoclassical assumptions and methodologies (as discussed in the next section) have persisted.

3.2.5 Mathematic details and the representation of technological change and growth in endogeneous growth theories (EGT)

Just like the Solow–Swan approach, endogenous growth theories assume a production function of multiple input factors that is used to derive an optimum point. Contrary to Solow and Swan they do not, however, require decreasing returns to the input factors, but regularly assume a term for technology or human capital to yield permanent non-vanishing positive growth by shifting the temporal equilibrium. Nevertheless, a neoclassical optimization is still possible, as the agents are perfectly informed about the shape of the production function and the result of investment in research and human capital formation. Thus, notwithstanding some minor flaws (like the suboptimal collective result of individual optimization with increasing returns as shown by Romer (1986)) endogenous growth theory fits very well into the neoclassical framework, much better even than the earlier Solow–Swan model that lacked the micro-level framework of individual optimization. Furthermore, somewhat unorthodox models applying the same mechanisms such as Aghion and Howitt's Schumpeterian–neoclassic synthesis (Aghion and Howitt, 1992, 2009) show the class of models to also be quite flexible. With an endogenous growth theory, individual optimization has come to replace the exogenous technology coefficient, completely endogenizing the cause of persistent economic growth. As this class of theories constitutes the state of the art approach for the vast majority of economics scholars, and specifically the neoclassical mainstream of the discipline, it is necessary to consider both the resulting dynamics and the underlying mathematics in detail in order to apply a profound critique before proceeding to other – less neoclassical – approaches to growth theory.

'Balanced growth' in EGT

The term 'balanced' growth has been used by some scholars (e.g. Aghion and Howitt, 1992) since there is no steady state equilibrium in the original sense, while the model develops along the lines of a sequence of often unique, mostly optimal, and usually deterministic (as long as no stochastic effects are considered) intertemporal equilibria.

Generally, the class of models assumes homogeneous agents whose decisions may be (and are in fact) aggregated to yield a macro-scale optimization problem. A resource[43] is divided into a part for current consumption (or immediate production of a consumption good in the current period) C_t and an investment in future resources or future production technology I_t. The future resources are produced from the investment according to a function $F(I_t)$. This condition is used as a constraint for the derived optimization problem

$$F(I_t) - C_t - I_t \geq 0. \tag{3.33}$$

Discounted utility of consumption serves as an objective function[44]

$$max \int_0^{\infty} U(C_t)e^{-\delta_t t}dt \tag{3.34}$$

An initial resource (or initial investment) $\overline{Y_0}$ is exogenous to the problem. The optimization yields the (current-value) Hamiltonian function

$$\mathcal{H}(C, I, \mu_h, t) = e^{-\delta t}U(C_t) + \mu_h(F(I_t) - C_t - I_t) \tag{3.35}$$

or equivalently the Lagrangian

$$\mathcal{L}(C_t, I_t, \mu_t) = \sum_{t=0}^{\infty} U(C_t)e^{-\delta t} + \sum_{t=1}^{\infty} \mu_t(F(I_{t-1}) - C_t - I_t) + \mu_0(\overline{Y_0} - C_0 - I_0) \tag{3.36}$$

giving as first-order conditions

$$\frac{\partial \mathcal{L}(C_t, I_t, \mu_t)}{\partial C_t} = e^{-\delta t}\frac{\partial U(C_t)}{\partial C_t} - -\mu_t = 0 \quad \forall t \tag{3.37}$$

$$\frac{\partial \mathcal{L}(C_t, I_t, \mu_t)}{\partial I_t} = -\mu_t + \mu_{t+1}\frac{\partial F(I_t)}{\partial I_t} = 0 \quad \forall t \tag{3.38}$$

as well as the initial constraints. This leads to

$$e^{-\delta t}\frac{\partial U(C_t)}{\partial C_t} = \mu_t \quad \forall t \tag{3.39}$$

$$\frac{\mu_t}{\mu_{t+1}} = \frac{\partial F(I_t)}{\partial I_t} \quad \forall t \tag{3.40}$$

and consequently

$$e^{\delta}\frac{\frac{\partial U(C_t)}{\partial C_t}}{\frac{\partial U(C_{t+1})}{\partial C_{t+1}}} = \frac{\partial F(I_t)}{\partial I_t} \quad \forall t \tag{3.41}$$

Finally the transversality conditions must provide that $lim_{t\to\infty}\mu_t I_t = 0$ or (from equation (3.39))

$$lim_{t\to\infty}e^{-\delta t}\frac{\partial U(C_t)}{\partial C_t}I_t = 0, \tag{3.42}$$

meaning that either there are no more investments ($I_{\infty} = 0$) or the utility is maximized by the investments($\mu_t = 0$), i.e. the present value of I_{∞} is zero.

Optimizing using the Hamiltonian yields the same results as will be given in the next section. Before focusing on that, and detailing a general critique of the neoclassical growth theory, the interpretation of the general result from equation (3.41) shall be discussed in the rest of this section. The method requires that the utility function $U(\cdot)$ exhibit decreasing returns $\frac{\partial^2 U(\cdot)}{\partial^2} < 0$; since the optimization does not have an interior solution for increasing return utility functions[45] and for constant return utility functions, the condition from equation (3.41) would yield a contradiction.[46] For regular decreasing returns utility functions, however, the development of the system can be observed from the equation (3.41)

$$\frac{\frac{\partial U(C_t)}{\partial C_t}}{\frac{\partial U(C_{t+1})}{\partial C_{t+1}}} = \frac{\frac{\partial F(I_t)}{\partial I_t}}{e^{\delta}} \tag{3.43}$$

since

$$\frac{\frac{\partial U(C_t)}{\partial C_t}}{\frac{\partial U(C_{t+1})}{\partial C_{t+1}}} \begin{cases} > 1 \Leftrightarrow & C_1 < C_2 \quad \text{C growing} \\ = 1 \Leftrightarrow & C_1 = C_2 \quad \text{C constant} \\ < 1 \Leftrightarrow & C_1 > C_2 \quad \text{C decreasing} \end{cases} \tag{3.44}$$

Where the first case (consumption grows and everything else grows in order for that growing consumption to be maintained) is equivalent to

$$\frac{\partial F(I_t)}{\partial I_t} > \frac{1}{e^{\delta}}. \tag{3.45}$$

This in turn is generally true for production functions with $\frac{\partial^2 F(I_t)}{\partial^2 I_t} > 0$ (increasing returns) and large values of I. Since the equations must be valid for the macro-level, an arbitrarily large I can be chosen. If the function is not homogeneous for small values of I, this simply means that there are multiple balanced trajectories, and only the most important one (the one for large Is) will be considered here. This is also case the deemed most likely by most scholars and analyzed in detail by Romer (1986). The case of general growth as given by equation (3.45) remains true for constant returns $\left(\frac{\partial^2 F(I_t)}{\partial^2 I_t} > 0\right)$ greater than $\frac{\partial F(I_t)}{\partial I_t} > \frac{1}{e^{\delta}}$. Constant return production functions smaller than that value and production functions with decreasing returns yield a contracting trajectory for the system's development. Thus, the conclusion that EGT models generally predict growth at the same or faster rates than their predecessors, the Harrod–Domar and early AK models, is justified. (See Figure 3.1.)

Critique of the neoclassical mathematics of growth

Returning to the neoclassical field theory of the economic equilibrium and Mirowski's (1989) criticism as discussed in section 2.1, this is equally or maybe especially true for neoclassical theories of economic growth. The Hamiltonian of a simple neoclassical model of endogenous growth as discussed in the previous section is (equation (3.35))[47]

$$\mathcal{H}(C,I,\mu_h,t)=e^{-\delta t}U(C_t)+\mu_h(F(I_{t-1})-C_t-I_t)$$

with conditions

$$\frac{\partial\mathcal{H}(C,I,\mu_h,t)}{\partial I}=-\frac{d\mu_h}{dt} \tag{3.46}$$

$$\frac{\partial\mathcal{H}(C,I,\mu_h,t)}{\partial C}=0 \;\Rightarrow\; e^{-\delta t}\frac{\partial U(C)}{\partial C}-\mu_h=0 \tag{3.47}$$

In the previous section the solution, further conditions for its validness and optimality (the transversality condition), as well as the interpretation of the results have been discussed using the equivalent Lagrangian method. The derived solutions are identical.

In order to have a unique equilibrium, the economic conservation law must also hold for such models. In effect, this means that the value $\mathcal{H}(C,I,\mu_h,t)$ of the Hamiltonian given above must be preserved under all transformations. (For details see section 2.1.)

Neoclassical economics is a theory of a scalar field of utility, which under certain additional required conditions, hopefully forms a differentiable utility function with a corresponding conservative vector field of the gradients of this function, by assumption equal to the prices. It is somewhat heroic to assume the existence of an appropriately shaped utility function, especially since it has to be derived by integration of not necessarily integrable conditions of substitution, that is, the isoclines of the utility function, e.g. the indifference curves in standard consumer theory (Mirowski, 1989; Hands, 1993, 2006). Furthermore, if the neoclassical equilibrium price that corresponds to the reservation prices of the last producer and the last consumer (the marginal utility and the cost of all producers and consumers) does not exist, this convention does, of course, not hold. There are a number of good reasons why this may be the case; among others from economic psychology the systematic bias of human decisions (Tversky and Kahneman, 1972; Kahneman and Tversky, 1973), falling marginal prices (Keen, 2009b), lagged price adjustment (Wheat, 2009), and nonlinear supply and demand functions caused by transaction costs, or income or price effects (Chen, 2005). Finally, the theory comes with unwanted side-effects. As assessed above, the optimization fulfills the law of the conservation of energy of the mechanical physics of the nineteenth century. The economic analogy in this case is a construction derived from additively connecting utility (potential energy) and expenditure (kinetic energy) that is rarely addressed, as Mirowski's (1989) detailed analysis of the history of the theory reveals, and not in general justified. It is possible to argue that any utility is also a potential source of expenditure (in order for the neoclassical agent to obtain this utility). However, this requires specifically shaped utility functions, complete knowledge, and is even then a heroically simplistic interpretation of economic behaviour (Mirowski, 1989).

Other shortcomings of the neoclassical growth theory as well as other variants of neoclassical perception of growth, specifically those related to growth cycles, will be discussed below in the section on growth cycles. Of course, these models are, though neoclassical in their assumptions, rather different from the class of growth models discussed in this section and are known as EGT models. Notwithstanding the sharp criticism of neoclassical models of endogenous growth as detailed above, the richness of models using this approach as well as their power to provide a simple theory of a system with roughly realistic macro-behaviour, yet including a micro-level, must be recognized. Attempts have been made to eradicate flaws such as the assumed homogeneity and perfect rationality and information of micro-level agents (not just in growth models, but from neoclassical economics in general) consequently making the micro-level (now non-homogeneous) much more complex. For examples see Föllmer (1974), Albin (1982), Kirman (1997) and Foley (1998). But, this has been found to be overly complex (Albin, 1982; Albin and Foley, 1992; Foley, 1998; Kirman, 1997; Keen, 2001) if theoretically possible at all. Neoclassical utility theory in general (Foley, 2010) alongside with other assumptions regarding price behaviour (Keen, 2009b; Wheat, 2009) as well as the role of space, distance and geography (Williamson, 1973; David, 2005) would have to be remodelled completely in a much more complex way. Before discussing the most manifest contrasting neoclassical endogenous growth models – the evolutionary growth theory – the next section will draw attention to some different models that added important new ideas and preceded the dawn of the evolutionary growth theory.

3.3 Empirics, final frontiers and the learning of technologies: some fundamentally different aspects

The approaches discussed in this section do not necessarily form a class of growth models with common properties, but they are similar in providing rigid observations of economic reality and drawing proper conclusions without trying to match the form of the standard models usual at the time. These models, being more radical in their changes to standard AK and EGT models, not only added some important empirical considerations, but shaped many of the ideas that subsequently became important for the subsequent evolutionary theory. This includes the empirically stable relations described by Verdoorn and Thirlwall (1980, 1986) as well as the consideration of demand-side effects on growth processes (and consequently also on investments in research and technological change) (Schmookler, 1966; Dixon and Thirlwall, 1975). This section also includes Arrow's introduction of 'human capital' in the form of a vintage learning model (learning by doing with depreciation); the idea of a connection between learning or 'human capital' and economic growth has proven pathbreaking: similar effects were considered by most of the later endogenous growth and evolutionary growth models. And finally it includes Georgescu-Roegen's (1975b) radical account of the interdependence of growth and manifest energetic and physical boundaries that unequivocally leads to the question whether the growing economy of the last decades and centuries is a giant Ponzi scheme – an idea that might in turn concur with Marx's nineteenth century account of the socio-economic system (Marx, 1963 [1885]).

3.3.1 Verdoorn's law, Thirlwall's law and the empirical approach to economic growth

While models giving exhaustive explanations of economic growth and technological change as a whole have not yet established many undisputed facts, some facts have become empirically apparent. Apart from the initially stated three facts (persisting growth, approximately regular cycles and lasting differences in both level and dynamics of growth between regions and sectors), two things that have come to be known as Verdoorn's Law and Thirlwall's Law are relatively well established (Thirlwall, 1980, 1986).

Verdoorn's law[48] (originally published by Verdoorn in Italian in 1949) states that the relation of the growth rates of labour (or labour productivity) $\dot{L}$ and output $\dot{Y}$ is approximately constant

$$\frac{1}{1-V} = \frac{\dot{Y}}{\dot{L}} \tag{3.48}$$

where V is the Verdoorn coefficient. Further, if a Cobb–Douglas or similar production function is implied (yielding $\dot{Y} = \alpha\dot{L} + \beta\dot{K}$), it follows that

$$\frac{1}{1-V} = \frac{\dot{Y}}{\dot{L}} = \alpha + \beta\frac{\dot{K}}{\dot{L}} \tag{3.49}$$

where α and β are labour and capital productivity respectively, and thus the technology coefficients of a Solow–Swan type model. It is estimated that $V \approx 0.5$, thus $2 = \alpha + \beta\frac{\dot{K}}{\dot{L}}$(Thirlwall, 1980).

Thirlwall's Law[49] states that the relation of export growth $\dot{Ex}$ (depending on the relative price levels and the income development of the rest of the world) and the elasticity of import demands π (which depends on the relative price levels and the price elasticity of demand) is approximately equal to the output growth

$$\dot{Y} = g \approx \frac{\dot{Ex}}{\dot{\pi}}. \qquad \text{(Thirlwall and Hussain, 1982, 7)}$$

Thirlwall and Hussain found strong evidence for this relation, incorporating also outstanding debts (the share of exports and receipts) and capital flows (Thirlwall and Hussain, 1982).

3.3.2 Marx and Georgescu-Roegen: unsustainable growth

Georgescu-Roegen (1975a,b) is one of the very few economists who took a fundamentally sceptical stand on the matter of economic growth. Being an ecological economist – which was more remarkable in the 1970s than it is today – Georgescu-Roegen stressed that on the one hand the necessity of permanent growth had never been sufficiently established, while on the other hand infinite growth is not possible. It confronts at some point physical and energetic boundaries that cannot be transgressed – the available energy from renewable sources (per timespan) and from non-renewable sources as well as the available labour and space is computable and finite. And so is the amount of waste products and pollution the earth (or the growing machine of production for that matter) can sustain without taking irreversible damage.

Though constantly denouncing the Marxian economists 'obsession' (Georgescu-Roegen, 1975b) with progress, Georgescu-Roegen's idea draws on Marx's idea. It is beyond the scope of the current study to discuss older ideas of economic growth, or specifically the theory of value of the nineteenth century. Still, as Georgescu-Roegen's argument is not particularly detailed at this point, the historical root of the idea shall briefly be discussed. It was Marx's observation that any production is accomplished by human labour, including capital goods used for further production. Neglecting non-linearities, specifically economies of scale, economies of scope and successful productivity increases through research (also a form of labour), Marx dismisses capital as a factor of production altogether and obtains a system with an entirely imaginary financial sector. If economic growth is measured using the monetary value of capital goods present, growth as well as perceived technological progress will also not be real, and the system will – as soon as this becomes obvious – collapse. Taking into account true technological change relying on research, which consumes large amounts of capital goods long before any results are to be expected, the system becomes vastly more complex and the argument as established by Marx may or may not continue to hold. What remains true, however, are the following two arguments applied to this idea by Georgescu-Roegen:

1. The amount of the available labour is finite and that labour and capital are not boundlessly substitutable (Georgescu-Roegen, 1975a).
2. The productivity of capital, hence in Marxian terms the productivity of technology used to increase labour productivity, is bounded by the amount of energy available (Georgescu-Roegen, 1975b).

Georgescu-Roegen thus not only concurs with Marx that the economic system is not sustainable and is bound for eventual collapse, he argues that economic growth is actually real but at the same time naturally limited (thus the system will eventually reach its limits), and even harmful in that it creates waste products and damages the environment. Others made similar arguments in favour of ideas similar to Marx, Georgescu-Roegen, or both, and recommending a 'non-growing' economic system, notably ecological economists (Daly, 1968) and neo-Marxians (Smith, 2010).

The Marxian case

To reformulate Marx's scheme of extended reproduction (Marx, 1963 [1885]), capital K_t is assumed to be invested in constant capital C or variable capital V (which virtually is labour). Returns to capital depend on the surplus rate p leading to a cash surplus of labour of $M = pV$ (thus $p = \frac{M}{V}$), hence generating a capital stock of $K_{t+1} = C + V + M = C + V + pV$ in the second period. The growth rate of labour $\frac{\dot{V}}{V}$ is considered relatively small compared to the growth rate of capital $\frac{\dot{K}}{K}$. Marx argues the growth rate of labour is close to zero, which for most countries is true at least as soon as the demographic transition[50] is complete. Letting the growth rate of labour be zero (labour is thus constant at $V_t = V_0$), Marx's dynamics of capital growth by $M = pV$ in every period yield

$$\frac{dK}{dt} = pV_0 \tag{3.50}$$

$$\int \frac{dK}{dt} dt = \int pV_0 dt \tag{3.51}$$

that resolves to

$$K(t) = pV_0 t, \tag{3.52}$$

which is a linear function of time t and thus growing towards infinity. The problem with this 'growth' is that it does not constitute genuine growth in the sense of technological, social or organizational progress, or qualitative or quantitative improvement. It is according to Marx's theory of surplus value a purely imaginary growth. Capital assets grow, but the actual productive capacity owned by these capital assets depends only on the labour force as labour is the only really productive input. Applying the laws of later modern macroeconomics, with perfect information, this should be purely monetaristic and should lead to an immediate devaluation of the capital assets. But, if information is imperfect and the agents are not informed about the fact that capital is in fact not productive and only constitutes the means of employing labour, the bubble will only be detected when the system finally collapses.

It should be pointed out that Marx probably erred in stating that capital does not matter at all for production. In Marx's own terms, capital is ownership of 'crystalized' past labour, which (and the structured, organized employment of which – taking the form or research, human capital, etc.) is crucial for technological progress. It might thus be deemed insufficient to reduce capital to the aspect of some special form of past labour. However, the argument is both disturbing and impressive. Marx explains that due to falling profit rates $\left(\frac{M_t}{C_t+V_t} = \frac{V_t}{K_t}\right)$ will eventually trigger the collapse of the whole system. This may be delayed by the expansion of the whole system, thus increasing the demand – here Marx is inconsistent, since his technical model as discussed above is not about demand but about production, productivity and ownership – but collapse will eventually take place. If it is indeed not recognized that the socio-economic system is not sustainable, there is no way to anticipate its eventual collapse which inevitably is brought about by its non-sustainability. Georgescu-Roegen developed a more modern version – taking into account ecological considerations as well as the physics of energy – of an interpretation of the economic system that has the same properties as Marx's nineteenth century model.

A Marxian system with growing labour force

Marx's model fitted with a non-zero constant growth rate of the labour force – a case that was not considered by Marx – leads to an exponential growth of the capital stock driven by the growth of the labour force. Thus, capital, output (following the development of the capital stock), and labour grow exponentially (with a constant growth rate). This case corresponds to Verdoorn's Law which was discussed in the previous section. While Verdoorn's Law predicts ($C_{Verdoorn}$ being a modified version of the Verdoorn coefficient indicating the linear relation of labour and capital)

$$L(t) = e^{\beta t}$$
$$K(t) = C_{Verdoorn} L(t) = C_{Verdoorn} e^{\beta t} \tag{3.53}$$

for Marxian capital growth, i.e.

$$\frac{dK}{dt} = pL(t) = pe^{\beta t} \tag{3.54}$$

which yields

$$\int 1 dK = \int pe^{\beta t} dt \tag{3.55}$$

$$K(t) = \frac{p}{\beta} e^{\beta t} \tag{3.56}$$

the system becomes

$$L(t) = e^{\beta t}$$
$$K(t) = \frac{p}{\beta} e^{\beta t} \tag{3.57}$$

where $\frac{p}{\beta}$ may be interpreted as a factor equivalent to the Verdoorn coefficient.

Georgescu-Roegen's theory

Georgescu-Roegen (1975b) develops an argument that leads to similar unsustainable dynamics, starting, however, not from the nature of labour and value but rather from the second law of thermodynamics. Well-established thermodynamic theory includes that the entropy – a function of the density of microstates of a physical system[51] (the higher the density of microstates, the higher the entropy) – tends to increase over time in closed systems. To locally counteract this process, energy is required. Energy may be interpreted as negative entropy and the process of collecting energy increases the entropy in other parts of the system. As the energy available to us is finite – for fossil energy sources absolutely, for renewable sources the quantity available within a certain timespan – the capacity of the economy to produce, to shape, to decrease entropy is finite as well. Though humanity uses only a tiny part of the theoretically available energy, the process depends also on the technological possibility and (and the physical and energetic requirements) of the exploitation of a certain energy source. Concluding that at some point, the cost of energy production increases in time, the productive capacity of an economy decreases. If the economy is closed (e.g. encompassing the whole world), this capacity to produce is the only source of revenue which in turn is the (discounted) current value of capital goods; its decay leads to a shrinking economy not only in terms of physical capacity but also in terms of capital – which will arguably give rise to turbulence or crisis if the mode of the economic system is not radically altered to match the energetic properties of the environment (Georgescu-Roegen's preferred suggestion).

Though Georgescu-Roegen did not formulate this explicitly, his theory leads to the suggestion that the requirements to produce a good should not be measured as its value, but rather the effort necessary to reacquire this good once it is consumed or decayed. This in turn is a usual public good problem including the known problems of free riding, moral hazard, etc., and, for 'rational' or 'heuristic rational' decision making without any countering mechanisms, results in the universal exploitation of the system and its eventual crash.

3.3.3 Arrow: learning by doing

Another influential idea was the description of learning by doing by Arrow (1962). At the time, few growth theories considered technology more than a possibility of interpretation of certain changes equivalent to others,[52] such as the quantity of capital goods. Recognizing the great importance of general knowledge of a technology in order to make use of it (as opposed to just having someone develop the technology), Arrow's idea introduced both human capital and a notion of technology with specific properties unique to it.

Specifically, Arrow uses a vintage model, separating the production into a number of plants of different quality, different productivity and different labour requirement. The older plants are out of use; only the most effective ones continue to be used.

Further a production function is assumed of

$$x = \Gamma(G) - \Gamma(\Lambda^{-1}(\Lambda(G) - L)) \qquad \text{(Arrow, 1962, 6)}$$

where L is, as usual, labour, G are capital goods, therefore more or less equivalent to the usual K, $\Lambda(G) = \int \lambda(G)dG$ is the aggregate learning function for the capital goods G, and $\Gamma(G) = \int \gamma(G)dG$ is the productivity (output capacity) of the capital goods G. The production function is derived from

$$(Y =)x = \Gamma(G) - \Gamma(G') = \int \gamma(G)dG - \int \gamma(G')dG = \int_{G'}^{G} \gamma(G)dG \qquad \text{(Arrow, 1962, 1')}$$

$$L = \Lambda(G) - \Lambda(G') = \int \lambda(G)dG - \int \lambda(G')dG = \int_{G'}^{G} \lambda(G)dG. \qquad \text{(Arrow, 1962, 2')}$$

Learning by doing is represented in this system by adding up the learning effects for the respective capital goods (the integrals) and subtracting the knowledge corresponding to the capital goods that are out of use G'.[53]

While the concept of human capital has been taken up by many of the later growth models (Cass, 1965, 1966; Uzawa, 1965; Romer, 1986; Lucas, 1988; Aghion and Howitt, 1992; Silverberg and Lehnert, 1993), Arrow's formulation of technological progress is, together with other vintage models, more or less isolated; although it has been widely recognized as a seminal work, first and foremost for its analysis of learning and knowledge as an effect (human capital) producing increasing returns and, hence, inherent instability (Antonelli, 2005, 2009; León-Ledesma, 2000). Later models chose to model human capital in a more controllable way in order to use it for optimization (the EGT approach as used from Cass through Lucas) or as a (sometimes stochastic) ad hoc term as a basis for evolutionary processes (Nelson and Winter, 1974; Conlisk, 1989; Aghion and Howitt, 1992; Silverberg and Lehnert, 1993). Of course, they are unable to retain a realistic model of the learning by doing as discussed by Arrow.

3.3.4 Schmookler: demand-side effects

Schmookler in 1966 was the first to call for an appropriate representation of demand-side effects in the theory of innovation. He showed empirically that research and innovation may be driven by demand. Increasing demand tends to lead to a boost in innovations in the relevant area (and vice versa) in the data Schmookler used. Certainly this also has a strong though indirect effect on economic growth. Until this time, virtually all models of economic growth relied heavily on production functions, and thus supply-side factors and for the analysis of technological change 'technology push' arguments. The approach Schmookler proposed was to 'reverse Say's Law': since the output of the economy must match the total income in equilibrium – according to Say's law income and demand is specifically generated by production. Hence, production generally does not exceed demand or vice versa. The reverse of Say's law will thus also be possible: if demand-pull effects are the stronger socio-economic forces, the production side will adapt and – considering general equilibrium properties – the prices will change accordingly. Schmookler did not argue that economic development, growth and technological progress in general were dominated by demand-side effects, merely that demand-side effects play an important role, which certainly seems Keynesian though Schmookler was probably much closer to Schumpeter than to Keynes.

The idea was taken up by: (1) neo-Schumpeterian evolutionary economists who use the demand side as a device of selection in an evolutionary model, e.g. Nelson and Winter (1982a); and (2) the (also neo-Schumpeterian) theory of long waves (Freeman and Perez, 1988), that finds the difference between the demand and the supply side to be the cause of catastrophic crises in the economic system. Some growth theorists from other traditions of thought also moved to apply demand-side factors, among others Kaldor (1972) and Dixon and Thirlwall (1975). All three schools have been criticized for the lack of more realistic mechanisms of the interrelation of demand-pull and technology-push factors, as well as the lack of empirical backing in their theories. For both an overview of the history of this school of thought and an example of heavy criticism of the neo-Schumpeterian[54] lines of reasoning, see Liagouras (2008).

3.3.5 Technological change in the light of empirical observations and the fundamental critique of the standard theory of economic growth

In this section, a number of approaches to economic growth have been presented that are fundamentally different from the orthodox (and mostly neoclassical) theory of the twentieth century. There are the empirical relations encountered by Verdoorn and Thirlwall; further, there are the writings by Arrow and Schmookler hinting at generally neglected patterns in the growth and development of real economic systems, the impact of knowledge and the effect of demand. Finally, there are 'radicals' and ecological economists who insist on discussing objective boundaries to growth, accusing the mainstream of carelessly neglecting these. In fact, Marx developed a philosophy of value, production, and capital in the nineteenth century that deemed only 'living labour' productive, which leads to the conclusion that most of economic growth – measured in capital growth[55] – is purely imaginary. Though there are good reasons to consider capital as a productive factor of it's own right (as discussed above), measuring the impact of productive factors on economic growth empirically is justified. Again, neglecting the attempts by Thirlwall and others to apply findings from Solowian growth theory to Verdoorn's Law, this empirical pattern (Verdoorn's Law) clearly implies that the relation of the growth rates of output and labour remains roughly constant – in other words the growth of production may be sufficiently well approximated by the growth of the labour force alone. In this way, Verdoorn's Law may be read as a late confirmation of Marx's theory.

Notwithstanding this consideration (which implies a 'true' growth process), there are also good reasons to reject the assumption of infinite growth. Among these are Marx and Marxians who argue that 'constant capital' is not productive and that, though the means of production (that is the constant capital) are real, the system is bound for crisis and growth is not sustainable. A newer argument that leads to the same conclusion is Georgescu-Roegen's discussion of the energetic boundaries of growth. Georgescu-Roegen is even very explicit in his accusations that mainstream economics deliberately uses a false and misleading system of accounting of economic growth (Georgescu-Roegen, 1975a) – an accusation that was only implicitly made by Marx. Solow was more careful with respect to the boundaries of growth than the proponents of AK and endogenous growth theories. He famously slammed the AK models and the endogenous growth theories in the tradition of the AK theory that 'exceed[ed] all boundaries before Christmas' (Solow, 1994: p. 50); still, the model he and Swan offered in the mid twentieth century resulted in a third dynamic (neither infinite growth nor no growth at all) with other disadvantages.[56] Figures 3.4 and 3.5 show the three concurring classes of theories of economic growth, one (3.4a) growing exponentially (AK and EGT), one (3.4b) converging (Solow–Swan) and one (3.5a) without any real growth. Note that the scale is logarithmic and the values are absolute, not per capita, thus the converging regime grows with the exogenous population growth rate. The most interesting part is the development of the capital–output rate: for AK–EGT-type models, capital and labour are strictly proportional, and in the converging (Solow–Swan) case, there are some minor adjustments as the system approaches the equilibrium in per capita terms (the equilibrium expansion path in absolute terms). In the Marxian case, output is bound to the labour force, capital however, is not. This leads to a falling output-to-capital ratio, which is generally a bad sign as it suggests an almost purely virtual growth and thus a falling real value of capital (i.e. price level). The same is true if not the labour force is not the unacknowledged limiting term that drags down real growth, but rather energy, etc., as in Georgescu-Roegen's analysis are.

Schmookler on the other hand hinted at the impact of then commonly neglected demand-side effects on innovation (and thus also on growth), calling for a more complex analysis considering both the demand and supply sides. Keynesian theory – including the model by Harrod (1939) – had been considering changes in investment and consumption behaviours, the famous Keynesian 'animal spirits' (Keynes, 1936), but growth theory had been largely confined to discussing growth as a change in the capacity of production, hence the supply side. It remained largely for the evolutionary growth theory first developed by Nelson and Winter (1974) and Nelson *et al.* (1976) (which will be discussed in the next section) to analyze the complex interdependence of demand- and supply-side effects on the micro-level.

Finally, Arrow (1962) was among the first to integrate radically new ideas on aspects of technological change in mainstream economic theory with a formal model (e.g. learning effects possibly implying increasing returns). Of course, 'classics' like Smith and Ricardo, the older institutionalists like Veblen, the first radicals like Marx and especially Schumpeter and his followers had a very clear

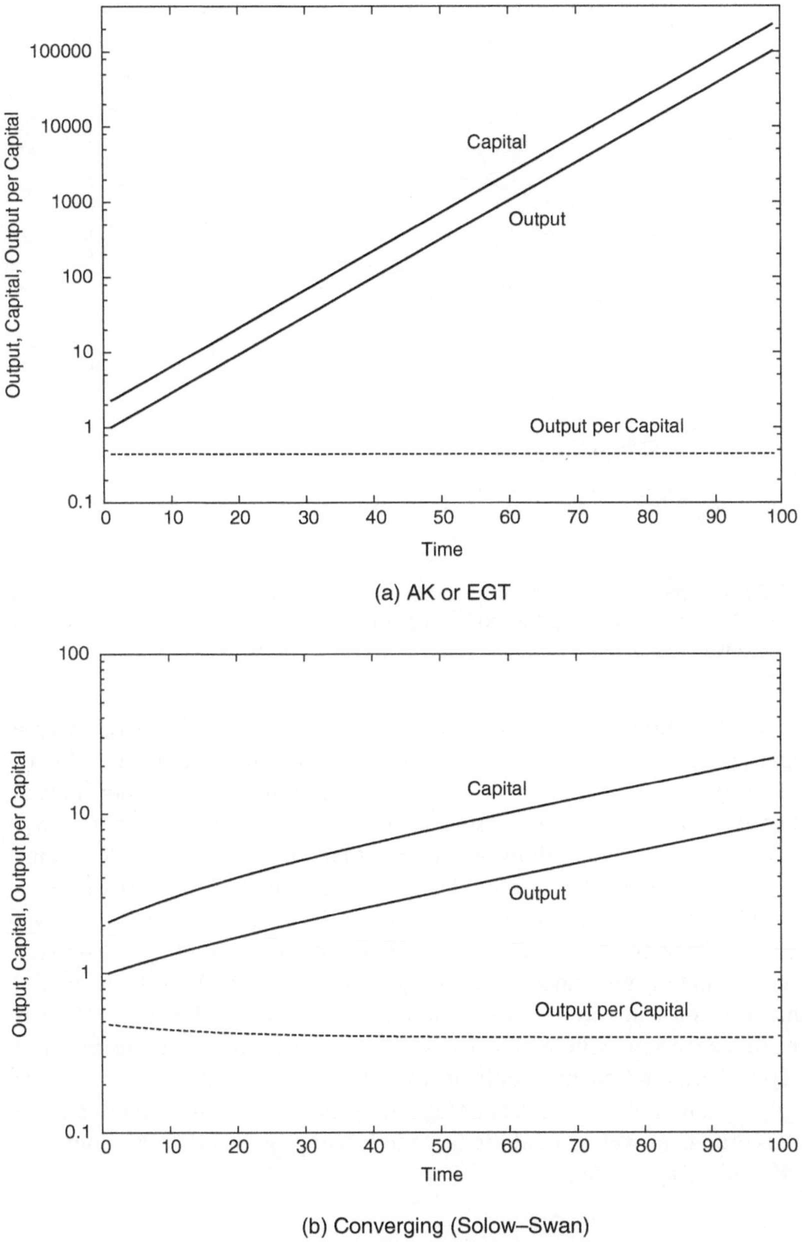

Figure 3.4 Growth regimes and their implications for the development of capital, output and the output per capital ratio (AK/EGT and Solow–Swan/converging).

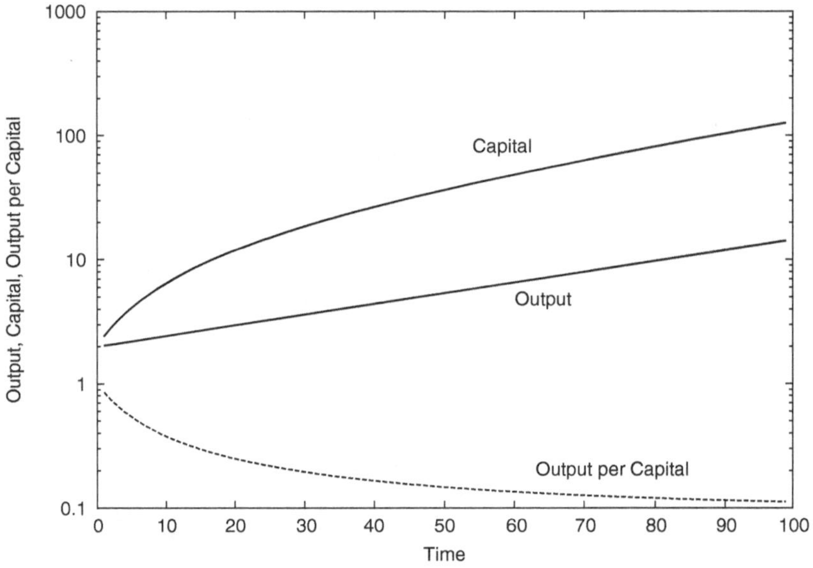

Figure 3.5 Growth regimes and their implications for the development of capital, output and the output per capital ratio (proportional to the rate of profit) in Marxian models.

idea of how technological progress works; what they lacked, however, was a formal approach and consequently a clear idea of how to incorporate a notion of technological progress into the theory of growth, and the empirical and formal approaches to economics in general. Solow (1956, 1957) as well as Kaldor (1957) included terms for the 'technology level' into their models; nevertheless, they were little more than residuals from theories of growth that were focused on macro variables like the capital stock and labour force. This is what made Arrow's model special. Drawing on a vintage approach to economic growth, he discussed the nature of human capital and of technological change as well as the necessity of learning, the construction of human capital for technological change. He was among the first economic scholars to discuss human capital and hence, as a central feature of growth, a relation that would be respected by most of the later influential theories of economic growth or technological progress, including the theory of endogenous growth as well as evolutionary models of growth and technology, the topic of the following section.

3.4 Evolutionary models: the nature of innovation and technology

The central claim of the neoclassical equilibrium theories is that there is a global attractor, that is a stable equilibrium or a stable trajectory. The central problem is to find that attractor and to prove its stability. Some models also aim to compare the stable path to the optimal one (e.g. the model by Romer (1986)). Modern neoclassical models go further and consider situations with several, and consequently non-unique, attractors (Sato, 1981). The neoclassical approach has been heavily criticised for its overwhelming inflexibility (Nelson and Winter, 1974; Kaldor, 1972, 1975; Mirowski, 1989; Foley, 1998; Chen, 2002; Antonelli and de Liso, 1997; Wheat, 2009; Keen, 2009b; Berger, 2009a; Foley, 2010). Notwithstanding the mathematical beauty of equilibrium models, they lack a justification for some or even all of their elements (as discussed above). As Winter *et al.* (2003) summarize, agents (firms, individuals) are in fact heterogeneous, and therefore, entry and exit matters,[57] heterogenity, complex interdependencies, skewed distributions and turbulence are empirically observable. And so is the fact that emerging and mature industries differ (Winter *et al.*, 2003).

Evolutionary models are a more or less direct answer to the persisting neoclassical incapability of including realistic microfoundations. As neoclassical models largely employ physical methods, and might be said to have more or less consciously copied them alongside physical models and metaphors that do not have meaningful economic interpretations (Mirowski, 1989), evolutionary economic theory tries to take advantage of methods and results from evolutionary biology. More concretely, the process observed from the evolution of species is adapted to explain the evolution of economic behaviour or organization. The process involves five elements:

1. some kind of 'memory' structure (the 'genome') capable of
2. reproduction by means of
3. an organism (or behaviour or organization) that is encoded in the mentioned memory structure and reproduces itself following the instructions given by the memory structure. Furthermore,
4. the reproductive success, i.e. the rate of reproduction and the degree to which the copy resembles the original, depends on the performance (the 'fitness') of the organism. This mechanism, called selection, provides some kind of heuristic optimization (of the reproduction process). Finally,
5. some diversity is needed in order to allow the optimization to occur efficiently, i.e. without diversity there would not be anything to optimize. In biological reproduction of organisms, the reproduction is never perfect, the copy never resembles the original perfectly, therefore some diversity is created as part of the process (mutation).

It may seem strange to tell the story from this point of view: an organism does not reproduce itself creating a new organism that inherits the genes from the original, instead the organism is constructed by the genes for the sole purpose of reproducing the genes. This idea by (Dawkins, 1976)[58] is, however, able to shed some light on the workings of evolutionary processes. They are not, as has been interpreted in earlier times, the selection and survival of the fittest, strongest and best. They are merely a very idiosyncratic mechanism of self-reproduction, that may favour the (whatever this means) 'fittest' individuals, that may work in favour of the greatest social good, but that may also on some occasions lead to socially or individually suboptimal or self-destructive behaviour, since the instruction-giving memory structure, the genome, is not capable of intelligent reasoning nor does it have a free will, but is merely a sequence of instructions that, when executed, will at some point reproduce the memory structure itself. Additionally, it may be necessary to add, that the memory structure by itself (without the organism) is of course not capable of productive or reproductive action; the memory structure and the organism are co-evolving.

A further very instructive concept from Dawkins (1976) is cultural evolution. Behavioural items (everything from trendy language and trendy clothing to attitudes and ideologies) that are copied in social systems, rise to global importance and fall out of use again. The analogies between cultural evolution and evolutionary biology are of course:

- the behavioural item as an organism
- memories of the behavioural item or of its observation (called 'memes' by Dawkins (1976)) as the genome
- the copying of behaviours as their reproduction
- 'coolness' as a fast and rapidly changing selective device
- mistakes in copying and creative alteration of behaviours as 'mutations'.

This analogy is occasionally also drawn in evolutionary economics; however, the arguably more powerful economic evolutionary concept is the evolution of organizations such as enterprises seeking competitive advantages[59] or seeking to adapt to their environment in terms of industry structure, the availability or scarcity of resources, etc., with powerful models created by Nelson and Winter (1974), Conlisk (1989), and Kauffman (1993).

An evolutionary process is thus capable of heuristic optimization, which is much more likely to be a justified modelling assumption for social or economic systems than the computation of a global, unique and stable equilibrium in the neoclassical theory. In short, the evolutionary view of economic growth sees entrepreneurship as the core element of both economic growth and technological progress; that is, innovation decisions under conditions of true uncertainty and motivated by the resulting comparative advantage should the innovation succeed. A successful innovation would yield a temporary monopoly of the new technology (which has resulted from the innovation) to the entrepreneur, who in turn can enjoy monopoly rents until another agent successfully imitates her technology. Evolutionary theory therefore does not add technological progress as a more or less welcome but necessary exogenous factor (as the Solow–Swan model does, see section 3.2.1), technology is one of the core elements. The idea is not new; it goes back to Hayek, Schumpeter and even Veblen.[60] Although Schumpeter especially added important visions to the economic theory: first and foremost the metaphor of the necessity of creative destruction of inefficient structures on the one hand and of creative entrepreneurship on the other hand.[61] More detailed evolutionary economic methods were not available until the emergence of game theory with the writings of von Neumann and Morgenstern (1944), Nash (1950) and others in the 1940s and 1950s.[62] Game theory put social scientists in a position to model direct interactions as such, as well as in repeated contexts or in structured populations. Game theory also easily allows for heterogeneous agents, as the level of modelling is the interaction level and not, as in the previous (neoclassical and non-neoclassical) models an aggregated level.

Growing in popularity since the Nelson and Winters first publication of an evolutionary model of economic growth (Nelson and Winter, 1974), that even claimed to explain Solow's (1957) original data better than Solow's original model, evolutionary economic models generally involve direct interdependence and often complicated mechanisms at the level of single heterogenous agents. This also implies that the behaviour of every individual agent has to be separately computed, increasing the computational requirements of the model considerably. From direct interdependence it follows further that the model dynamics of both agent-level and aggregate variables may be inhomogeneous. This results in a multidimensional model with a high number of dimensions and a non-trivial inhomogeneous structure. The models often exhibit persistent disequilibrium properties or do not even theoretically permit equilibria. This raises major problems, since it is not immediately clear how the model can be 'solved' in order to derive predictions about its behaviour and, eventually, about real economic systems that are to be modelled. For a discussion of methods from evolutionary game theory that are applied here, see section 2.3. Due to its power and relatively modest requirements, simulation is the most widely used method to work with evolutionary and complex economic models.

3.4.1 Nelson and Winter: the evolution of technology

In the 1960s Nelson started to research technological change, more specifically he published a technology diffusion model in 1968 (Nelson, 1968). It was not designed as a growth theory and therefore did not account for economic growth as contemporary neoclassical models did. It was designed to address a very different point, a mechanism that is assumed implicitely (Solow, 1956; Swan, 2002 [1956]) or modelled as a simple optimizing decision (Cass, 1965, 1966; Uzawa, 1965). Nelson's model was a seminal work in that it succeeded in formalizing a proper microfoundation, matching the driving force of technological progress as described by Schumpeter and his followers, a microfoundation for the gradual improvement of technology through entrepreneurship. Targeted at temporary first comer advantages, putting a technological improvement into practice may or may not be worth the risk. While, as both Nelson and other Schumpeterians before him saw, this constitutes an evolutionary process of technological transformation, it has further and more severe consequences on the macro-level of the system. A better technology leads (assuming that it does not just crowd out other input factors) to economic growth, increased production, possibly greater consumption, greater investment and possibly further technological advance. These consequences and the corresponding co-evolution of the micro-layer of the system were explored by Nelson and others during the next 15 years.

Nelson, Winter and Schütte (Nelson *et al.*, 1976; Nelson and Winter, 1974) constructed the first influential evolutionary growth model. It was published in 1973 as a University of Michigan Discussion Paper. This largely unnoticed publication gained more attention when Nelson and Winter summarized its main results again in another paper mainly dealing with the failure of neoclassical growth theory to reflect the economic reality in its models in 1974 (Nelson and Winter, 1974). Two years later, a 'progress report' (as the paper labels itself) bearing the same title as the initial discussion paper appeared in the *Quarterly Journal of Economics* (Nelson *et al.*, 1976). It details the mathematical and empirical properties of the study that were also explained in the discussion paper.

Subsequently, Nelson and Winter continued to explore the possibilities and properties of evolutionary models. Another, in some respects more elaborate, model to explore the micro-level effects in detail[63] (Nelson and Winter, 1982b) was presented alongside the one from 1973 and Nelson's diffusion model from 1968 in their seminal book *An Evolutionary Theory of Economic Change* 1982 (Nelson and Winter, 1982a).

The 1982 model was primarily designed to investigate the inner structure of an industry, its development in time, and the process of searching for and research of new technologies, rather than merely determining the consequences of an evolutionary simulation at the micro- for the macro-level as the 1973 model does. The authors themselves (Winter *et al.*, 2003), as well as other scholars, have been working to advance these techniques and to develop further evolutionary models that may help to develop an integrated and robust evolutionary theory of technological change (Conlisk, 1989; Arthur, 1989; Dosi *et al.*, 1994; Kirman, 1997; Kauffman *et al.*, 2000; Silverberg and Verspagen, 2005, 2007). For an overview discussing Nelson and Winter's approach and its impact on evolutionary economics, see e.g. Antonelli and de Liso (1997) or Nelson (1995).

Technology diffusion model (1968)

Nelson's (1968) diffusion model is clearly recognizable as the precursor to the later 1973/1974 model by Nelson and Winter (Nelson and Winter, 1974; Nelson *et al.*, 1976), that in turn was one of the first formal applications of evolutionary theory to economics. It is itself not evolutionary in that sense; however, two of the elements of an evolutionary system are present: an 'organism' is present in both phenotype (the firm, the technology) and genotype (the encoded structural properties of the organism, the production costs) and a mechanism for selection (the expansion of the technology, by assumption following the unit profits) is present as well. What is missing is just a device to generate diversity.

With w and r denoting the factor costs (wage and interest rate) and $h(w, r)$ being the cost function for some benchmark technology, the function

$$C_i = \frac{1}{\alpha_i} h(w, r) \qquad \text{(Nelson, 1968, 16)}$$

denotes the unit cost for technology i depending on its productivity α_i. The price, however, is assumed as a global decreasing function of the output – thus introducing demand-side effects.

$$P = P(Q) \quad \text{with } \frac{\partial P}{\partial Q} \leq 0 \qquad \text{(Nelson, 1968, 1a)}$$

Global price and technology specific costs generate the 'proto-evolutionary' equation for the technologies' rate of expansion or decline (the development of its share of the industry)

$$\left(\frac{\dot{Q}_i}{\sum Q}\right) = \lambda(P - C_i) \qquad \text{(Nelson, 1968, 17)}$$

where λ is some positive constant. It is easy to see that the technology with the highest α will – having the lowest cost and thus the highest profit – gradually come to dominate the industry – it generates a clean s-shaped curve as shown in Figure 3.6. (The dynamic of Nelson's model can be reproduced as shown in Figure 3.6; the simulation code is available from the author on request.) Further, if diversity is introduced ('mutation'), the development becomes truly evolutionary and dynamic, generating particular temporary growth effects, cycles, particular patterns in the industry structure and turbulence. This, however, has only been added by Nelson and Winter's as well as other collaborator's subsequent research till 1982.

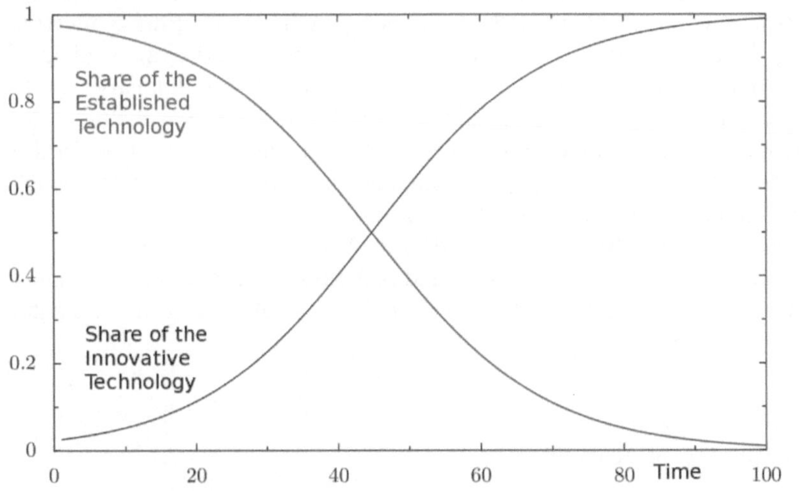

Figure 3.6 Technological change induced by the diffusion of an innovative technology in Nelson's diffusion model.

Growth model (1973–4)

The centrepiece of the theoretical debate that built up around the 1973 model was, that the model was able to fit the empirical data famously used by Solow in 1957 (Solow, 1957) better than Solow's original model (Nelson *et al.*, 1976; Nelson and Winter, 1974), while Nelson and Winter consider their model's assumptions to be vastly better justified (Nelson and Winter, 1974). This impressive feature has been subject to some discussion. It has been argued that that claim is just valid for this particular dataset – in other words that the model has been constructed to fit these exact data (Antonelli and de Liso, 1997). The canny answer Nelson and Winter's original 1974 paper has to offer to counter this argument is that the very same argument applies to Solow's neoclassical growth theory. 'To a considerable degree the success … has been due to the ad hoc flexibility of the theory' (Nelson and Winter, 1974: pp. 887–88).

The model was revolutionary in two respects. First, the agents are not homogeneous and do not even engage in maximization calculations, as in the few neoclassical models that allowed for heterogeneity.[64] Second, the economy is not assumed to be in equilibrium, but rather to constitute an environment of dynamic selection in the strict evolutionary sense. The latter difference especially raises issues of how to 'solve' this model and how to compute it as discussed above.

The model is agent-based with interdependent agents in the sense that they engage in imitation of technology. Furthermore they compete globally for labour according to the mechanisms of a neoclassical labour market with a global price derived from a strictly growing labour supply function. The labour demand of the enterprises is determined by the evolutionary part of the model. The goods market is supply driven; the price elasticities are infinite, that is, demand follows supply and the price p does not react to any changes in the extent of the market. The evolving memory structure is the technology, in the terms of the productive factors a_K for capital productivity and a_L for labour productivity; the vehicle of evolution, the evolving organism, is, in effect, the firm.[65] Evolutionary variety is introduced as a stochastic search function for research into new technologies additionally to imitation of existing ones. The selection mechanism defines a system where enterprises with better technology tend to own more capital, which in turn is required for both imitation and innovative research.

The state of the industry is represented by an $N \times 3$ matrix, N being the finite and fixed number of firms in the industry and each firm being represented at time t by a vector of its factor productivities for capital[66] $a_{K,i}$, labour $a_{L,i}$ and its capital endowment K_i, thus[67]

$$(a_{K,i,t}, a_{L,i,t}, K_{i,t}).$$

Production at time t is given by

$$Q_{i,t} = F(a_{K,i,t}, K_{i,t}) = \frac{K_{i,t}}{a_{K,i,t}}$$

for the individual firm, which is aggregated over all i for the total production of the industry. The development equations for the state variables are

$$K_{i,t+1} = K_{i,t} + pQ_{i,t} - \delta K_{i,t} - R(K_{i,t}) - w\left(\sum_i \frac{a_{L,i,t}}{a_{K,i,t}} K_{i,t}\right) \frac{a_{L,i,t}}{a_{K,i,t}} K_{i,t} \quad \forall i$$

$$(a_{K,i,t+1}, a_{L,i,t+1}) = \begin{cases} (a_{K,i,t}, a_{L,i,t}) & \text{if } \frac{Ki,t+1}{Ki,t} \geq 1+\tau \\ r((a_{K,i,t}, a_{L,i,t}); (a_{K,1,t}, a_{L,1,t}), \ldots, (a_{K,N,t}, a_{L,N,t})) & \text{if } \frac{Ki,t+1}{Ki,t} < 1+\tau \end{cases}$$

An increase in capital stock is by definition equal to the net investments, which in turn in this model is equal to the total revenues[68] $pQ_{i,t}$ minus the total costs including depreciation at a rate[69] δ, labour[70] and capital costs $R(\cdot)$.[71] Technology, represented by factor productivities may be changed by innovative and imitative research,[72] but only if the success of the firm (its capital increase) falls below a margin of[73] $\frac{Ki,t+1}{Ki,t} = 1 + \tau$. This threshold condition (dropped in Nelson and Winter's later models (Nelson and Winter, 1982a)), termed 'satisficing', is somewhat implausible, since successful enterprises, and sometimes especially the more successful ones, should also engage in research and development. The condition might, however, be necessary to 'keep the model alive'; that is, prevent it from rapidly developing a performance gap between the centre of the industry and the leading enterprises, which might then swiftly discover the technologies with values a_L and a_K close to 0 followed by the explosive expansion of production and the monopolization of the industry structure. Similar to the preceding deterministic models, the firms engage in selling their products in a largely neoclassical market, thereby taking part in forming or finding the current period's equilibrium price and managing their capital stock according to the given investment function. And much like in these deterministic models, this process does not offer the firms any opportunity for strategic behaviour or true decision-making, as they are completely deterministic and pre-specified in the model. However, firms act in another way. They may choose whether to engage in innovative and imitative research and they may set their R&D expenditure r for both types of research. This in turn is a strategic decision, the success or failure of which is dependent upon the simultaneous decisions of the other firms. For most settings[74] this changes the sector's industry structure in terms of the individual capital stocks, production and technology levels.

It is obvious that this model will behave more dynamically than the pure technology diffusion version Nelson had designed in 1968 (Nelson, 1968); the model includes a complete evolutionary process including stochastic introduction of diversity (quasi-'mutation'), and thus the best technology is not necessarily always the same. Therefore, the dominance of a particular firm may or may not be persistent – though it is likely that at some point only one firm will be able to afford research in sufficient quantities. For the first part of a simulation, growth is cyclical with always (with any true innovation) one firm gaining a comparative advantage.[75] Furthermore, turbulence and a moderate overall growth may realistically be expected. A simple Python simulation as depicted in Figures 5.12–5.17 (source code available from the author on request) confirms this expected behaviour. The depicted system turns into a monopoly after about 150 iterations. Until then, periodic alternation of fast and slow growth can be seen; it is likely that Nelson and Winter's (1974) claim to match real data better than Solow's neoclassical framework (while also including well-justified microfoundations) may indeed be justified, as the proximity of the resulting pattern to real data of economic growth[76] suggests.

Co-evolution of the industry structure (1982)

Nelson and Winter continued their work on evolutionary economic models. In 1982 they published an influential work on the prospects and implications of an evolutionary economic theory in general (Nelson and Winter, 1982a). The models from 1968 (Nelson, 1968) and 1973 (Nelson and Winter, 1974; Nelson *et al.*, 1976) were presented again, complemented by more theoretical considerations of a Schumpeterian economic theory and once again a critique of the neoclassical growth theory. The volume also includes (Chapter 12) an extended version of the evolutionary model of technological change, a model that resembles the 1973 model in some features but differs from it in others. Nelson and Winter state explicitly that the two models serve different purposes, namely the investigation of micro–macro relationship in the context of technological change (the 1973 model) and the analysis of the research process in the process of technological change (Nelson and Winter, 1982a). Contrary to the 1973 model, the 1982 model is designed to cover a range of different settings, also generating diverse dynamic patterns.

What the model retains from its more macroeconomic predecessor is mostly the theoretical approach, not the particular dynamics. The evolving object – in terms of the equivalent to the biological gene – is still the technology of production, which still determines the production capacity of the individual firms (still the 'vehicles' of the evolution mechanism) together with the firm's capital stock. It is still changed by innovative and imitative research with stochastic success. The capital stock and, more importantly, the increase or decrease of the capital stock remain the model's primary measure of the firm's fitness. Therefore, even though the technology is no longer represented as a multiple of capital and labour productivity but as a single capital productivity level $A_{i,t}$, the appropriate and complete representation of the industry may still be written as an $N \times 2$ (with N firms) matrix as above $(A_{i,t}, K_{i,t})$.

The goods market is now demand and supply driven. Labour demand and labour costs as separate variables have been dropped,[77] the capital stock development has been changed, the upper boundary for engaging in R&D (denoted above as τ) has been removed, and the critical influence on the profit margin as a quasi selection mechanism has shifted from the labour costs to the demand price function $P_t = D(\sum_i Q_{i,t})$ for the produced goods, which is constant in the 1973 model, but may in the 1982 model have a decrease in the quantity of goods $\sum_i Q_{i,t}$. The mechanism is arguably quite similar, since the effect is the same: increases in production lower the profits; in the 1973 model, the cost side depended on the state of the industry in a nonlinear manner[78] while the profit side was linear; in the 1982 model, the cost side is a simple multiplication of the individual firms capital and technology level, while the profit side depends on the state of the whole industry.

According to the available technology and capital stock, each firm produces an output of

$$Q_{i,t} = A_{i,t}K_{i,t} \qquad \text{(Nelson and Winter, 1982b, 1)}$$

and generates profits of

$$\Pi_{i,t} = \pi_{i,t}K_{i,t}$$

where $\pi_{i,t}$ are the profits per unit of output[79]

$$\pi_{i,t} = P_t A_{i,t} - c - r_{i,t} \qquad \text{(Nelson and Winter, 1982b, 3)}$$

The future characteristics of the firm, i.e. its technology and capital stock, are determined from the current ones as

$$K_{i,t} = IK_{i,t} + (1-\delta)K_{i,t}, \qquad \text{(Nelson and Winter, 1982a, Ch.12, Eq.7 (Part))}$$

where δ is the depreciation rate and I is an investment function dependent on the ration of revenues to costs, the current sector share, the current return on capital and the depreciation rate[80]

$$I = I\left(\frac{P_t A_{i,t}}{c}, \frac{Q_{i,t}}{Q_t}, \pi_{i,t}, \delta\right).^{81}$$

(Nelson and Winter, 1982a, Ch.12, Eq.7 (Part))

As in the earlier model, the technology level $A_{i,t}$ may be improved by innovative research resulting in a randomly chosen new technology level, and by imitative research making the industry's 'best-practice' technology available to the firm. Both types of research require a monetary expenditure $r_{i,t}$. Both succeed only with a certain probability; afterwards the best technology level available to the firm is chosen.

Nelson and Winter discuss three simple benchmark cases:

- No research is allowed ($r_{i,t} = 0 \quad \forall i \forall t$). In this case, the 'mutation' mechanism is not present. The resulting dynamics are perfect selection dynamics as in Nelson's earlier model (Nelson, 1968). There is a unique equilibrium, with only the firm(s) that possesses the industry's best-practice technology (and possess it initially) remaining and all others driven out of the market.[82]

- No innovative research; imitation is allowed. Here, 'selection' works initially, but collapses as soon as all firms have, by imitation, caught up with the technology leaders. The 'mutation' mechanism is not present.
- Constant imitative and innovative research for each firm is independent of productivity. In this case the pattern of the industry dynamics is more complex; productivity increase will be stochastic but will, according to the law of large numbers, approximately match a deterministic function of the (constant) research expenditure.

Independent innovative and imitative research leads to more complex patterns. Innovative research leads to better performance, better prices and a higher sector share for those firms that succeed in developing better technology (higher A_i, leading to higher Q_i, higher π_i and higher $K_{i,t+1}$). If the performance gap in terms of per unit of capital profits is large enough, some less lucky firms will be gradually driven out of competitiveness. Differences in the initial technology levels may result in the same effect. The average productivity and performance gap decreases with growing imitative research or growing success rate of imitative research. Nelson and Winter report quite complex results for which the success or failure of particular strategies is not ex ante predictable, as the chances of imitative and innovative research, as well as research in general, depend on the differences in productivity and sector share among the firms.

Nelson and Winter were among the first to investigate the evolution of an industry structure. Since the model is microfounded, in that it develops with entities representing the individual firms, it permits this kind of analysis, which has not been possible for earlier purely macroeconomic models of economic growth. Nelson and Winter consider the number of firms initially present in the sector and study its effect on several variables after 101 iterations. No significant difference exists in terms of the best-practice productivity; however, the average productivity and the total net worth tend to fall with the initial number of competitors. Furthermore, different investment regimes[83] seem not to cause a large difference, though Nelson and Winter elaborate at length on this aspect and find some minor effects. In a setting with very few firms (an oligopolistic case, more specifically up to four firms), all the firms tend to retain their position in the sector and to keep their technology close to the industry's best practice. For greater numbers of competitors at the initial stage, the industry structure becomes asymmetric within 101 iterations, indicating a tendency torwards monopolization. Nelson and Winter suspect – probably correctly – that in settings with few firms, the already large industry shares for several competitors give them greater opportunity (capital for R&D investment) to defend their position. However, the dynamic does still tend towards monopolization, though this may take time.

Relevance for the development of evolutionary growth theory

The mechanism that Nelson and Winter detected is simple yet powerful; it is highly likely that real-world systems do in fact follow similar patterns in the short run, especially as long as the carefully modelled micro-level of the model shows non-trivial dynamics. Though Nelson and Winter were especially interested in developing an evolutionary theory of technological change (as their 1982 contribution (Nelson and Winter, 1982a) and other later research (Nelson and Winter, 1982b; Nelson, 1995; Winter *et al.*, 2003) indicates), the model they developed is capable of reproducing real-world economic growth patterns including growth cycles. The cycle pattern their mechanism generates may also be much more realistic than other growth cycle models (see section 3.5) in that the waves (rather than cycles) it generates are irregular and microfounded. For details, see section 5.3.2. In turn, the model has been an inspiration for many subsequent evolutionary growth models including those discussed in this section. Some of the models did indeed employ the property of the Nelson–Winter model to generate cycle patterns to model growth cycles and technological paradigm change (i.e. 'long waves') (Silverberg and Lehnert, 1993, 1994). After some time, however, Nelson and Winter's mechanism tends towards monopolization, which in turn strips the system of its evolutionary selection mechanism; this is also when the system stops showing cyclic motion. Other evolutionary simulation models (Silverberg and Verspagen, 1996) show similar results; a plausible model which leads to perpetual non-trivial motion driven by evolutionary mechanisms remains to be found. This will, with the additional inclusion of network effects and a slightly different model (but inspired by this one), be attempted in Chapter 5.

3.4.2 Conlisk: a stochastic replicator model of technology change

Nelson, Winter and their collaborators (Nelson, 1968; Nelson and Winter, 1974; Nelson *et al.*, 1976; Nelson and Winter, 1982a) delivered a considerable blow against the complete dominance of rather simple traditional neoclassical or Keynesian models of economic growth, with no account for the micro-layer of technological change as one of the most important driving forces of economic growth. They showed that evolutionary models with an explicit and well-justified notion of the micro-layer are equally able to explain growth as measured empirically. Instead of computer simulation as in Nelson and Winter, Conlisk (1989) intended to develop a simple stochastic replicator dynamic model, which is similar to Nelson and Winter's model but can be analytically solved using stochastic methods without requiring computer simulation.

Conlisk uses a vintage model (similar to the one by Arrow (1962)) with stochastic technological progress. The macroeconomic framework is rather simple. The labour force N_t develops according to

$$N_{t+1} = (1 + n)N_t \qquad \text{(Conlisk, 1989, 1)}$$

with the constant growth rate n. Production takes place in 'plants' according to a vintage production function. Any plant yields an output according to its technological efficiency, but requires the same normalized amount of labour input, 1. Hence, at time t, exactly N_t plants can be operated; the most efficient ones are chosen. The production function results as

$$Y_t = \sum_{i=1}^{N_t} y_{i,t} \qquad \text{(Conlisk, 1989, 2)}$$

where y_t is the infinite dimensional vector of the plants ordered by productivities (denoted by the '*Rank*[·]' function)

$$y_{t+1} = Rank[x_t, (1-\delta)y_t]. \qquad \text{(Conlisk, 1989, 4)}$$

δ is the depreciation of existing capital goods (the plants), x_t is a vector of m_t new plants[84] with technological productivities $x_{i,t}$

$$x_{i,t} = e^{\mu_t + \sigma\varepsilon_{i,t}} \quad \text{with: } \varepsilon_{i,t} \sim IID(0,1). \qquad \text{(Conlisk, 1989, 5)}$$

The logarithm of $x_{i,t}$ is hence stochastically distributed around a value μ_t – and thus $x_{i,t}$ is an exponential function of a random value that roughly follows the expression e_t^{μ}. The crucial question for the extent and development of economic growth is therefore which form $\Delta\mu = \mu_{t+1} - \mu_t$ takes and how it evolves.

It is easy to see that Conlisk is able to reproduce the typical AK model using this specification (by simply assuming a constant growth rate for μ) as well as the Arrow model. The evolutionary growth that forms the core of his argument, however, specifies μ as the average of the k_t most efficient production technologies

$$\mu_t = \frac{1}{k_t} \sum_{i=1}^{k_t} log z_{i,t} \qquad \text{(Conlisk, 1989, 7)}$$

with z_t following the same *Rank* function as y_t only without depreciation

$$z_{t+1} = Rank[x_t, z_t]. \qquad \text{(Conlisk, 1989, 6)}$$

and the number of the relevant technologies growing with a constant (zero or positive) growth rate β.

Equipped with this model specification, Conlisk proceeds to derive results for different population growth (n) and productivity development (k_t,m_t). If the population size remains constant, technological progress ($E(\Delta\mu)$) becomes a function of σ, m_t and k_t.[85] In particular, if $n = 0$ (no population growth), the progress function ($E(\Delta\mu)$) behaves roughly like a moving average of $\Delta log \frac{Y_t}{N}$, which leads to the Slutzky–Yule effect generating cyclic behaviour,[86] a development very similar to the result of Nelson and Winter's first evolutionary model of economic growth. Still, Nelson and Winter's model tends to simultaneously produce monopolization effects, and thus in their system the Slutzky–Yule effect (resulting from spill over effects) will vanish as soon as the system has transformed into a monopoly. In Conlisk's model, cycles are persistent. For a growing population, Conlisk's system might either approach a constant growth rate $g = E(\Delta\mu) < g_{max} = \sigma\varepsilon_{max}$ (when $\frac{m}{k}$ is fixed) with $\frac{m_t}{k_t} = \frac{s}{\beta}$ or $\frac{m_t}{k_t}$ grows without bounds pushing the growth rate to its maximum $\sigma\varepsilon_{max}$. In this case, growth might start slowly (with low growth rates) to subsequently accelerate rapidly to the maximum in a short transition period, labeled a 'big push' in Conlisk's paper (Conlisk, 1989), which is, however, deemed unreasonably large by Conlisk.

3.4.3 Silverberg and Lehnert's model: the evolution of technology paradigms

Silverberg and Lehnert, drawing on earlier work by Silverberg, provided a model and empirical evidence in long-term time series of energy sector technologies in order to explain long cycles, that is Kondratieff cycles (Silverberg and Lehnert, 1993, 1994). The basic idea is that superior technologies appear at certain rare points in time and will slowly take over their sector. The theory of technological paradigm changes that this model follows had been discussed for some time when Silverberg and Lehnert published their model, especially by evolutionary economists. The quailitative historical basics of this theory will be discussed in more detail in section 3.5.6 in the context of Freeman and Perez's work (Perez, 1983; Freeman and Perez, 1988).

The model is constructed along the lines of Goodwin's (1967) Marx–Keynes–Goodwin model (see section 3.5.3), defining the factor productivities of capital c and labour a, as well as the relative growth of the capital stock $\frac{\dot{k}}{k}$, the employment share v and the wage rate w as crucial variables to construct a Marx–Keynes–Goodwin-type Phillips curve[87]

$$\dot{w} = -mw + nwv \qquad \text{(Silverberg and Lehnert, 1993, 1)}$$

with m and n being constants. The differential equation for the development of the capital share for any technology i out of N technologies, however, also resembles features of Nelson and Winter's (1974) model

$$\frac{\dot{k_i}}{k_i} = \frac{1}{c_i}\left(1 - \frac{w}{a_i}\right) - \gamma. \qquad \text{(Silverberg and Lehnert, 1993, 2)}$$

where γ is the physical deterioration of the capital stock. The technology specific employment share v_i with $\sum v_i \leq 1$ in turn depends directly on the capital stock in use for the technology[88]

$$\frac{\dot{v_i}}{v_i} = \frac{1}{c_i}\left(1 - \frac{w}{a_i}\right) - \beta \qquad \text{(Silverberg and Lehnert, 1993, 4)}$$

with $\beta = \alpha + \gamma$ being the decrease of the per capita capital employment by deterioration γ and growth of employment α. Silverberg and Lehnert add that the model may be extended to include gross technology investment flows in order to allow investment in superior technologies.

In order for the technology to grow and eventually take over the sector, it is required that $\frac{\dot{v}_i}{v_i} > 0$, thus

$$\frac{1}{c_i}\left(1-\frac{w}{a_i}\right)-\beta>0 \tag{3.58}$$

which leads to

$$w < a_i(1-c_i\beta) = T_i \qquad \text{(Silverberg and Lehnert, 1993, 5)}$$

Silverberg and Lehnert term $T_i = a_i(1 - c_i\beta)$ the evolutionary potential of the technology i. With the steady state independent from the distribution of the technologies requiring $w = T_i \forall_i$, the above condition translates to T_i, the evolutionary potential for any new technology to be higher than the system average and, for the rise of this technology to last, higher than any of its competitors' evolutionary potential. The relation of the evolutionary potential further governs the speed of the diffusion of superior technologies. The interpretation provided by Silverberg and Lehnert is that a higher T is equivalent to a negative eigenvalue associated with the v_i of the new technology in the vicinity of the current state of the system, allowing the technology to diffuse (Silverberg and Lehnert, 1994). Silverberg and Lehnert finally test their model with data on the development of technologies in the US energy sector since 1850, their conclusion being: 'Our high-dimensional, nonlinear dynamic system thus appears to a certain extent to be imitating a moving average process in converting the white noise innovation input into a cyclical macro-variable (output). It is well known that a moving average process will extract a cyclical structure from white noise (the Slutzky–Yule effect)' (Silverberg and Lehnert, 1994: pp. 90–91).

A later similar model developed by Silverberg and Verspagen (1996) was studied using computer simulation, and yielded a similar result to what was discussed above with respect to Nelson and Winter's model. The model shows strong oscillations or cycles in the rate of technological change during the first timesteps (in this case the first 1000–2000 timesteps), which later cease as the industry sector begins to monopolize as a result of some firms being systematically more successful than others (Silverberg and Verspagen, 1996).

3.4.4 Kauffman's technology landscapes

A further major step in evolutionary modelling of growth and technology was taken by Kauffman (1993) (a biologist and familiar with state of the art theories of biological evolution) and collaborators (Kauffman and Macready, 1995; Kauffman *et al.*, 2000). In this case, the approach was radically different from the earlier (largely Schumpeterian) evolutionary theory of economic growth. Kauffman dived deeper into the biological root of the metaphor: the genotype in evolving populations, the argument proceeds, while simply being comprised of some single value representing a strategy, a technique or a technology level in earlier evolutionary economics, is rather a sequence of values, in biology called a gene. Any value of the sequence may change independently, altering the entirety of the phenotype, the evolutionary fitness of the resulting subject determines the success or failure of the evolving entity.[89] A more sophisticated picture of genotypes (in general) as proposed by Kauffman (1993) is an n-dimensional space of possible configurations for a gene sequence (or its economic or social equivalent) consisting of n entities.[90] Furthermore, the effect of each of the n dimensions ('genes') is assumed to interfere with k other dimensions (but is independent of the $n-k$ remaining dimensions). More specifically, each gene assumes a specific value out of A possible values. The complete sequence is represented by

$$w_i = \{w_{i,1}, w_{i,2}, \dots, w_{i,n}\} \quad \text{with } w_{i,j} \in \{a_1, a_2, \dots, a_A\} \quad \forall j \in 1, \dots, n$$

(Kauffman *et al.*, 2000, 3 and 4)

where w_i is a vector representing the complete genotype including all genes, each of which assumes one out of A possible states. The fitness contribution of a specific gene j – depending on k other genes $w_{i,e} = \{w_{i,j_1}, w_{i,j_2}, \dots, w_{i,j_k}\}$ – is therefore a function $\phi_{i,j} = \phi_j(w_{i,j}, w_{i,e})$ and the total fitness value of the 'organism' is assumed to be the arithmetic mean of all contributions

$$\theta(w_i) = \frac{1}{n}\sum_{j=1}^{n}\phi_{i,j} = \frac{1}{n}\sum_{j=1}^{n}\phi_j(w_{i,j}, w_{i,e}) \qquad \text{(Kauffman \textit{et al.}, 2000, 7)}$$

To derive an *economic* model of a technology NK-landscape, the biological terms can simply be replaced. Kauffman and his collaborators (Kauffman *et al.*, 2000) used a labour-driven model of growth and progress with the production q_t (at time t) being a function of the labour force l_t and the labour productivity θ_t.

$$y_t = q_t = F(\theta_t(w_{i,t}), l_t) \qquad \text{(Kauffman \textit{et al.}, 2000, 1)}$$

where θ_t is the renamed fitness value of the 'organism' which now is an elaborate technology of production (a 'production recipe') $w_{i,t}$ with n separate elements as defined above in biological terms as the 'genome', the complete sequence of genes of a genotype. The values that an element of the technological specification of a mode of production may take, the 'alleles' $w_{i,j} \in a_1, a_2, \dots, a_A$, are the possible methods that may form a part of the technology at this particular point j in the specification of the technology.

Having defined such a technology landscape (or more generally NK-landscape), Kauffman *et al.* proceeded to analyze the features of this landscape, namely the number of local optima. For simplicity and without loss of generality, the dimensions are assumed to be binary values, A thus being limited to 2 and the value space of the $w_{i,j}$ being reduced to $w_{i,j} \in 0, 1$. The neighbourhood structure is defined as follows: all those technologies $\widetilde{w}_i$ are neighbours of w_i that differ in only one of the n elements from w_i – hence with $A = 2$ each technology has n direct neighbours. If 'mutation' or, in economic terms, innovation is allowed only to take place in one gene at a time, only in those cases where there are no other local optima, is there certainty to reach a global optimum. This happens for $k = 0$; under this condition, the contributions of the elements would simply be additive, either optimal or not, and would be switched to the optimum one by one until the global optimum is reached. For larger k, the number of local optima rises and for $k = n - 1$ the landscape finally reaches a completely random form with a large number of local optima. In fact, with searching for better technologies with more distant configurations allowed, Kauffman and his team found that the 'optimal' search distance – the one most rapidly leading to the global optimum – is neither the very small one (direct neighbourhood) nor the whole population, and that it depends on the value of k.

Characterising technological change as an analogy to natural biological evolution, Kauffman, Macready and collaborators (Kauffman and Macready, 1995) arrived at the conclusion that these two systems are similar in more than one way. First, considerable novelties occur rarely. Second, if they do, major improvements occur quickly at first, which corresponds to the first steps taken after starting at a random point on an NK-landscape; the rate of improvements, however, slows down roughly following a power law distribution until finally only some major designs remain. Kauffman's technology landscapes are also capable of providing an explanation for technological lock-ins: an industry may be trapped at a local optimum without the capacity to innovate such radically different designs to reach a distant global optimum. Certainly, this explaination is radically different from the ones provided by other economists who were working on the topic of lock-ins (Katz and Shapiro, 1985; Farrell and Saloner, 1986; Arthur *et al.*, 1987; David, 1985).

Kauffman and his team have arguably established a new class of evolutionary models in economics, a class the scope of which is not necessarily already fully explored nor even fully anticipated. It is a class of models that differs from the earlier Nelson–Winter-type approaches, that employs more advanced mechanisms of evolutionary processes, like complex interdependences, redundancy, etc. The growth model that has been developed by Kaufmann *et al.* (2000) to apply their earlier results to the economics of technological change was not extensively tested using empirical data and it may in fact be too general for this. It is, however, a concept of another class of models that in some ways surpasses the limits of the early evolutionary economic theory.

3.4.5 Technological change and evolutionary growth theory

Other than in more macroeconomic approaches like those discussed in the sections 3.1, 3.2 and 3.3 evolutionary growth theories not only acknowledge that technology and technological change play a crucial part in economic growth, they also attempt to include this explicitly. This coincides with the expressed intention of developed proper microfoundations for macro-level growth and is greatly facilitated by the use of agent-based modelling (Pyka and Fagiolo, 2005). Indeed, evolutionary growth theory succeeds not only in reproducing growth patterns and even cycles (for instance in Nelson and Winter's case (Nelson and Winter, 1974, 1982a)) but also in establishing a variety of further important aspects. This includes the suggestion that innovation may be clustered in time (especially in Kauffman *et al.*'s (2000) theories), and some insight in to the driving mechanism of economic growth and technological change. Evolutionary economists did for instance recognize very early the importance of innovation diffusion (Nelson, 1968) very early and have continued to pursue the analysis of the effects of innovation and innovation diffusion on economic growth. This discussion is not the focus of the current thesis; however, this line of research is discussed in a number of works, for instance in Silverberg *et al.* (1988), de Liso and Metcalfe (1996), Winter *et al.* (2003) as well as Saviotti and Pyka (2008).

It should, however, be noted that the evidence generated by evolutionary economic models has as of today not been striking enough to convince a broader group of scholars. In fact, two decades after the publication of Nelson and Winter's model and 12 years after that of their widely recognized monograph, in 1994, neoclassical economists could simply dismiss Nelson and Winter's model in a footnote as 'a complement to formal theory' with the formalisms of the model yet to be made (Romer, 1994). Though this opinion is breathtakingly unjustified, it is a sign of how much remained and still remains to be done to develop an integrated and convincing theory with striking evidence. Nevertheless, the method is sound and the models are much richer in ideas and flexibility than their neoclassical counterparts; the development of an integrated model of economic growth and technological change with network effects later on in Chapter 5 will rely heavily on evolutionary agent-based modelling and draw important lessons from Nelson and Winter's technology diffusion mechanism.

3.5 Economic cycles: no such thing as an equilibrium

The most important general phenomena that must be explained by economic growth theory number two: the – on average – constantly high growth rates of the economies and the relatively regular cyclical disturbance especially in the growth of output. Cycles of different length (and thus frequencies) have been or are claimed to have been identified. While seasonal cycles are obvious, and most scholars are still confident that there are longer-than-seasonal cycles of some kind, it remains unclear, which kind of cycles these may be and how they are caused.[91] Following the taxonomy of de Groot (de Groot, 2006; de Groot and Franses, 2008) the most prominent suggestions are

- 4–5 year cycles, the well-known business cycle, also called Kitchin cycle, proposed in the 1920s, supposedly caused by important inventions with a certain impact on the economic structure.

- 7–11 year cycles, the usual economic investment cycle, also called Juglar cycle. This cycle was discovered in the mid nineteenth century with later explanations and models suggesting investment and financial markets as a probable cause. Scholars are not generally convinced that this cycle is actually different from the Kitchin cycle; in fact, business and economic investment cycles are commonly treated as the same phenomenon.[92]
- 15–25 year cycles, the Kuznets cycle, empirically identified in 1930 for economic data of the USA. It is likely that waves of immigration from Europe caused this phenomenon and that this pattern does not exist any more, although some economists argue that empirical evidence can be found in a spectral analysis of GDP fluctuations in OECD countries (Diebolt and Doliger, 2006). The proponents of this suggestion seem, however, to argue that these cycles should be interpreted as longer but otherwise ordinary business cycles.
- The 48–60 year long-wave or Kondratieff wave, claimed by Kondratieff and the conjuncture institute in Moscow in the 1920s, though explicitly and only for capitalist economies (as de Groot quotes Kondratieff (de Groot, 2006: p. 98)). The long-wave theory has received moderate to high attention from multiple generations of economists; however, it remains uncertain whether the wave does actually exist in this regularity, and by what it might be caused.[93] The most convincing interpretation, if the cycle does exist, however, is the slow technological and organizational change in rigid, locked-in systems; while, once the rigidity is broken, a rapid revolutionary change in both technology and organization may take place (Perez, 1983). De Groot and Franses add testable hypotheses and empirical investigation to Perez's theory, arguing that the cycle of innovations correlates with economic cycles and that basic innovations do not occur uniformly distributed but clustered in times of rapid advance[94] (de Groot and Franses, 2009). This process may or may not be confined to a sector or an economic region.
- Even longer waves than the Kondratieff wave. de Groot names them 'hegemony cycles' (de Groot, 2006) since the suggestion of such a wave is based on the argument that political–economic hegemony periods of national industries last about 150 years (roughly: Netherlands 1600–1750, Great Britain 1750–1900, USA since 1900). It should be added, however, that the justification of the term 'cycle' (suggesting a regular phenomenon) is highly questionable in this case, since the suggested wavelength of the cycle does not permit more than two or three observations since the dawn of modern economies, since economic systems are arguably subject to heavily path dependent influences given the timeframe of the cycle, and since against the extent of random disturbances the value of this long-term trend would be vanishingly small and impossibly difficult to measure and prove. This argument, often also applied to the 60-year Kondratieff wave,[95] is even more striking as cycles are not perceived as regular patterns but as disturbed patterns that may be stretched and compressed or otherwise altered.
- Irregular and non-global waves. For many of the other suggested cycles it is argued that they are specific to certain policy regimes or systemic conditions

and thus not universal, but temporarily and possibly also regionally limited.[96] Other systematic wave patterns caused by interlacing negative and positive feedback mechanisms have been assumed to exist at industry levels, the earliest being financial market models in the spirit of Keynesian animal spirits (Samuelson, 1939a,b). Models in the Marxian tradition suggest 'bootstraps', the use of output as a production factor (in the Marxian tradition namely capital goods). Demand or supply shocks will in this case directly translate into turbulences in the sectors' production (more specifically, rising demand translates directly into shortages in supply) (Forrester, 1977). Others argue that mere time lags that are usually ignored in standard neoclassical theory lead to this kind of turbulence (Wheat, 2009); Schumpeterians often argue that innovation motivated by temporary monopoly revenues and the subsequent diffusion of technology may lead to these patterns. This property is present (and explicitly intended) in a number of the models (Nelson and Winter, 1974; Silverberg and Lehnert, 1993; Conlisk, 1989) discussed in section 3.4 on evolutionary growth models. This mechanism is also said to correspond to the periodic change of technology paradigms (and smooth development within the technology paradigms); for a more detailed account of the different interpretations of circular patterns in evolutionary growth theory, see de Liso and Metcalfe (1996).

Looking at output data from a random sample of countries or regions – as a simple example using data from Germany and the USA from 1950 to 2009 (see Figures 1.1 and 1.2) – it becomes obvious that short cycles in seasonally adjusted time series are indeed present, but in both length and amplitude still heavily perturbed, subject to influence by random historic events and differing between countries. Indeed, the spectrum of autocorrelations of the USA and Germany (see Figure 3.7) is entirely different. The autocorrelation coefficient $AC(LAG)$ of the GDP growth rates (seasonally adjusted GDP) $\dot{Y(t)}$ has been methodically computed for time lags of 1 through 41 years.

$$AC(LAG) = \frac{1}{(n-LAG)Var(\dot{Y(t)})} \sum_{t=1}^{n-LAG} (\dot{Y(t)} - Mean(\dot{Y(t)})) \times (Y(t)(t \dot{+} LAG) - Mean(\dot{Y(t)})) \quad (3.59)$$

where n is the total number of observations (60), and $Var(\dot{Y(t)})$ denotes the variance of the distribution of $\dot{Y(t)}$.

While for Germany the influence (autocorrelation) is high for a short cycle (4 years, or probably slightly more than 4 years as long-term data reveal) and multiple short cycles; that is, the current value is correlated not only to the value one cycle (4 years) ago, but also to the values two (8 years), three (12 years), etc., cycles prior to the present period. Longer differences are difficult to recognize and, if present, probably not significant. For the USA, the short influence is present for the last cycle (4 years) but for longer time lags, its influence vanishes rapidly.

A longer cycle of about 20 to 25 years length is clearly visible as well; that is, the current growth of output is correlated with the data in the last few years, and with those of roughly 25 years ago. Interestingly, the short cycle with a stable wavelength is arguably still visible over the 25 year lag as a disturbance of the long cycle. The difference between Germany and the USA is not only that a long cycle does not appear to exist in German data, but that the wavelength of the short cycle is, while stable over time in the countries time series, not the same for both countries, as the peak autocorrelation values of both countries do not coincide for longer time lags. Though this does not rule out that it might be the same 'true cycle', if the process was a 'true cycle' with additional random elements, it suggests that the influence of biases is strong. To support this argument, a tradition in economic literature must be mentioned that tries to estimate cycle lengths macro data of different countries, using not only the GDP but a much broader set of variables and much longer time series. One of the most recent examples by de Groot and Franses with the results published in 2008 estimates very different cycle lengths for the US, Great Britain and the Netherlands (de Groot and Franses, 2009).

It is well known that moving average filters applied to time series with cycles can generate time series with a different cyclic pattern, specifically longer cycles (Slutzky, 1937 [1927]). Moving averages may occur naturally if random or cyclic disturbances do not only affect a time series at one point in time, but retain a slowly decreasing influence over several periods. This may indeed be true for shocks in variables with positive feedback loops which must be assumed for much of the economic system.[97] It may thus be suggested, that

1. different regions and probably also different sectors may experience cycles of different amplitudes and wavelength
2. an aggregated level (such as a country or a group of countries) might show another different pattern with longer cycles caused by resonance and dampening in a system of time series with different and asynchronous cycles
3. in more homogeneous systems (Germany in the example above), in contrast to more heterogeneous ones, a cycle might be quite strong, thus its effect may be clearly measurable over several wavelengths.

For an empiric account of this, see the GDP growth rates of Germany and the USA in Figure 1.2, the autocorrelation spectrum of those growth rates in Figure 3.7 and the R^2 of the autocorrelations (as a direct comparison of the strength of the respective autocorrelations) in Figure 3.8.

Consequently, following this argument, another interpretation becomes viable: cycles, at least the longer waves, may not have an underlying 'true cycle' process, may in reality just be random shocks. With a wave pattern caused by naturally occurring moving average filters and similar patterns, they are thus truly independent in amplitude and wavelength at the sectoral or regional level. This in turn would create longer cycles at more aggregated levels.[98] Even random biases generate strong cyclic patterns when transformed by a moving average filter. This finding by Slutzky (1937 [1927]), was not only one of the first insightful

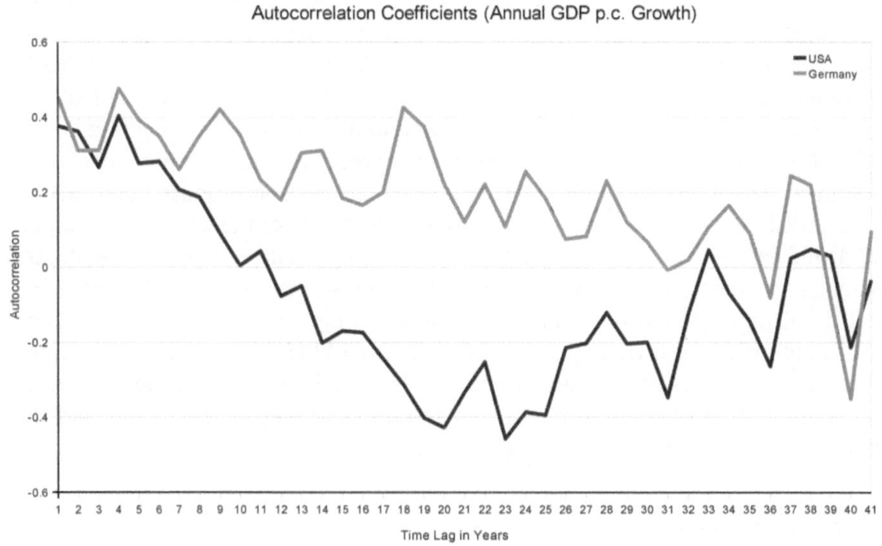

Figure 3.7 Autocorrelation spectrum of the GDP per capita growth rates of Germany and the USA. The growth rates are shown in Figure 1.2 (with a moving average filter applied for convenience; data from the Bureau of Economic Analysis (2010) and the German Federal Statistical Office (2010).)

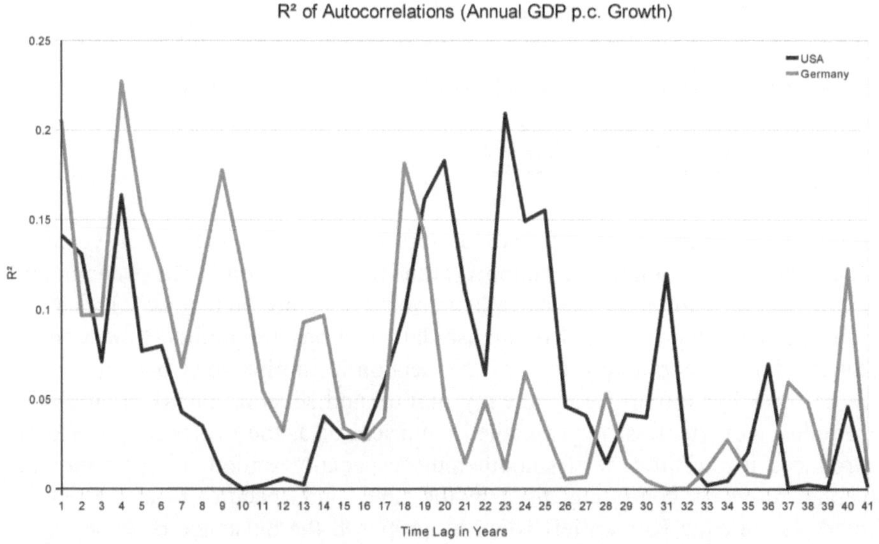

Figure 3.8 R^2 of the autocorrelations of the GDP per capita growth rates of Germany and the USA. (Data from the Bureau of Economic Analysis (2010) and the German Federal Statistical Office (2010).)

contributions to this area, it also came from the same research institute as the very idea of long waves, Kondratieff's conjuncture institute in Moscow. Sadly, however, the conjuncture institute was closed down shortly thereafter, arguably marking the end of great economic science in the eastern hemisphere in that era. Several other approaches have been followed to develop a theoretical foundation of economic cycles. Another of the earliest approaches was to explain cycles by nonlinear dependencies in the financial sector. Samuelson proposed that private investment decisions depend on the current growth rate of the economy's output, mediated by naive growth expectations (i.e. the expectation that the growth rate remains constant) (Samuelson, 1939a,b). This leads to temporary self-reinforcing effects of high growth or shrinking of the economy. Another approach, which was more or less already implied in Marx's (1963 [1885]) analysis of 'extended economic reproduction' and was formalized by Goodwin (1967) using Keynesian elements, is that of the interplay of wage rates and employment rates and the resulting development of the output growth.

3.5.1 Slutzky: natural moving averages

Drawing on earlier findings by Yule, Slutzky (1937 [1927]) showed that a moving average filter of length L, applied to a stationary[99] random time series $X(t)$, that is

$$\widetilde{X(t)} = \frac{\sum_{\tau=1}^{L}(X(t+\tau))}{L}, \tag{3.60}$$

or equivalently a dampened motion

$$\begin{aligned}\widetilde{X(t)} &= \frac{1}{L}(\text{current random shock} + (1-L)\text{ past random values})\\ &= \frac{X(t)}{L} + \frac{\sum_{\tau=1-L}^{-1}(X(t+\tau))}{L}\end{aligned} \tag{3.61}$$

produces, within limited time frames, almost regular cycles, with the wavelength depending on the original distribution $X(t)$ and the filter length L. The phenomenon is illustrated in Figure 3.9 (code available from the author on request). This cycle motion can be approximated by sinusoidal functions for limited time frames. Slutzky did not explicitely discuss the economic implications but mentioned the business cycle (and also planetary motion and several others) as empirical examples. Nevertheless, for smoothed empirical data, the cycles might also be introduced in the process of smoothening, especially smoothening by moving average filters as commonly used in the time, and specifically also used by Slutzky's collegue Kondratieff when he proposed the existence of long waves (Kondratieff waves) in economic data.[100]

For time series satisficing the properties above required for $X(t)$, two successive values are below (or above) the mean value with probability 0.5, three successive values with probability 0.25, etc., thus approximately very short but rather

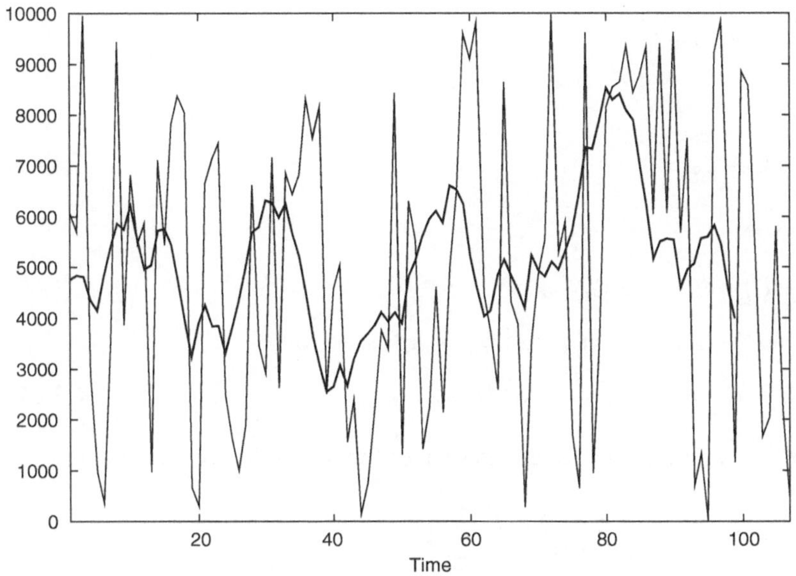

Figure 3.9 Random data (uniform distribution) and nine-period moving average filter (bold line).

irregular 'cycles'.[101] Furthermore, the second order difference of two successive values (the difference of two successive differences of successive values) tends to be negatively correlated, which, with dampening motion, translates to smooth cycles of a longer wavelength than one or two periods. Generally, moving average filters produce longer cycles from basic data with short cycles, which is called the Slutzky–Yule effect.

3.5.2 Samuelson's oscillator: cyclic, equilibrium and disequilibrium regimes

Keynesian scholars have argued that macroeconomic cycles, boom and bust, might be caused by excessive reactions of the financial sector to changes in the real economy. The most well-known example is Samuelson's 1939 model (Samuelson, 1939a,b). Assuming that economic output at time Y_t is composed of an autonomous term (imagined as government expenditure) G_t, linear private consumption

$$C_t = \alpha Y_{t-1}, \qquad \text{(Samuelson, 1939a: p. 76 [without number])}$$

with a being the constant marginal rate of consumption, and non-linear private investment

$$I_t = \beta(C_t - C_{t-1}) = \alpha\beta(Y_{t-1} - Y_{t-2})$$

(Samuelson, 1939a: p. 76 [without number])

the composed output function reads

$$Y_t = G_t + C_t + I_t.$$

Defining

$$D_t = C_t - C_{t-1} \tag{3.62}$$

yields

$$Y_t = G_t + C_t + I_t = G_t + \alpha Y_{t-1} + \beta D_{t-1} \tag{3.63}$$

$$\begin{aligned} D_t &= \alpha Y_t - \alpha Y_{t-1} = \alpha G_t + \alpha^2 Y_{t-1} + \alpha\beta D_{t-1} - \alpha Y_{t-1} \\ &= \alpha G_t + (\alpha^2 - \alpha) Y_{t-1} + \alpha\beta D_{t-1}. \end{aligned} \tag{3.64}$$

Thus the system is represented as a system of first-order difference equations

$$\begin{pmatrix} Y \\ D \end{pmatrix} = \begin{pmatrix} \alpha & \beta \\ \alpha^2 - \alpha & \alpha\beta \end{pmatrix} \begin{pmatrix} Y \\ D \end{pmatrix} + \begin{pmatrix} G_t \\ \alpha G_t \end{pmatrix}. \tag{3.65}$$

The behaviour of this system is governed by its eigenvalues. Specifically, the system generates spiral motion, growing, constant or shrinking cycles, if the eigenvalues are complex. The eigenvalues are computed from the corresponding homogeneous system's characteristic equation[102]

$$\lambda_{1/2} - a(1+\beta)\lambda_{1/2} + \alpha\beta = 0 \tag{3.66}$$

which resolves to

$$\lambda_{1/2} = \frac{\alpha(1+\beta)}{2} \pm \frac{\sqrt{\alpha^2(1+\beta)^2 - 4\alpha\beta}}{2}. \tag{3.67}$$

The eigenvalues λ are complex if and only if the term in the square root is negative, hence if

$$\alpha < \frac{4\beta}{(1-\beta)^2}. \tag{3.68}$$

Samuelson (1939a) considers five cases, two of which are stable and do not exhibit cycles ((A) $\alpha = 0.5$, $\beta = 0$ and (D) $\alpha = 0.8$, $\beta = 4$) while three do ($\alpha = 0.5$, $\beta = 1$, (B) $\alpha = 0.5$, $\beta = 2$ and (C) $\alpha = 0.6$, $\beta = 2$).[103]

In mathematical terms, Samuelson used one of the most easy approaches to generating a dynamic system that results in cyclic behaviour: a relatively easy oscillator with complex eigenvalues. A Keynesian interpretation of a dynamic macroeconomic system applied to this oscillator yields the extremely flexible (in terms of including different regimes of cyclic, growing and shrinking behaviour) theory that forms the core of Samuelson's papers of 1939. Of course, later theorists such as Goodwin (1967) and Day (1994) would add more complex dynamic models that are also capable of explaining cycles while involving a more sophisticated representation of the economic reality. These models, however, also increase the mathematical complexity; Samuelson's work is probably the most concise demonstration of the power that dynamic systems may have if used in a theory of economic growth.

3.5.3 Goodwin: a cyclic predator–prey model of capital and labour

Goodwin (1967) developed another, more Marxian, model of growth cycles. The mathematical approach is similar to Samuelson's but more complex, using a system of non-linear differential equations. It follows that the eigenvalues depend on the coordinates and thus the type of dynamics, and the stability changes with the point where the system is currently located in its phase space.[104] In effect, the system resembles the Lotka–Volterra predator–prey model[105] with the workers' share of profit (u) and the employment share (v) taking the role of predator and prey respectively: the predator variable increases with increasing numbers of prey animals (the higher the employment share, the higher the workers share of profit[106]); however, more predators tend to decimate the prey (the higher the workers' share of profit and therefore the wage, the fewer workers are employed). This yields the Lotka–Volterra system (with A,B,C and D being positive constants):

$$\frac{du}{dt} = Auv - Bu$$
$$\frac{dv}{dt} = -Cuv + Dv \tag{3.69}$$

Goodwin developed his system according to the following assumptions:

- The growth rates of both population n (identical to the labour force) and labour productivity a are constant $\frac{da/dt}{a} = \alpha$ and $\frac{dn/dt}{n} = \beta$.
- The workers share of profit is $u = \frac{w}{a}$ (w being wage); it follows that $\frac{du/dt}{u} = \frac{dw/dt}{w} - \frac{da/dt}{a}$.
- The wage rate w in this equation follows the Phillips curve assumption of a wage function that rises with the employment share $w = -\gamma + \rho v$. Together with the first point, this yields the first of Goodwin's Lotka–Volterra equations (see below).
- As the workers' share of profits is $u = \frac{w}{a}$, the capitalists' share is $1 - u = 1 - \frac{w}{a}$, the overall production is $q = al$ ($l < n$ being effective employment) and the growth rate of the capitalists profit is $\frac{dq/dt}{q} = \left(1 - \frac{w}{a}\right)\frac{1}{\sigma}$ ($\frac{1}{\sigma}$ being the capital productivity).

- The employment share is $v = \frac{l}{n}$; from the production function $q = al$ follows $v = \frac{q}{an}$, and thus $\frac{dv/dt}{v} = \frac{dq/dt}{q} - \left(\frac{da/dt}{a} + \frac{dn/dt}{n}\right) = \left(1 - \frac{w}{a}\right)\frac{1}{\sigma} - (\alpha + \beta)$. This is the second of Goodwin's Lotka–Volterra equations.

Thus, Goodwin's Lotka–Volterra-type system follows as

$$\frac{du}{dt} = \rho uv - (\alpha + \gamma)u \qquad \text{(Goodwin, 1967, 1 and 2)}$$
$$\frac{dv}{dt} = -\frac{1}{\sigma}uv + \left(\frac{1}{\sigma} - \alpha - \beta\right)v$$

Substituting

$$A = \rho$$
$$B = (\alpha + \gamma)$$
$$C = \frac{1}{\sigma}$$
$$D = \left(\frac{1}{\sigma} - \alpha - \beta\right) \tag{3.70}$$

the system is exactly identical to the Lotka–Volterra system above.

To analyze the system, it is not sufficient to evaluate the eigenvalues (as they are not static); rather the fixed points in phase space must be found first, to proceed with investigating the stability in the vicinity of these equilibria by computing the eigenvalues at these positions. Once this is accomplished, general conclusions concerning the behaviour of the system and its economic interpretation become attainable.

The necessary and sufficient conditions for a fixed point in phase space are $\frac{du}{dt} = 0$ and $\frac{dv}{dt} = 0$. There are two solutions:

$$u_1 = 0, \quad v_1 = 0 \tag{3.71}$$

which is the trivial solution with zero employment, production and wages; and

$$Av = B, \quad -Cu = D \tag{3.72}$$

The Jacobian matrix of the system (from which the eigenvalues are then computed) is

$$\mathcal{J} = \begin{pmatrix} \frac{\partial du/dt}{\partial u} & \frac{\partial du/dt}{\partial v} \\ \frac{\partial dv/dt}{\partial u} & \frac{\partial dv/dt}{\partial v} \end{pmatrix} = \begin{pmatrix} (Av - B) & Au \\ -Cv & (D - Cu) \end{pmatrix} \tag{3.73}$$

For the trivial solution, the Jacobian is reduced to

$$\mathcal{J} = \begin{pmatrix} -B & 0 \\ 0 & D \end{pmatrix} \tag{3.74}$$

which yields the eigenvalues

$$\lambda_{1/2} = \frac{D - B \pm \sqrt{(D-B)^2 + 4BD}}{2} \tag{3.75}$$

Since the radicand is always positive and has a larger absolute value than $D-B$,[107] the dominant eigenvalue (the eigenvalue with the largest real value) is always positive and the system is therefore always diverging in the vicinity of the first fixed point (the trivial zero production fixed point), hence the fixed point is an unstable hyperbolic fixed point.[108]

For the other fixed point, substituting B and D in the Jacobian

$$\mathcal{J} = \begin{pmatrix} 0 & Au \\ -Cv & 0 \end{pmatrix} \tag{3.76}$$

This implies a pair of purely imaginary eigenvalues

$$\lambda_{1/2} = \pm\sqrt{-AC}. \tag{3.77}$$

Thus in the vicinity of the fixed point the system is neither contracting nor expanding, and must therefore have stable (closed) limit cycles orbiting the fixed point. In fact, there is an infinite number of such limit cycles distributed infinitely densely around the fixed point. (See Figure 3.11.) As has been demonstrated, the 'Marx–Keynes–Goodwin' system is not converging nor does an equilibrium exist in this setting; hence, it is a genuine model of persisting disequilibrium. However, the far-from-equilibrium situation is non-chaotic (there is no chaos in two-dimensional systems) and easily computable. From the closed limit cycles, it follows that the development of the system is cyclic with strictly regular cycles (3.11). It is sensitive to shocks, since the limit cycles are closed: once the system has been moved to another limit cycle by a one-time shock, it will remain on the other limit cycle. It is thus obvious that the system will transform into a potentially chaotic system if further variables are added.[109] The model adds another important feature: it combines persistent growth with cycles, the growth properties of the resulting system (Figure 3.10) getting rather close to the empirically observed patterns. The technological progress in the model is, however, exogenously fed into the system by means of the constant growth rates of productivity. Notwithstanding, Goodwin's model (or as it later came to be called, the 'Marx–Keynes–Goodwin' model) was an early promising framework for the explanation of growth cycles, which was in place before the advent of evolutionary economics and also before many of the great advances in chaos

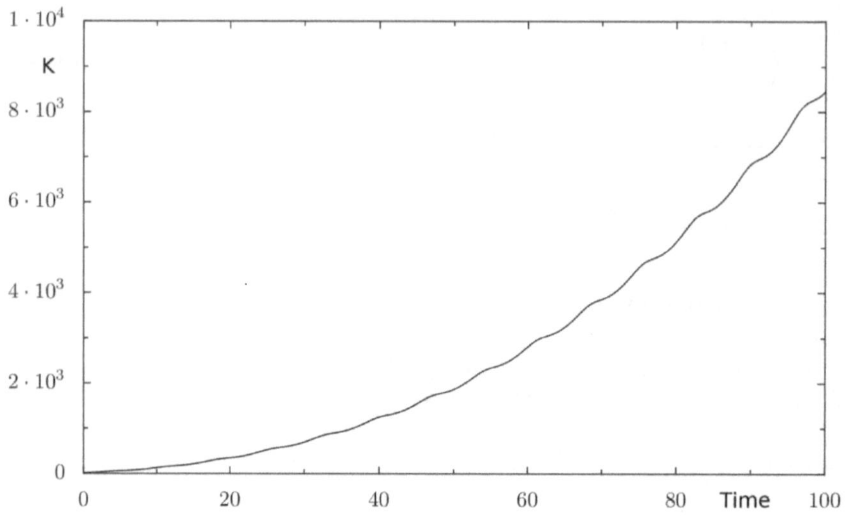

Figure 3.10 Capital growth in Goodwin's growth cycle model as simulated using computer simulation (source code is available from the author on request).

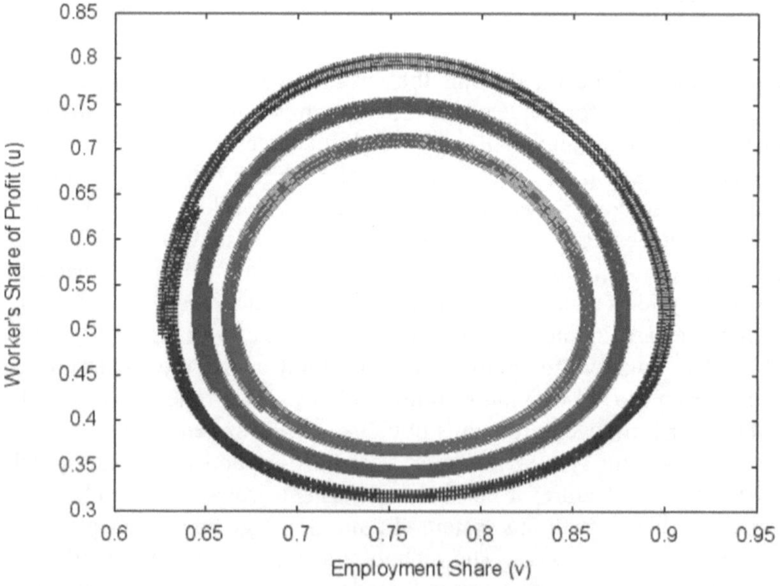

Figure 3.11 An approximation of Goodwin's limit cycle around a neutrally stable equilibrium in employment rate and wage rate generated by computer simulation.

theory of the 1970s. Both its mathematical framework (the use of a Lotka–Volterra system) and its theoretical idea (Keynesian macro-dynamics with a Marxian labour market) proved to be seminal advances for modern economics and have been used in similar ways in a number of later models (e.g Silverberg and Lehnert, 1993; Keen, 1995).

3.5.4 *Kaldor and Lorenz: the transition to chaos*

Kaldor is probably best known for his work on empirical relations of economic growth rates and other quantities (such as labour productivity and output growth rates) and his theory on the divergence of geographic regions caused by inherent disequilibrium patterns in economic growth (Kaldor, 1972). The foundation for this theory was, however, his earlier formal work on business cycles (Kaldor, 1940). Together with the extensions developed by Dixon and Thirlwall (1975), León-Ledesma (2000) and Lorenz (1987), the approach proved most valuable for analyzing far-from-equilibrium patterns that are believed to be observed in growth data (Kaldor, 1972; Nelson and Winter, 1974; Lorenz, 1987; Keen, 1995; Chen, 2002, 2005; León-Ledesma, 2000).

Kaldor (1940) started with a Keynesian capital market imbalance: investment $I(Y)$ does not necessarily (and will usually not) equal savings $S(Y)$, but both are positively dependent on the 'level of activity' which behaves more or less similarly to the output Y, hence[110]

$$\frac{dI}{dY} > 0, \quad \frac{dS}{dY} > 0. \qquad \text{(Kaldor, 1940: p. 79 [without number])}$$

Furthermore it is assumed that the level of acivity depends positively on the difference between investment and savings, hence (assuming a linear function)

$$\frac{dY}{dt} = \alpha(I(Y) - S(Y)) \quad \text{with} \quad \alpha > 0. \tag{3.78}$$

Fixed points are defined as

$$I(Y^*) = S(Y^*). \tag{3.79}$$

The stability is given by the single eigenvalue of the difference equation (3.78)

$$\lambda = \alpha(\frac{dI}{dY} - \frac{dS}{dY}), \tag{3.80}$$

which is negative if the slope of the savings function exceeds the slope of the investment function

$$\frac{dS}{dY} > \frac{dI}{dY} \tag{3.81}$$

in the vicinity of a fixed point. This fixed point is thus stable. An unstable fixed point is characterized by the investments rising faster in its vicinity than the savings, i.e.

condition (3.81) does not hold. Kaldor notes that the functions $I(Y)$ and $S(Y)$ may be complex and have a number of intersections, which generates a landscape of stable and unstable fixed points between which the system may move when fueled with random shocks. Kaldor and Mirrlees (Kaldor, 1957; Kaldor and Mirrlees, 1962) extended the model to account for a number of phenomena, including technological change – in this case a 'technical progress' function defines a technological boundary to investment[111] such that the model shows cycles around a zero growth rate, but includes a drift representing technological progress.

The system is a very simple notion for disequilibrium in economic systems. Later, with demand-side effects as an additional aspect, Kaldor argued vehemently that equilibrium theory and especially the equilibrium models of Solow–Swan as well as the endogenous growth type are misleading (Kaldor, 1972, 1975). Demand-side effects may induce persisting different growth rates as Thirlwall was later able to show (Thirlwall, 1980, 1986), basically since the growth rate does not only depend on quantities in the growing region or country but also on the rest of the world. Though this gained much attention for explaining persistent differences in national growth rates, it is possibly equally applicable to different economic sectors. Increasing returns may work in the same direction (León-Ledesma, 2000).

Still, Kaldor's original system is unable to sustain persistent cycles. Without shocks, the system would of course sooner or later reach one of the stable fixed points. This, however, is changed by a few minor modifications as Lorenz (1987) showed. If used to model a system that is composed of multiple parts (sectors in Lorenz's example, but possibly also regions), it generates chaotic patterns and strange attractors.

Again, investment $I(Y,K)$ (depending on output Y and capital K) does not necessarily (and usually does not) equal savings $S(Y)$ (also depending on output). The output changes according to the difference between investment and savings multiplied by a factor (multiplicator) $\alpha > 0$

$$\frac{dY}{dt} = \alpha(I(Y,K) - S(Y)) \qquad \text{(Lorenz, 1987, 12, part 1)}$$

while the change of capital is simply the difference between investment and depreciation (with a depreciation factor $1 > \delta > 0$)

$$\frac{dK}{dt} = I(Y,K) - \delta K. \qquad \text{(Lorenz, 1987, 12, part 2)}$$

The equilibrium follows as

$$\frac{dY}{dt} = 0 \quad \frac{dK}{dt} = 0, \tag{3.82}$$

hence

$$I(Y,K) = S(Y) = \delta K \tag{3.83}$$

Following Kaldor, Lorenz assumes $\frac{\partial I}{\partial Y} > 0$, $\frac{\partial I}{\partial K} < 0$, and $\frac{\partial S}{\partial Y} > 0$.

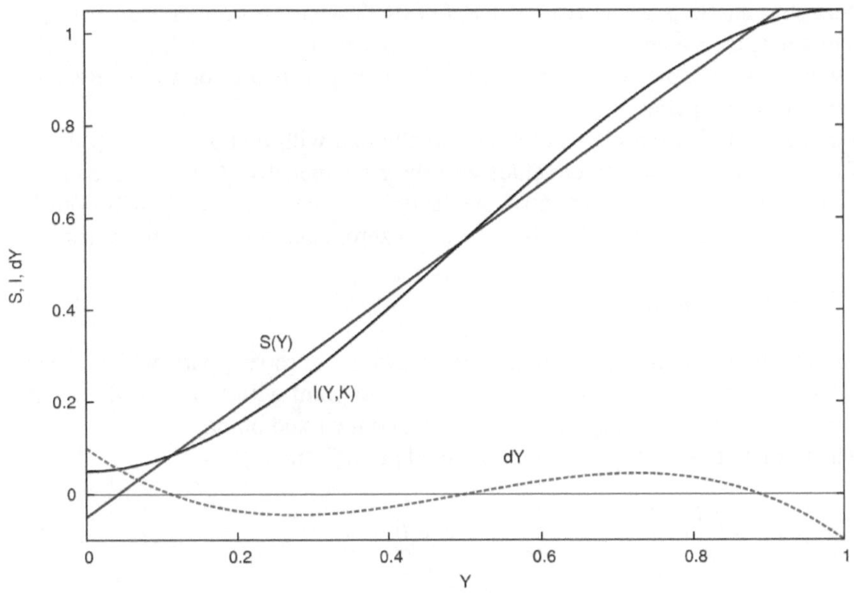

Figure 3.12 Investment, savings and output change in Lorenz's model.

The capital part of the equilibrium $\frac{dK}{dt} = 0$ exists by definition for any valid Y and is (given Y) unique, since $I(Y,K)$ is strictly falling in K while δK is a simple increasing linear function in K. The output equilibrium is more complex. $S(Y)$ is again by definition a strictly increasing function in Y. However, $I(Y,K)$ is assumed to follow an s-curve shape, which means that there are up to three equilibria even for any given K (as shown in Figure 3.12), which may or may not be stable.

The stability has to be assessed by evaluating the eigenvalues of the system's Jacobian matrix.[112]

$$\mathcal{J} = \begin{pmatrix} \alpha(\frac{\partial I}{\partial Y} - \frac{\partial S}{\partial Y}) & \alpha \frac{\partial I}{\partial K} \\ \frac{\partial I}{\partial Y} & \frac{\partial I}{\partial K} - -\delta \end{pmatrix} \tag{3.84}$$

using the characteristic equation (where $\mathcal{I}$ is the identity matrix)

$$0 = det(\mathcal{J} - \lambda \mathcal{I}). \tag{3.85}$$

As the system is only two-dimensional, it comes down to ($tr(\mathcal{J})$ denoting the trace of the Jacobian)

$$\lambda_{1/2} = \frac{tr(\mathcal{J})}{2} \pm \sqrt{-det(\mathcal{J}) + \left(\frac{tr(\mathcal{J})}{2}\right)^2}. \tag{3.86}$$

As detailed in section 2.4, the interesting part in eigenvalues of systems of difference equations is the real part $Re(\lambda)$. If this part is smaller than 0 for all eigenvalues, the system is stable; if it is larger than 0 for at least one eigenvalue, the system is unstable; and, particularly, if it is larger than 0 for all eigenvalues, the system is a repeller.

Lorenz identifies the interesting case as the one with $det(\mathcal{J}) > 0$[113] (Lorenz, 1987). For this case, the eigenvalues will (both) be negative (and the fixed point stable) if, and only if, the trace of the Jacobian is negative. They will both be positive, if the trace is positive. If the trace is zero, equation (3.86) becomes

$$\lambda_{1/2} = \pm\sqrt{-det(\mathcal{J})}. \tag{3.87}$$

These eigenvalues are purely imaginary, leaving a zero real part, which means that the fixed point is neutrally stable. This fixed point will in turn be the centre of (infinitely many) limit cycles circling around the fixed point.

The trace in turn is negative (i.e. the fixed point is stable), if

$$tr(\mathcal{J}) = \alpha\left(\frac{\partial I}{\partial Y} - \frac{\partial S}{\partial Y}\right) + \frac{\partial I}{\partial K} - -\delta < 0 \tag{3.88}$$

hence, the critical value computes as

$$\alpha_{Hopf} = \frac{\delta - -\frac{\partial I}{\partial K}}{\frac{\partial I}{\partial Y} - \frac{\partial S}{\partial Y}} \tag{3.89}$$

For smaller values $\alpha < \alpha_{Hopf}$, the system has a stable fixed point.[114] At α_{Hopf}, a Hopf-bifurcation occurs, the fixed point becomes unstable and a limit cycle around the fixed point emerges which becomes the attractor for $\alpha > \alpha_{Hopf}$ (i.e. trajectories converge to the limit cycle).

When three or more systems that undergo a bifurcation of this kind are coupled, a strange attractor emerges; the resulting integrated system is then essentially unpredictable (without exact knowledge about all parameters), i.e. chaotic. The coupling is achieved by defining the underlying simple (uncoupled) system as a sectoral model and allowing intersectoral trade. Lorenz runs numerical simulations and shows that deterministic chaos does indeed emerge.

3.5.5 Minsky and Keen: the growing disequilibrium and the catastrophic collapse

Minsky (1980) offered a different, less technical, institutionalist model for business cycles. Minsky observed that the boom and crash cycle is accompanied (especially the more modern the economy in question is) by expanding and contracting volume of overall debt. Debt expands during the upswing – and it probably expands beyond all reasonable limits – only to eventually crash followed by a much more careful (and indeed much too careful) lending regime. Technically, the performance is greatly underestimated in times of slow but steady growth. Enterprises react by reinvesting internally and taking out very few loans (not least because of the unwillingness of capital owners to lend). Thus, profits start to rise as well as the interest rates the firms are willing to pay; likewise general confidence in the firms success grows as well. Now, the performance of the firms' is greatly overestimated, with the Keynesian 'animal spirits' driving everyone to take part in order not to lose the opportunity of fantastic profits. This leads to a short-lived positive feedback loop; the volume of overall debt rises exponentially until eventually the general misestimation is detected, the system crashes, and the cycle returns to a mode of slow growth.

Being an institutionalist, Minsky contributed the idea and the theoretical framework without bothering to develop a sophisticated model framework. This was finally accomplished by Keen (1995), a Marxian mathematical economist who shared many of the ideas of the institutionalists but argued that the widespread rejection of mathematical methods was to 'throw the child out with the bath' (Keen, 2001). Keen started with a 'Marx–Keynes–Goodwin' model (Goodwin, 1967) (see section 3.5.3) and introduced debt as a further variable. This solves one of the major problems of Goodwin's model: it is not entirely convincing that investment should equal profits as assumed in Goodwin's model. Keen loosened this assumption, introducing a net investment function depending on the profit rate, output, and capital depreciation. The dependence on the profit rate π is such that net investment is zero for negative profits and then increases superlinearly (matching profits when the profit rate is equal to 10% of the output but continuing to increase beyond that value). While this specification seems to be arbitrary, filling the gap between profits and investment with loans and adding a share of profit for the owners of the external financing (interest) leads to a more complex dynamics that behaves as discussed in Minsky's theory.[115] The limit cycle in the zero-debt-plain (Goodwin's case) narrows with increasing debt (i.e. with increasing possibilities of financing by taking loans), then widens again at some level of external financing, and then vanishes completely. The orbits move up to a larger debt level (during this time all variables are relatively stable), then employment share and wage rate grow increasingly unstable, and finally the system collapses when the attractor leaves the normal limit cycle. This creates a catastrophic situation with debt dropping to low levels again.

Government policy had already been one of Minsky's (1980) major concerns – according to Minsky one of the main tasks of a government is to prevent both the excessive overestimation of economic capacity (and thus the exponential increase in the debt level) and especially to prevent the eventual crash. Keen (1995) includes a 'government' actor with counter-cyclic behaviour in his system, and finds that the crash may indeed be prevented; doing so, the 'government' (adding further dimensions to the system) renders the system's behaviour even more complex (Keen refrains from using the term 'chaotic' (Keen, 1995)) – the system the develops in irregular waves.

3.5.6 Perez and Freeman: long waves of technological paradigms

Another theory of economic cycles resulting from the technological development of the employed methods of production is due to Freeman and Perez (Perez, 1983; Freeman and Perez, 1988). Their approach is to depart from the historic development of the economic system and apply Keynesian and Schumpeterian arguments to the derived findings. In particular, and notwithstanding all empirical difficulties, they focus on long (Kondratieff-type) waves rather than the short business cycles. According to their theory, innovations may be separated into four types (Freeman and Perez, 1988: pp. 45–7):

- Incremental innovations constituting continuously emerging minor improvements of known technologies
- Radical innovations – the development of fundamentally new technologies, that may or may not take off economically, depending on the 'technology system'
- Changes of the 'technology system' breaking the way for radical innovations in the social, economic and organizational system
- Changes in the 'techno-economic paradigm', changes of the industry structure of the economy as a whole – probably resulting from sufficiently many radical innovations and subsequent changes of the organizational system. These 'technological revolutions' alter the mode of production, exchange the core industries of the economy, and influence the social and economic structure deeply. These types of innovation constitute changes of the techno-economic (and implicitly social) system that are far-reaching enough to correspond to philosophical concepts like Kuhn's (1996) changes of scientific paradigms or Foucault's technologies of knowledge and power (Foucault, 1976 [1978]).

Providing a Schumpeterian microfoundation, Freeman and Perez's 'technological revolutions' were among the first economic theories to give a detailed and consistent explanation of long waves.[116] They stressed that an innovation – and much less an invention – is not enough to cause technological progress; the techno-economic environment must be ready to incorporate radical changes. This, Freeman and Perez (1988) argue, occurs regularly but only every several decades; each wave has therefore identifiable core technologies that are new in this period of time, that drive the initial boom, and that tend to play only minor roles during subsequent waves. They discuss the history of modern economics in some detail and identify five Kondratieff cycles since the end of the eighteenth century. The key sectors of these waves are, according to Freeman and Perez (1998, pp. 50–3):

- textile and iron working (first wave; 1770–1830, the industrial revolution)
- steam powered industry and transportation (second wave; 1830–1880)
- heavy industry (third wave; 1880–1930)

- synthetic and petrochemical industry, automobiles, and air transportation (fourth wave; 1930–1980, 'golden age of growth')
- computers, electronics, and capital goods (current wave, 1980–?).

Freeman and Perez add that the organizational changes each wave requires rely on a set of 'key factors' to be used for production with low (and, with larger amounts, falling) relative cost, good availability over the foreseeable future and the potential to be incorporated in multiple sectors, products, technologies, etc. They analyze these key resources to have been cotton and pig iron in the first cycle, coal and transportation in the second, steel in the third, energy and oil in the fourth and microelectronic devices ('chips') in the current wave. It might be added that the key factors may not always be as easy to separate from intermediate products of the emerging industries as Freeman and Perez may have assumed. This is most clearly visible from the contemporary IT economy, which relies heavily on software, services and especially digital communication that is of course impossible without 'chips', but requires much more both in intermediate products and in the organization of the industry sector. To a lesser extent, the same is true for the key factor of transportation in the second (steam power) Kondratieff wave.

Adding a Keynesian argument, Freeman and Perez emphasize that the cycle structure is reinforced by mechanisms in the financial sector. First, during the upswings, in the time of the greatest boom, the potential of the new technologies tends to be overstated due to insufficient data on these young sectors, which only show very rapid growth in the early phase of their life cycle. Second, the actors in the financial sector overestimate the capacity of the declining industries, thereby increasing the severity of the ensuing crash. And finally, industries need time to mature, as long as the standard setting process is not finished, these industries are subject to increased risk and the product life cycles tend to be very short.[117]

Freeman and Perez's 'technological revolutions' are close to the concept of lock-ins: in diversified (modern and pre-modern) economies, the economic sectors do not exist independently from one another; the structure of the economies is heavily integrated; therefore, the technologies must be consistent as well. Though this is principally not more than a multi-actor coordination game,[118] once established, a regime must be overcome as a whole in order to replace outdated technologies – an important and typical mechanism of technological lock-ins, which will be discussed in more detail in Chapter 4.

3.5.7 Lucas' neoclassical vision of growth cycles: a perturbation in equilibrium

The idea of persistent cycles has considerable fascination for those believing in the general-equilibrium nature of the economic system. The traditional approach is to concentrate on the macro-level and the long run – disregarding any cycles. On average, everything runs smoothly and is pretty stable, usually (except for occasions of extraordinary disaster) even in the long run. Hence, the general equilibrium is seemingly almost within reach – if only a good explanation for the cycles is found. However, neoclassical economy did not have much luck so far in accomplishing this. And so it is with both eloquence and cynism that Chen (2005), an evolutionary economist with a focus on complex methods, tells the history of the repeated attempts and failures of neoclassical economics to derive an equilibrium theory of disequilibrium (cycles), starting with Frisch's dampened motion, to Friedman's arbitrageur to Lucas' neutral money. This section will focus on the most modern and most elaborate of these approaches, Lucas' studies of business cycles (Lucas, 1972, 1975, 1981).

Lucas' hypothesis

Drawing on earlier neoclassical theory, namely Phelps and Friedman, Lucas (1972, 1981) (whose later EGT model was discussed in section 3.2.2) developed his theory of a business cycles: with incomplete information agents will react uniformly as soon as they detect a (random but not just short-term) disturbance, thus creating a coordinated movement. Lucas argues that the agents themselves will perceive these ups and downs as business cycles. The agents are assumed to have strictly rational expectations – expectations that will not defeat themselves by offering arbitrage opportunities.[119] Lucas' model proved to be one of the most influential contributions to new classical economics, as Chen puts it the 'counter Keynesian revolution' (Chen, 2005) of the 1970s, notably adding an elaborate microfoundation with theoretical, deterministic mathematical and econometric parts. (The econometric components will, however, not be detailed in this section.)

Lucas considers a two-sector economy, about which the non-mobile labour force is randomly distributed: a share $\frac{\theta_t}{2}$ is in one sector, the remainder $(1-\frac{\theta_t}{2}=\frac{2-\theta_t}{2})$ in the other one. x_t indicates the (nominal) demand for goods at the macro-level and influences the money supply $m' = mx_t$ of the economy. With supply following demand at the aggregated level (by definition) the contribution per worker (in nominal terms) is $\frac{x_t}{\theta_t}$ (and $\frac{x_t}{2-\theta_t}$ in the other sector). The workers will optimize their working time (or labour productivity) $n\left(\frac{x_t}{\theta_t}\right)$ (and $n\left(\frac{x_t}{2-\theta_t}\right)$) with $\frac{\partial n(\cdot)}{\partial \cdot} > 0$ accordingly, thus generating the total (real) production of the economy

$$Y_t = \frac{\theta_t}{2} Nn\left(\frac{x_t}{\theta_t}\right) + \frac{2-\theta_t}{2} Nn\left(\frac{x_t}{2-\theta_t}\right) \qquad \text{(Lucas, 1981: p. 80, 28)}$$

The price level is determined sector-wise but with the demand side using arbitrage possibilities until the price is equal on both sectors.

$$p(m, x_t, \theta_t) = m\phi\left(\frac{x_t}{\theta}\right) \qquad \text{(Lucas, 1981: pp. 73 and 79 [without number])}$$

where ϕ is a function $\phi > 0$ with elasticity greater than 0 but smaller than 1. Thus, while the labour allocation parameter θ_t induces temporary changes ('cycles'), the money shocks x_t have an – again temporary – positive impact on both price level p and real output Y_t. That results, in Lucas' model, in an empirically observable correlation of inflation $p_t - p_{t-1}$ and output Y_t which is nevertheless only temporary and not direct but induced by a common cause. The implicit conclusion is that: (1) monetary changes usually have no real effect, (2) the relation of nominal wages and employment known as Phillips curve is an illusion and (3) cycles may be real (or partly monetary[120]) but still – given rational expectations – nothing more than temporary fluctuations around equilibrium values.

Chen's evolutionary critique

Lucas' way to approach the phenomenon is – as with many neoclassical economists – econometric. Indeed, as mentioned, he was also prone to use computer simulation several times throughout his work (Lucas, 1975, 1981). So, however, are his critics: full of doubt about his conclusions, disapproving of his assumptions and eager to put his theories to a test. The most vivid of Lucas' recent critics is Chen (2002, 2005) an evolutionary economist who did much research on complex systems, chaos theory and the empirics of cycles in economic data.

Chen argues that Lucas' theory has flaws in both the assumptions and the statistical properties of the model. The theoretical construct – isolated individuals making the same decisions at the same time according to temporary imbalances in the input variables thus creating fluctuations and amplifying existing ones – is, as Chen argues, not stable (Chen, 2002). In particular, it is not immune to Lucas own critique of expectations being self-defeating if arbitrage is possible. Lucas does not consider the price of leisure, leisure is merely included implicitly as the price of work is defined and the amount of work (as opposed to leisure) is chosen according to this price (the wage) and its relation to the price of goods. Yet, there should also be a price of leisure and prices of leisure relative to goods and to work; the price of leisure should also be influenced by the amount of labour invested in the current period which in turn creates opportunity for arbitrage and smoothens the fluctuation towards the general equilibrium level. Further, and more generally, Chen argues that Lucas' microfoundations are a delusion – as all agents are identical there are no complex interferences to be studied as a true micro-level. In Chen's words, the multi-body system of an economic micro-level would be reduced (with all the bodies having the same position, direction and speed) to a one-body system (Chen, 2002).

Finally, at the statistical level, Lucas' system as well as earlier approaches by Frisch and others consider either random walk, continuous random walk (Brownian motion) or geometric Brownian motion processes as the processes underlying macro-level cycles at a micro-level. These, however, do not match the data and are unable to produce persisting fluctuations, the first two (random walk and Brownian motion) being inherently dampening, the latter (the geometric Brownian motion) escalating. To capture this property, Chen defines the relative deviation of a stochastic process X with the expectation $\overline{X}$ as

$$\psi = \frac{VARIANCE}{MEAN} = \frac{\sqrt{(X-\overline{X})^2}}{\overline{X}}. \qquad \text{(Chen, 2002, 1)}$$

This value decreases with the mean value for the random walk and the Brownian motion as the disturbances are linear and independent from the current scale of X; for the geometric Brownian motion, ψ increases with the mean value. Empirical macro data, however, show the relative deviation (ψ) to be constant; a condition that is not satisfied by any of these processes – it is, however, satisfied by a simple birth–death process[121] (Chen, 2002, 2005). Chen concludes that this stochastic formulation may thus be the preferable way to establish economic micro-foundations which in turn leads in the direction of evolutionary economics again.

3.5.8 Growth cycles and technological change

For models intended to explain not just growth but growth cycles or, more generally, lasting disequilibrium development it is perhaps more obvious than for the other classes of growth models. Some approaches start by analyzing the micro-relations of the system and will only then (if at all) give the model a formal mathematical representation to be used for econometric testing and prediction. Minsky (1980) is clearly an exponent of this approach as well as Freeman and Perez (1988). Another way to construct a testable formal model is to start with the mathematics and add the economic interpretation. With some econometric or statistic exceptions (an early one being Slutzky (1937 [1927])), the use of this method will generally not be documented; rather it will often remain obscure how a model was derived. Still, for many theories it is obvious that they employ previously known mathematical systems (of which the authors must have been aware) only adapting them to the requirements of a justified economic interpretation. Samuelson's oscillator (Samuelson, 1939a,b) being a rather simple linear system of difference equations was such a theory,[122] though it was doubtlessly a brilliant idea to use difference equation oscillators to explain growth cycles. The same is true for Goodwin's use of the Lotka–Volterra model, and to a lesser extent Keen's generalization of Goodwin's model to higher dimensional systems as well as Lucas' attempt to model cycles as synchronous reactions to random shocks.[123]

Considering the theories of the analytical-theoretical approach again, for both Minsky (1980) and Freeman and Perez (1988) the second step – to derive a formal model – has been complemented by other scholars (Keen (1995) in the case of Minsky's theory, Silverberg and Lehnert (1993) as well as de Groot and Franses (2008, 2009) for Freeman and Perez) who in turn were much more prone to explore with promising mathematical systems possible ways to reproduce the theoretical frameworks of Minsky and Freeman and Perez respectively.

It should be added that there is nothing wrong with exploring possibilities of identifying mathematical mechanisms able to generate certain kinds of macro-behaviour; the exact interpretations of these models as well as possibly resulting policy implications and scientific standards, however, should be viewed with caution. They constitute mere possibilities in a huge space of potential explanations. Furthermore, there is a good reason why these approaches are especially numerous in the theory of economic cycles: as discussed in the first part of this section, there is not much at all that is certain about growth cycles. Specifically, the longer cycles are subject to discussion. Empirical studies (de Groot and Franses, 2008, 2009) in this field established that there are several possible cycle lengths, not necessarily the same for different sectors, different economies and at different times; they seem to be related to other time series like the number of innovations (de Groot and Franses, 2009) but details are largely either unknown or disputed.

One thing, however, remains clear: it is undisputed that real economic systems are not in a smooth near-equilibrium development but rather subject to persistent turbulence which takes the form of cycles of different lengths. The exact mechanisms causing this behaviour at a micro- or intermediate (meso-) level are unclear, but the history of the theoretical research on growth cycles provides a diverse pool of different models.

Most models of growth cycles dismiss the idea of taking technology into consideration when trying to identify what causes cyclic perturbations completely. Slutzky (1937 [1927]) does not speculate about the cause of natural or artificial moving average filters that he deems responsible for cycles (or possibly the delusion of cycles); for Samuelson's (Samuelson, 1939a,b) model a lagged feedback loop of consumption and income suffices – no mention of technology so far. Goodwin (1967) has a static concept of technology rather similar to AK and Solow–Swan models; technology is an exogenous (and growing) factor. For Kaldor (1940), Lorenz (1987) and Minsky (1980) again, technology is not explicitly relevant while Keen's (1995) mathematical formulation of Minsky's model follows Goodwin (1967) in every aspect not touched by Minsky's theory. Finally, Lucas' temporary distortion cycles do not need technology either. Possibly the inclusion of technological change in this model might sweep the stable equilibrium away or add further arbitrage opportunities; at best, technology could substitute the effects already included. The notable exceptions are the deliberately evolutionary theories discussed in this section and the previous section: Freeman and Perez (1988) understand technological change, innovation and progress as the driving force of growth and the cyclic growth pattern in particular; Silverberg and Lehnert's (1993) approach might be seen as a formalization of Freeman and Perez. Nelson, Winter and collaborators (Nelson and Winter, 1974; Nelson *et al.*, 1976; Nelson and Winter, 1982a,b) though not explicitly discussing growth cycles consider innovation (which at a macro-level condenses to technological change) as the main factor in economic growth. Kauffman *et al.* (2000) finally also outline a relation between technological progress (on their technology landscapes) and growth cycles.[124] In the following chapters it will be argued that cycles and technological change are probably linked, that the micro-level as proposed by evolutionary economics and particularly the approach of agent-based simulation and its macro-level consequences (namely lock-ins) may be able to contribute the means to explain this connection.

A further interesting question is whether cycles are a natural phenomenon that will never change, a natural process of periodic creative destruction as the Schumpeterians might argue. As long as there is no sufficient explanation of what causes these cycles, this is not conclusively established. There are also only limited data on how long the cycle pattern has been present in the economic data and for how long it will remain. Several theories have accused capitalism of causing the cycles, especially the downswings, starting with Marx (1963 [1885]), likewise Goodwin (1967) and Minsky (1980) as well as reportedly, the discoverer of the long waves, Kondratieff; see de Groot (2006)

Though much has been written on growth cycles, the issue remains a mystery. It is rather undisputed that at least short (Kitchin) cycles do exist; it has also been established that a number of autocorrelations with longer time lags can be found in growth rates and that such differing positive and negative autocorrelations also exist for innovation data (de Groot and Franses, 2008, 2009). A relation between those two variables, in fact technology and growth, has been suggested (Nelson and Winter, 1982a; Perez, 1983; Freeman and Perez, 1988; Silverberg and Lehnert, 1993; Kauffman *et al.*, 2000; de Groot and Franses, 2009). However, the theoretical explanation for this remains disputed and the empirical evidence might not be that surprising since in a complex socio-economic system many macro-variables are probably related for very different reasons.

Nevertheless a variety of mechanisms have been suggested including:

- Keynesian 'animal spirits' or other investment fluctuations
- A direct or indirect labour market bargaining struggle between employers and employees
- Instability generated in the financial system
- lock-in into Technological paradigms and occasional paradigm changes
- Simple random shocks, a hypothesis that fits well with the neoclassical economic model (though this hypothesis has been strongly criticized)
- 'Natural' moving average filters. This also suggests that growth cycles are not regular cycles but merely recurring perturbation; a pattern that is consistent with empirical findings of different cycle length for different countries (de Groot and Franses, 2008, 2009). An explanation for the occurrance of implicit moving averages will be suggested in Chapter 5.

3.6 Growth theory: a pool of models

There are several approaches to modelling. Georgescu-Roegen slammed mathematical models in one of his papers (Georgescu-Roegen, 1975b) with the following analysis. Economists employ three different methods: the analytico-physiological, the mathematico-imaginative and the mechanico-descriptive.

The first one – the one that Geogescu-Roegen himself clearly preferred – is to analyze the observed phenomenon staying close to the facts and without using specific mathematical models. The second one (the mathematico-imaginative) is to construct elaborate but purely imaginary models without any relation to reality, as the name chosen by Georgescu-Roegen for this group cynically makes clear. The final one (the mechanico-descriptive – also a derogatory title) describes the approach of empirically backed model-building; 'mechanico' specifically refers to the simplification of the system under observation to some few of its properties; Georgescu-Roegen complains that 'some of them may do even more harm than the … mathematico-imaginative exercises' (Georgescu-Roegen, 1975b: p. 766). The question whether the analytico-physiological modelers themselves may be simplifying and whether their flaws are – compared to their mathematical counterparts – sometimes just hidden behind equivocal poetic elaborations is not addressed. Still, Georgescu-Roegen's analysis of the ways scholars in economics work, is sound and incisive. Returning to growth theory, this is an even more striking description. Some of the models are clearly part of the mathematico-imaginative branch – not caring at all about empirical data: Samuelson (1939a,b), Harrod (1939), Domar (1946), Uzawa (1965), Goodwin (1967), Romer (1986, 1990) and Conlisk (1989) among others. Others try to support their assumptions or results with empirics – Georgescu-Roegen could possibly describe that as mechanico-descriptive – but fail to avoid simplifications that are not backed empirically: among others Solow (1956, 1957), Nelson and Winter (1974, 1982a,b) and Lucas (1972). Some few including Georgescu-Roegen (1975b,a) himself but also Minsky (1980) and Freeman and Perez (1988) live up to what Georgescu-Roegen christened the analytico-physiological approach – though not necessarily with better results.

Empirical data for different countries, different sectors, and at more or less aggregated levels show rather similar pictures (see Figure 1.2 for an example). Usually the economic entities are growing in terms of production as well as income and capital. The growth occurs at a faster than linear – possibly exponential – rate and is subject to cycles. 'Cycle' usually refers to alternating periods of slower and faster growth, not actual shrinking of the economy.[125] The different classes of growth models discussed in this chapter try to approximate this process and are successful to vary different degrees. The AK and later EGT models generally achieve a good reproduction of the long-term average growth process – an exponential growth path. The short-term does, however, not match reality (cycles); even less are these models able to model the corresponding micro-level of this economy and the possibility of more severe disturbances and crashes. A final flaw of AK models is, as famously noted by Solow (1994), that they predict growth without bounds – however, immediately obvious from physical and energetic boundaries, there are real bounds that in reality can never be transgressed. The second class of growth models, the Solow–Swan models, still a textbook standard of the growth theory, predicts convergence to a global equilibrium (in per capita terms). Facing the exponentially rising time series of economic growth before 2008, the only answer from the perspective of Solow–Swan models could be that most of the empirically observed growth – most notably the convex shape of the growth data series – is accounted for by technological change exogenous to the model. Though one might thank this theory for reminding the profession that technology is indeed related to economic growth, this is, as Nelson and Winter famously argued, to derive exactly the results from the model that were previously fed into it as exogenous variables (Nelson and Winter, 1974). A completely different branch of growth models are the theories concerning growth cycles. Providing numerous mathematically elaborate models of cyclic patterns in (more or less) complex systems, this class of models has sought to explain growth cycles. Doing so, they showed that relatively simple mathematical methods from dynamic systems theory are able to accomplish this task; the growth process, however, is usually introduced separately – in Goodwin's case as an exogenous variable. This kind of combination in turn results in a rather good reproduction of observed real data,[126] however without real turbulence; with turbulence added the cycles generated by e.g. the Marx–Keynes–Goodwin model are not necessarily stable. Evolutionary growth theory applies exemplary microfoundations retaining the growth and the cycles as explained by rival theories. The driving force of economic growth of course is different in this class of growth models. In accordance with Schumpeterian theory, entrepreneurship arises from comparatively better performance and the resulting higher profits; better performance in turn is generated by better technology, which is attained stochastically depending on the R&D expenditures. Nelson and Winter's (1974) model results in growth data (see Figure 5.12) resembling real data series for the first couple of periods. The simulation of such a system reveals, however, that this dynamic regime collapses after a certain number of periods, giving way to a monopoly with only sudden step-wise growth (that occurs whenever the monopolist gains access to a new technology).[127] It is probable that a good and specific microfoundation is necessary to finally obtain a theory capable of powerful predictions. Out of all growth theories only the evolutionary approaches (Nelson and Winter, 1974, 1982b; Conlisk, 1989; Silverberg and Lehnert, 1993; Kauffman, 1993; Kauffman *et al.*, 2000) are remotely close to providing such a micro-foundation; the meso- and macroformulations of these theories, however, are still not convincing in the light of both what may be seen in the macroeconomic data series and what is known about of the meso- and macro-levels the economy. Specifically, the industry structure, the network effects, and the economic-organizational bounds to growth and technological and structural change – most importantly lock-ins – are mostly not accounted for. These mechanisms and their interrelations with growth theory will be discussed in the following chapters, particularly Chapter 4.

4 Network economies

The phenomenon of network externalities as a subject of interest to economics emerged in the 1980s. Within a few years, several seminal contributions by two to three[1] different groups of rather closely linked researchers were published. First were David and his group (David, 1985; David and Bunn, 1988) analyzing a number of historical cases of technology standards that may have had an inferior outcome; the approach was, however, purely qualitative. Then a group of mostly evolutionary and complex modelling economists around Arthur, David, Kaniovski and Dosi[2] (Arthur *et al.*, 1982, 1987; Arthur, 1989; Dosi and Kaniovski, 1994; Dosi *et al.*, 1994) and a group of more neoclassical game theorists[3] (Katz and Shapiro, 1985, 1986; Farrell and Saloner, 1985, 1986) independently formalized the theory. Both groups contributed a great variety of important findings, in turn developing a rich and accurate terminology for this field of study. The term *network externality* which was coined by Katz and Shapiro, (Katz and Shapiro, 1985) was intended to describe a special case of Coase's externalities.[4] This was a failure of a theoretical neoclassical market to attain the theoretical optimum resulting in one actor losing utility (compared to the optimum) due to actions of another actor, a loss that can for some reason not be traded, i.e. reduced by compensating the other agent for not taking the harmful action. In the present case of network externalities, this reason is generated by the specific property of a good to gain value for any individual user the more users subscribe to the usage of this good (or service). In the absence of an almighty planing entity the optimum state will probably not be attained. This effect is redoubled if the possibility of switching between goods with network externalities is restricted or if there are even switching costs. David (1985) and Arthur and his collaborators (Arthur *et al.*, 1987; Arthur, 1989) extend this argument to a technology (or an industry) being completely dominated by one standard (or product) that will continue to dominate this technology (or industry) even if the dominating standard (or product) is at some point in time not any more (or has never been) the 'best' one – purely by the means of an installed base of users. They call this effect a technological *lock-in*.

The vivid scientific discussion starting in the 1980s sparked continued development of this field of research with an increasing variety of models and applications notably by Klemperer (1987), Economides and White (1994), Choi (1994, 2004), Shy (1996, 2001, 2002), León-Ledesma (2000), Manenti and Somma (2002),

Sundararajan (2008), Kretschmer (2008), Leoncini *et al.* (2008), Stango (2004), Stango and Greenstein (2008) and Dolfsma and Leydesdorff (2009).

Though the broader interest in network externalities and the emergent field of research centring on this topic is rather new, the idea is not. It is generally agreed[5] that both the working of networks alongside its implications for industry dynamics and the impact of network effects on the development of the economy as a whole were recognized much earlier. The implications of network effects on industrial competition, namely entry-deterrence strategies in order to collect monopoly rents, had been implied in the distinction between 'competition in the field' and 'competition for the field' by Chadwick in 1859 (see Demsetz, 1968) and later Demsetz (1968). Network effects as such had already been discussed by Veblen[6] and the implications for economic growth have more or less explicitly been present in some heterodox growth theories, namely Kaldor (1957, 1972), Dixon and Thirlwall (1975), Forrester (1977) and later León-Ledesma (2000). The analysis of economic growth and network effects is connected to the theories of *Circular Cumulative Causation*[7] most of which draw largely on approaches by Veblen, Kapp, Myrdal, Georgescu-Roegen and others (Berger, 2009a,b). Nevertheless, the great advances of the theory of network externalities from the 1980s and onward have been neglected in the recent analysis of far-from equilibrium models of economic growth and technological change, with the notable exceptions of León-Ledesma (2000), Shy (2002) and Silverberg and Verspagen (2005, 2007).

It is this impact of network externalities on the nature of technological change, and consequently also on the progress of economic growth, that is the central topic of the current work. Having discussed many major theories of economic growth of the recent decades extensively in Chapter 3, the argument will proceed with a discussion of the approaches to network externalities from the 1980s onwards and their effect on economic growth. This will provide the basis for the development of a generalized model of growth and technological change with network externalities, and for empirical considerations and computer simulation in the following chapters (Chapters 5 and 6).

The current chapter will briefly discuss two technical issues that will be important later on. It will then proceed by discussing both the orthodox economic perspective (developing out of standard neoclassical theory) and the complex systems view on network externalities in sections 4.2 and 4.3 respectively. These two sections detail the core and many variants and aspects of the current state of the art technology of network externalities. Three more brief sections on less widely discussed questions follow with section 4.4 on the possible shape of network externality functions, section 4.5 on qualitative and quantitative evidence of network externalities, and section 4.6 on the extent to which network effects and network externalities are present and prevalent in different economic systems, industry sectors, and beyond. Section 4.7, immediately preceeding the chapter conclusions (section 4.8), on the possible impact of network externalities on technological change, is a theoretical bridge to the model in the chapter that follows.

4.1 Two technical preliminaries

Increasing returns are relevant for the subsequent analysis of network effects and network externalities in order to point out in which way they relate to the standard approach to microeconomics (the neoclassical 'market' perspective) and in which way they do not. In fact, there is no disagreement that a set of tools different from neoclassical equilibrium models is required to analyze network effects; even fairly neoclassical approaches (in the following called 'mixed neoclassical and game theory approaches') employ game theoretic models.

Furthermore, as both increasing returns and network effects work toward monopolization (i.e. large vendors, widely used technologies and standards, etc., are favoured), a few remarks on the neoclassical theory of monopolies and oligopolies are in order as well (section 4.1.2).

4.1.1 The trouble with increasing returns

In order for the method of general equilibrium calculus as discussed in section 2.1 to function properly, the objective function and the restriction function (specifically the isoclines of those functions) must not only be continuous (such that the problem is integrable) but also shaped in such a way that the equilibrium is unique. This is best accomplished by concave isoclines of the restriction function and convex isoclines of the objective function for maximization problems (and vice versa for minimization problems); for one of the two, the curvature may be zero (i.e. the isocline is linear).[8] It will still work while the curvature of the isoclines of the restriction function is smaller than that of the isoclines of the objective function but this is an inconvenient case.

Considering a simple maximization problem, concave isoclines of the restriction function correspond to decreasing returns: the more of a resource or product is required, the more expensive it becomes per unit. It was known from the earliest days of neoclassical field theory that these assumptions are rather heroic. Marginal production costs for instance may decrease with quantity thus generating increasing returns. In particular, Marshall (1990 [1890]) was aware of the problem and considered external effects as a concept to deal with it. Young (1928) discussed the problem at length and pointed out that increasing returns would in turn lead to convex or partly convex (non-monotonic curvature) isoclines. The optimization method that focuses on identifying identical slopes of the isoclines of restriction and objective function (i.e. tangential points), works well for monotonic curvature. With Young's case (non-monotonic curvature), the optimization may also yield saddle points and, of course, also multiple local equilibria.

That, however, is not all. Consider a case with the marginal costs $c'(x)$ increasing linearly with quantity x until reaching a natural boundary of some kind (at which point the marginal cost curve becomes vertical).[9] At $x = 0$, marginal utility $u'(x)$ is assumed to be larger than marginal costs, hence production (or trade, whatever the object of the maximization problem is) of the first units is feasible and profitable. As long as marginal utility is not increasing or increasing slower than marginal

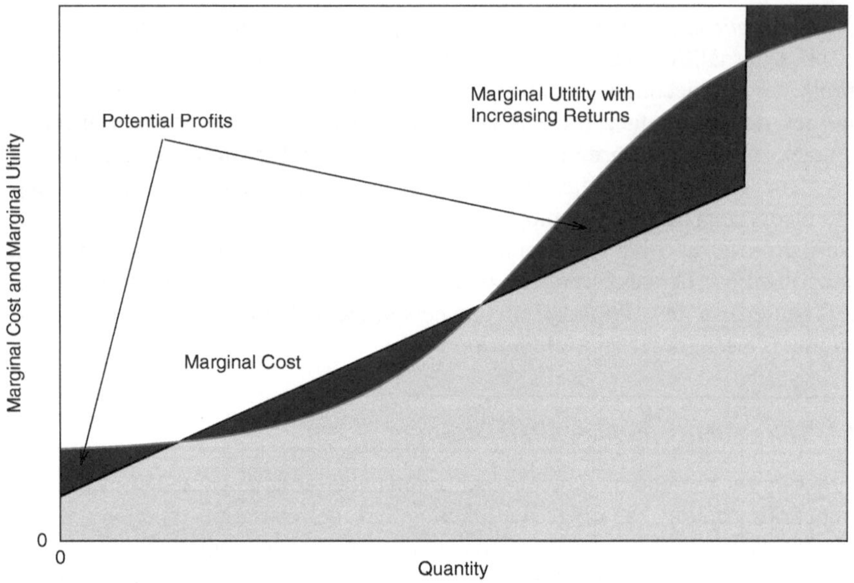

Figure 4.1 Increasing returns to quantity and feasible (profitable) modes of production.

costs (i.e. in the case of decreasing returns), a unique equilibrium

$$x^* > 0: \quad c'(x^*) = u'(x^*) \quad \text{with} \quad \frac{\partial c'(x)}{\partial x} > \frac{\partial u'(x)}{\partial x} \tag{4.1}$$

exists. For increasing returns (i.e. marginal utility increasing faster than marginal costs), this is not true any more. Figure 4.1 illustrates the situation. The areas with marginal utility greater than marginal cost (the lighter areas) indicate that production is feasible and profitable in this area. There are two locally stable equilibria where the two functions intersect and the marginal costs grow larger than marginal utility, i.e. the marginal cost function crosses the marginal utility function from below. One equilibrium sustains a much higher quantity of production than the other. Additionally, with distributed decision making without perfect foresight and perfect information – conditions that are quite likely in technology choice and other situations with network effects – this might even lead to catastrophic events. Should, for a situation starting in the higher equilibrium, the quantity fall for some reason (e.g. some kind of random shock) below the value that makes high production feasible (i.e. the system is then in the section in the middle of Figure 4.1), production will immediately fall to the lower equilibrium.

For a two-dimensional system with two resources or goods both of which are subject to increasing returns, the usual picture of monotonic isoclines and indifference curves becomes one with a heavily irregular shape. As shown in Figure 4.2, the cost levels may increase and then decrease again for a certain

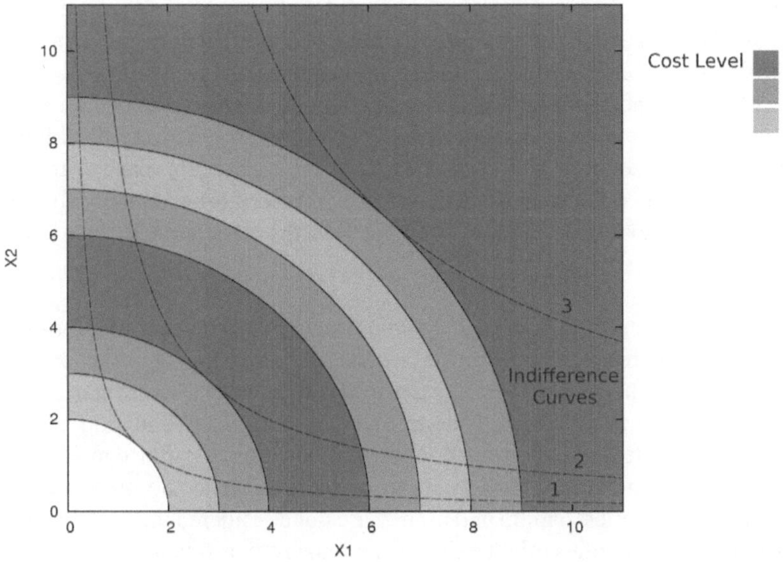

Figure 4.2 A system of cost levels (stylized to be discrete, darker shades indicate higher cost level) and indifference curves with increasing returns.

higher quantity. The 'pocket' of higher costs between the lower and higher cost-efficient quantities would never be chosen, the system would 'jump' between the lower and higher cost-efficient quantities (if budget changes or something similar require that).

Increasing returns have been studied by a number of scholars with respect to growth theory (Kaldor, 1972, 1975; Romer, 1986; León-Ledesma, 2000, 2006), development theory (Matsuyama, 1991), network effects (Arthur, 1989, 1996), and economic geography (Krugman, 1997). Network effects are one of the primary examples of increasing returns: The size of a network, and hence the total quantity in use, make the product more valuable, thus increasing either the revenue of the producer side, the consumer's utility, or both. Either way, at some point, the growing size of the network increases profitability or utility, thus inducing further growth. It is clear that the neoclassical equilibrium approach is insufficient to address such systems properly. Most formal models of network effects have therefore incorporated elements and methods of game theory or complex systems theory.

4.1.2 Neoclassical monopolies: mighty market-leaders and powerful price-makers

Several sections in the first part of this chapter employ a mixed game theory and market approach to analyze network externalities. This is convenient as the scholars referred to who identified the phenomena under consideration have used this approach. The neoclassical model of monopolized sectors as developed by Cournot (1897 [1838]) and integrated into a general economic theory by Marshall (1990 [1890]) as a general approach to industry economics has, however, some flaws. Therefore, it is necessary to challenge and amend the microfoundations of the model at this point and to include elements of game theory. Strictly speaking, none of the following sections uses a pure neoclassical market approach without game theory, so this is done rather implicitly. Thus, before proceeding with the analysis of network externalities some remarks on the neoclassical theory of monopolies are in order.

- First, with well-behaved neoclassical demand and cost functions, the former being strictly decreasing, the latter constant or increasing, it is generally obscure why the monopoly should exist in the first place.[10] Information is assumed to be perfect, thus the mode or technology of production is no secret in this model; specifically the information regime is not considered the cause of the monopoly. Willingness for investment in the presence of profitable opportunities, capital and money are not considered unavailable, to the extent that monopolies might exist because no entity is financially capable of challenging it. A newcomer would turn a monopoly into a Cournot duopoly with positive net profits; the next agent to enter the market would still be able to appropriate profits by shifting the situation a little more towards perfect competition until perfect competition is finally reached (or in the mathematical model rather asymptotically approximated).[11] Therefore, the question must arise how neoclassics can explain monopolies at all. As mentioned above, the classic example both Cournot (1897 [1838]) and Marshall (1990 [1890]) used to justify and introduce the monopoly model in their respective works was a unique natural resource the monopolist owns or controls and that is not artificially reproducible.[12] For this case the neoclassical system is a perfectly concise model; however, it is not applicable to generally producible and reproducible goods and services.
- Second (with respect to the textbook presentation of the first problem), the neoclassical monopoly and oligopoly theory it usually either introduced using this example (as Cournot and Marshall did themselves) or the model is just presented as a generalized model of monopolies without reference to the perfect competition model (Varian, 1992; Mas-Colell *et al.*, 1995). By abstractly defining the model's assumptions, the price setting power of the monopolist and the restriction of entry into the market, they avoid the problem of explaining the relation between this and the neoclassical perfect competition model, and why the former does not just devolve into the latter. In fact, the usual reasoning with respect to the role and purpose of the monopoly model is in the other direction: why the perfect competition equilibrium does not devolve into a monopoly solution in the presence of – empirically observed – increasing returns, an issue that soon came under heavy attack from less neoclassical economists such as Young (1928).[13]

- Third, returning to the question of how to apply the neoclassical monopoly model to a general setting, a setting without natural monopoly or perfect control of the earth's complete supply of a particular natural resource. There are two possible answers. On the one hand, a neoclassical explanation can operate with fixed costs – in effect the definition of a 'natural monopoly' (the theoretically possible monopoly rents in the neoclassical market are lower than twice the fixed costs of production, leaving room for just one producer to work profitably). On the other hand, game theory can help neoclassics to derive another solution[14]: If a newcomer enters the market, the monopolist will drive the price below marginal costs, incurring heavy losses both for herself and the newcomer until the newcomer gives up. While the latter explanation does also change the general approach of neoclassical theory, both explanations are in essence insufficient to explain the wide varieties of monopolies that may be observed in reality. The monopolies present in real-world economic systems include sectors which dynamics neoclassical theory would probably describe as 'monopolistic competition', one quasi-monopolist, many small competing suppliers. Many of the sectors of the new economy displaying heavy network externalities show such a structure.[15] This certainly does not involve natural monopolies, as obviously the many small competitors were able to pay the fixed costs of production. However, network externalities and the competition of standards may be a possible explanation for this (both with neoclassical and heterodox models).
- Fourth, with network externalities of technologies and the products employing them, the emergence of monopolies becomes logical, an idea that has been developed using mixed game theory and neoclassical market approaches by several scholars (Katz and Shapiro, 1985, 1986; Farrell and Saloner, 1985, 1986; Klemperer, 1987). As will be detailed below, (Bertrand-type) price competition is the rational strategy for the competitors to allow their own technology to build a network that will be able to both attract further adopters and to reduce the temptation for current adopters to defect to the competing standards. Even a model based principally on the neoclassical market setting will yield monopoly outcomes in this case.

Network effects add another aspect: increasing returns with the number of consumers, i.e. in effect, with every consumer who bought the product (or access to the network) the reservation price of the next buyer is decreased and she is willing to pay a little more. With increasing returns to the number of buyers, the monopolist will thus try to appropriate the increased utility using price discrimination. She restricts the number of goods sold at a time and raises the price frequently to match the reservation price of the customer who will be next to buy (among the remaining customers the one with the lowest reservation price). At this point, however, this mode of production becomes vulnerable to imitation and thus 'product piracy'. A 'product pirate' imitates the monopolist's product or technology and sells at a considerably lower price. Note that product 'piracy' does not make sense in normal perfect markets, since the 'pirate' would face the same production costs as the vendor of the original. In a monopoly with network externalities, however, the monopolist loses the sales the product 'pirate' has realized and also a part of the monopoly rents due to the substantially lower price. Thus, the monopolist will be able to establish a monopoly position for the network technology; it is, however, unlikely that she will be the single vendor of this technology and even if she is, she will not be able to appropriate large shares of the monopoly rents. (For details see the section on 'piracy' below.)

4.2 Modelling frameworks of network economies: the traditional perspective

Including interactions and interactive decisions of agents causes considerable changes to the development of the aggregate level (the macro-system) even if typical neoclassic assumptions such as perfect rationality and complete information are retained. This is because a rather simple isomorphic system – the micro-layer identical to the macro-layer (the neoclassical view) – is replaced by a large number of inherently different agents (network view). This inherent difference emerges through their distinct places in the network, their neighbourhood, etc. Chen compares this to the difference of a single body system and a multi-body system in mechanical physics, the former being trivial in nature and movement while the latter may give rise to chaos (Chen, 2002). It should also be noted that a number of other scholars, including many geographical economists, pointed out this fact and reasoned that the neoclassical view is a tremendous oversimplification of reality (Williamson, 1973; Krugman, 1997; David, 2005; Kirman, 1997). The theorists discussed in this section were somewhat more reluctant to a voice general and fundamental critique of the orthodox economic theory but they presented important insights in the functioning of networks and their consequences for economic theory, and especially for the theory of technological change.

4.2.1 Network utility and network externality

A utility an agent experiences by using a particular product, standard or technology that does not arise from the properties inherent to this product, standard or technology themselves, but rather from the number of other agents that use the same product, standard or technology shall – following Katz and Shapiro (1985) – be called this agent's *network utility*. Consider a – rather traditional – utility function

$$u(r,n) = r + v(n) \qquad \text{with: } \frac{\partial v(n)}{\partial n} > 0 \tag{4.2}$$

where u is the utility, r is that part of the utility that arises purely from the physical and technological properties of the product (or standard or technology) itself,[16] and n is the number of subscribers to the usage of this product (or standard or technology). As determined by the derivative of the function $v(n)$, u rises strictly in n and indicates therefore a positive network utility $v(n)$ in addition to its physical and technological utility for the user r. The effect of a user subscribing to a particular product (or standard or technology) on another user's utility (thus $v(n) - v(n-1)$) shall be called the *network externality* (see below for a more detailed analysis).

Again, following Katz and Shapiro (1985), indirect effects leading to the same economic consequence as the direct effect shall be called *indirect network externalities*. A simple example are economies of scale in the production of the product in question which leads to the product being cheaper the more users are willing to buy it.[17] In this case the utility function would have to be constructed as containing a term for the purchasing costs c.

$$u(r,n) = r - c(n) \quad \text{with:} \quad \frac{\partial c(n)}{\partial n} < 0 \tag{4.3}$$

This is mathematically equivalent to equation (4.2) above: It consists of an internal physical utility $(r - c(0))$ and a network utility $(c(0) - c(n))$ as well.

Katz and Shapiro published their analysis of this phenomenon in two subsequent papers (Katz and Shapiro, 1985, 1986) the first of which introduces the general terminology and proceeds to derive the phenomenon from a simple microeconomic model, while the 1986 paper analyzes the game theoretic properties of the model and their implications in detail. They published the first paper earlier in 1983 as a discussion paper of the Woodrow Wilson School (Princeton), which had been known to other researchers working in the area (Farrell and Saloner, 1985), though only the 1985 publication gained widespread interest. The model and notation of network externalities developed by Katz and Shapiro (and similar models) have subsequently been used by many of the scholars concerned with this matter including Arthur (1989), Klemperer (1987), Choi (1994) and Shy (2001).

In the following, a modified version of Katz and Shapiro's (1986) model will be considered. This model shall furthermore be extendable to cover some of the phenomena discussed by other contributors to the economic literature on network externalities, which will facilitate using the same model in subsequent sections on bandwagon effects (section 4.2.2) and public and private sponsoring of standards (section 4.2.3). The most basic case is a two-standard two-period case (the case that was discussed by Katz and Shapiro). To define this model, the following assumptions are made:

1. There are two mutually exclusive standards i; $i = A, B$.
2. The model comprises two time periods, $t = 1, 2$, with a number of consumers to choose in each of the two periods between A and B and to adopt exactly one of the two standards. Call the number of consumers in period 1, N_1, and the number of those in period 2, N_2, the total number of consumers being $N = N_1 + N_2$. The model may be extended to assume the divide between the two time periods is arbitrary, defined by the number of consumers that have already chosen and those that are still to chose $D = \{N_1, N_2\}$, and apply the

derived theorems to all possible divisions.

$$D_\Xi = \{\Xi, N - \Xi\} \quad \Xi = 0, 1, 2, ...N$$

3. The utility of the consumers is comprised of an *intrinsic utility*[18] $r_{i,t}$ (where i denotes the standard as above and t is the time period) and a network utility $v(n_i)$ depending on and increasing steadily with the number of consumers using the same standard[19] n_i. Hence the total utility of a consumer is

$$u = r_{i,t} + v(n_i)$$

4. As the unit price of a standard is equivalent to a negative intrinsic utility, the intrinsic utilities should be assumed to be net utilities (to include the unit price for the purchase of the standard).
5. Agents are – for now – assumed to be homogeneous.
6. No utility is generated even for period 1 users before period 2 users have made their choices. The only effect of the division into two periods is thus that period 2 users are informed about period 1 users' choice while period 1 users can only anticipate what period 2 users might rationally do. Nevertheless, for certain cases (see below) the outcome of this model depends on which users have the more *credible thread* not to follow the other's choice. As shown below, in the basic model this advantage rests with period 1 users.
7. It is assumed that the usual assumptions of game theory hold: all agents are defined as rational with a well-defined preference ordering according to their utility as well as well-informed about the rationality of all other agents (*common knowledge of rationality*).
8. All agents are assumed to be perfectly informed about the rules of the game and about the preferences of all the participants (perfect information).

The discussion of this model follows again the analysis by Katz and Shapiro (1986) in their slightly different model which forms the basis for the current one. Where possible, the equations will also follow the notation used by Katz and Shapiro.

As all agents are perfectly rational and homogeneous, all agents choosing in the same period are completely equivalent and will thus always make the same decision. Further, as it is known when the game ends, the game may be solved using backward induction. The second group knows the choice of the agents that chose in period 1; without loss of generality it may be assumed they chose A. In this case, consumers in period 2 face the following condition.

$$r_{A,2} + v(N_1 + N_2) \geq r_{B,2} + v(N_2) \tag{4.4}$$

If this condition holds true they will make the same decision as their first period counterparts (choose A). Call the difference of the net intrinsic payoffs in period 2 $\Delta_2 = r_{B,2} - r_{A,2}$. The condition becomes (corresponding to Katz and Shapiro (1986), equation (1))

$$v(N_1 + N_2) - v(N_2) \geq r_{B,2} - r_{A,2} \equiv \Delta_2 \tag{4.5}$$

The opposite inequality

$$r_{A,2} + v(N_1 + N_2) \leq r_{B,2} + v(N_2)$$

would imply that they choose *B*, the competitor standard to the first period consumers' choice.[20] Period 1 consumers do, however, foresee the result of the second period. If they foresee period 2 consumers will choose *A* (that is, if inequality (4.5) holds) they face the decision to choose *A* if

$$r_{A,1} + v(N_1 + N_2) \geq r_{B,1} + v(N_1)$$

hence (defining $\Delta_1 = r_{B,1} - r_{A,1}$)

$$v(N_1 + N_2) - v(N_1) \geq r_{B,1} - r_{A,1} \equiv \Delta_1 \tag{4.6}$$

Otherwise, if they foresee the following (period 2) consumers will choose *B* (that is, if inequality (4.5) does not hold), they ignore the anticipated action of the period 2 consumers and nevertheless choose *A* if

$$v(N_1 + N_2) - v(N_1) \leq r_{A,1} - r_{B,1} \equiv -\Delta_1 \tag{4.7}$$

These two cases are the only relevant ones: either inequalities (4.5) and (4.6) hold resulting in all agents choosing standard *A*, or inequality (4.5) does not hold but inequality (4.7) does; consequently, first period consumers would decide for standard *A* while second period consumers opt for standard *B*. The remaining cases (all agents choose *B*, and agents choose *B* in the first and, *A* in the second period) would defeat the underlying assumption, that *A* is the standard chosen by first period consumers. In this case, as the system is otherwise perfectly symmetric between the two competing standards, *B* would simply take the position of *A* directly resulting in an equilibrium with fulfilled underlying expectations.[21] All possible cases are documented in Table 4.1.

To analyze the particular consequences of this pattern it shall first be assumed that the intrinsic preference ordering for the two technologies does not change over time (this case is shown graphically in Figure 4.3), that is

$$\begin{aligned} &\Delta_1 > 0 \Leftrightarrow \Delta_2 > 0 \\ &\Delta_1 < 0 \Leftrightarrow \Delta_2 < 0. \end{aligned} \tag{4.8}$$

Table 4.1 Cases and outcomes in two-period network-choice games with perfectly rational agents

	Second period	*First period*	*Result, first and second period*
	Inequality (4.5)	*Inequality (4.6)*	
Case 1	true (⇒ *A*)	true (⇒ *A*)	First: *A* and second: *A*
Case 2	true (⇒ *A*)	false (⇒ *B*)	Self-defeating expectations
		Inequality (4.7)	
Case 3	false (⇒ *B*)	true (⇒ *A*)	First: *A* and second: *B*
Case 4	false (⇒ *B*)	false (⇒ *B*)	Self-defeating expectations

In this case, the first and second period consumers always make the same choice (case 1 from Table 4.1). This is obvious from the fact that different choices for different periods are only possible if inequality (4.5) does not hold but inequality (4.7) does. This implies (as the left side of both inequalities is positive) that $\Delta_2 > 0$ and $-\Delta_1 > 0$ which violates condition (4.8). Indeed, if for a standard A either inequality (4.5) or inequality (4.6) do not hold, the agents choosing in the respective period will not only opt for the competing standard B but are also able to force their counterparts choosing in the other period to match their choice. As the intrinsic preference ordering remains the same for both periods, this choice is always optimal (both Pareto optimal and socially welfare optimal). However, if this is not the case, it is possible that

$$\begin{aligned} v(N_1+N_2)-v(N_2) &\geq \Delta_2 > 0 > -\Delta_2 \\ v(N_1+N_2)-v(N_1) &\geq \Delta_1 > 0 > -\Delta_1. \end{aligned} \tag{4.9}$$

In this case there are two equilibria fulfilling their underlying expectations: both groups of consumers choosing A and both groups of consumers choosing B. Out of these two equilibria, only one can be optimal while the other is strictly inferior. In this example, given by conditions (4.9), B is strictly superior since

$$\Delta_i = r_{B,i} - r_{A,i} > 0.$$

Still, if both groups believe the other group will choose A, they refrain from pursuing the better option B, resulting in a system of self-fulfilling prophecies of the dominance the inferior standard not entirely different from Cooper and John's (1988) idea of the coordination failure.

Figure 4.3 shows the resulting situation: the areas along the coordinates (BB) are necessarily superior results; the area with stronger network effects in both periods (AA or BB) that corresponds to conditions (4.9) is not. In this case, first period consumers still possess the more credible threat. In fact, using ordinary backward induction, first period actors know that second period consumers will follow their decision. In turn, they choose according to their net intrinsic utility (B) and still attain the optimal result.

This picture changes dramatically when considering changes in the intrinsic preference ordering (that is, conditions (4.8) do not hold). Consider $\Delta_1 < 0$ and $\Delta_2 > 0$: this means the first standard (A) is intrinsically preferable in the first period, while, by the time the second period consumers choose, standard B has become the better one in terms of net intrinsic preferences. Reversing the positions of A and B (and hence of Δ_1 and Δ_2) leads to a case exactly symmetric to this one (therefore this one shall suffice). Changing intrinsic preferences is also a central focus of what Katz and Shapiro discussed when they first used the term network externality (Katz and Shapiro, 1985, 1986).

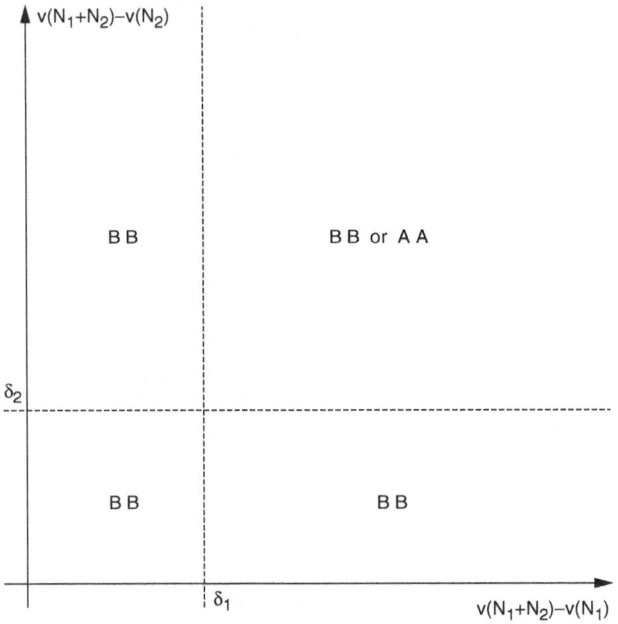

Figure 4.3 Standard adoption by the first and second groups with time-constant preferences (i.e. equation (4.8)) depending on the network utility differencees $v(N_1+N_2)-v(N_1)$ and $v(N_1+N_2)-v(N_2)$. The superior technology in this case is B; if the other technology A is superior, the picture is symmetric (see the lower left graph in Figure 4.4.)

Again, if the intrinsic utility difference is greater than the network utility (the areas along the axes in Figure 4.4), that is

$$v(N_1+N_2)-v(N_2) \leq \Delta_2$$
$$v(N_1+N_2)-v(N_1) \leq \Delta_1, \tag{4.10}$$

at least one of the two groups decides autonomously from the other. In this case the result is still optimal with respect to general welfare; Pareto optimality is usually not attained. If the network effect does, however, exceed the effect of the intrinsic utilities, the result is not necessarily optimal any more. Here again, for both A and B, inequalities (4.5) and (4.6) hold, expectations are self-fulfilling, and the two equilibria are AA and BB (as seen in Figure 4.4). On which one of the two equilibria the actors coordinate depends on which of the two groups has the more credible threat to enforce their preferred option. Applying backward induction, this is always the first mover group as they cannot reconsider their decision. On the other hand, for real systems the second group would have every reason to somehow commit to choosing their preferred standard before the first group gets the chance to choose. Either way, one of the solutions is welfare optimal (though

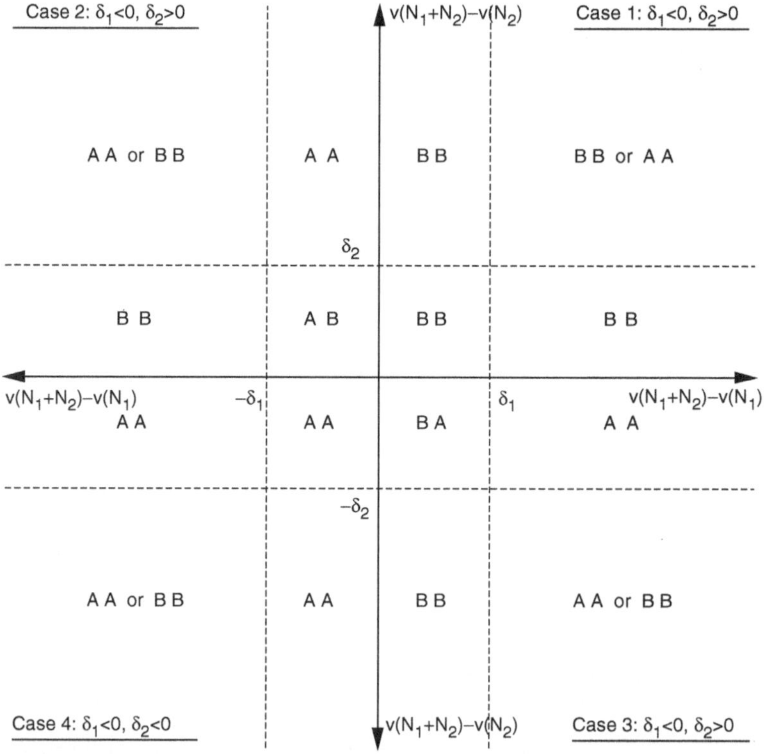

Figure 4.4 Standard adoption by the first and second group with changing preferences (i.e. equation (4.8) does not hold) depending on the network utility differencees $v(N_1+N_2)-v(N_1)$ and $v(N_1+N_2)-v(N_2)$. Note that these are four different graphs (four cases), each axis shared by two of them; that is, both axes are positive in both directions (negative network utility differences are impossible by definition). The upper right graph (case 1) is identical to Figure 4.3.

both are Pareto optimal); which one that will be is absolutely independent from the question of which one wins the bid (as this question only depends on the more credible threat).

Hence, the described situation is the easiest example of what is called a (negative) network externality, an effect by which agents lose utility through actions of otherwise unrelated third agents. This effect is beyond their control. Moreover, they do not have any immediate possibility of bribing other agents into enforcing the superior equilibrium, even though this would generate additional utility (or additional 'welfare'[22]).

It should, however, be noted, that this term is inherent in a rather traditional benchmark approach to the economy. It only makes sense when compared to a first-best optimum (in this case the welfare optimum), a consideration that indeed is central to Katz and Shapiro's writings on network externalities. Of course, this comparison is also present in more heterodox theories on network externalities (if not, the term network externalities would be misplaced), but for David and collaborators (David, 1985; David and Bunn, 1988) as well as Arthur and his co-authors (Arthur *et al.*, 1987; Arthur, 1989), this is more another consequence of a lock-in situation, while Katz and Shapiro and other economists in the neoclassical tradition[23] conduct extensive and exact computations of the social welfare in first-best (benevolent dictator) and other solutions.

4.2.2 Bandwagon effects

Consider the distribution of the adopters on the two periods $D(\Xi) = \{\Xi, N - \Xi\}$. Shifting Ξ towards lower values (that is shrinking the first and expanding the second group) leads to increasing $v(N_2)$ and – $N_1 + N_2$ being unchanged – to smaller $v(N_1 + N_2) - v(N_2)$. This is identical to the left side of inequality (4.5); thus the condition (4.5) tends to be violated more easily. As the condition defines when second period consumers are forced to follow the first period consumers' choice, the second group gains more independence. On the other hand $v(N_1 + N_2) - v(N_1)$ – the left side of inequality (4.6) – increases. As a consequence, condition (4.6) tends to hold true more often, which means that first period consumers are more dependent on the second period actors choices. The modification shifts power from the first to the second group, as the first group shrinks and the second group increases in number. The opposite is of course true if Ξ is shifted towards higher values closer to N – moving agents from the second to the first group.

Modifying the model such that consumers choose one by one and not simultaneously in groups, the dynamics change slightly. Intuitively, those who choose earlier gain an even stronger power as they remain a credible threat, while later consumers choose in small groups as well as without any certainty about the choice of other late-choosing agents. Suppose the first agent has a slight preference for product A ($r_{A,1} > r_{B,1}$) while all others would – without network effects – prefer B ($r_{A,i} < r_{B,i}$ with: $i \neq 1$). With the network externality resulting from the first agents' choice, they choose, however, A against their 'intrinsic' preferences. This leads the second agent to choose A because the first agent did, the third agent to choose A because the first two did, etc. If the second one would have chosen B instead, the network utility from A and B would have been equal for the third agent as they both would have had an installed base of one agent – however, the second agent might not know or anticipate or trust the following actor's preferences. This obviously leads to a – possibly vastly – inferior result. The effect is implied in Veblen's Veblen (2008 [1899]) writings on conspicuous consumption and trends; since the mid twentieth century, and with the rise of dynamic theory, the phenomenon has received more attention as a central and illustrative example for positive feedbacks. It has been termed 'bandwagon effect' (Leibenstein, 1950) and has played and continues to play a major role in the literature on network effects (Farrell and Saloner, 1985, 1986; Banerjee, 1992; Kretschmer, 2008).

The model cannot be described as the group-wise decision problem any more. Consider the decision of a consumer $\kappa \in N$. Call the number of consumers that have chosen before κ (the consumer κ) $N_{ch(\kappa)}$ and those among them that chose the first and the second product $N_{ch(\kappa),A}$ and $N_{ch(\kappa),B}$ respectively. Inequality (4.5) becomes

$$v(N_{ch(\kappa),A}+1)-v(N_{ch(\kappa),B}+1)\geq r_{B,\kappa}-r_{A,\kappa}\equiv\Delta_{\kappa}. \tag{4.11}$$

If strictly true, that is if

$$v(N_{ch(\kappa),A}+1)-v(N_{ch(\kappa),B}+1)>r_{B,\kappa}-r_{A,\kappa}\equiv\Delta_{\kappa},$$

κ chooses A, if the inequality (4.11) is not fulfilled, that is

$$v(N_{ch(\kappa),A}+1)-v(N_{ch(\kappa),B}+1)<r_{B,\kappa}-r_{A,\kappa}\equiv\Delta_{\kappa},$$

κ chooses B. If both sides of the inequalities are equal, κ is indifferent. If there exists a $\kappa^*\in N$ such that, given that all prior consumers $N_{ch(\kappa)}$ and κ^* chose rationally, for all later-choosing agents ($\kappa>\kappa^*$) the absolute value of the network effect is always larger than the absolute value of the difference of the net intrinsic payoffs

$$|v(N_{ch(\kappa),A}+1)-v(N_{ch(\kappa),B}+1)|>|r_{B,\kappa}-r_{A,\kappa}|\equiv|\Delta_{\kappa}| \qquad \kappa>\kappa^*, \tag{4.12}$$

all following agents $\kappa>\kappa^*$ must take the same decision, either all of them choose A or all of them choose B. If further, there is at least one among the agents κ choosing after κ^* with a net intrinsic preference for the action not taken, $r_{B,\kappa}-r_{A,\kappa}>0$ if these agents choose A or $r_{B,\kappa}-r_{A,\kappa}<0$ if they choose B, the situation is said to exhibit a *bandwagon effect*. The same argument still holds if κs are considered not to stand for single agents but for small groups of agents.

In addition, a variant of the theorem of Arthur *et al.* (1982) may be applied to this model.[24] Let $N\to\infty$ and assume that the difference of the net intrinsic payoffs Δ_{κ} of all κs are bounded, that is the function has an absolute minimum $\Delta_{\kappa,min}$ and an absolute maximum $\Delta_{\kappa,max}$ while the network effects are cumulating over time, then the boundaries $\{\Delta_{\kappa,min},\Delta_{\kappa,max}\}$ may be considered as absolute[25] *absorbing barriers*. Once the network effect $v(N_{ch(\kappa),A}+1)-v(N_{ch(\kappa),B}+1)$ transcends the boundaries in one or the other direction, all following agents will choose the dominating network technology, A (or B respectively) if the lower (or upper) boundary had been transcended (see Figure 4.5). As N tends to ∞, the probability that the process remains in the area of 'intrinsic' choice between the two boundaries tends to zero. The bandwagon effect is another notation for the phenomenon of network effects. If a specific network of users of a particular product, standard or technology is large enough, more and more agents decide to use this network. The bandwagon effect captures specifically a situation of subsequent decisions of different agents, where later decisions are influenced by earlier choices that in turn were taken only because of still earlier choices to join the same network. This may be the case to the extent that a network may always be just large enough to incite just one more agent to join the network, in effect turning the whole population into adopters.

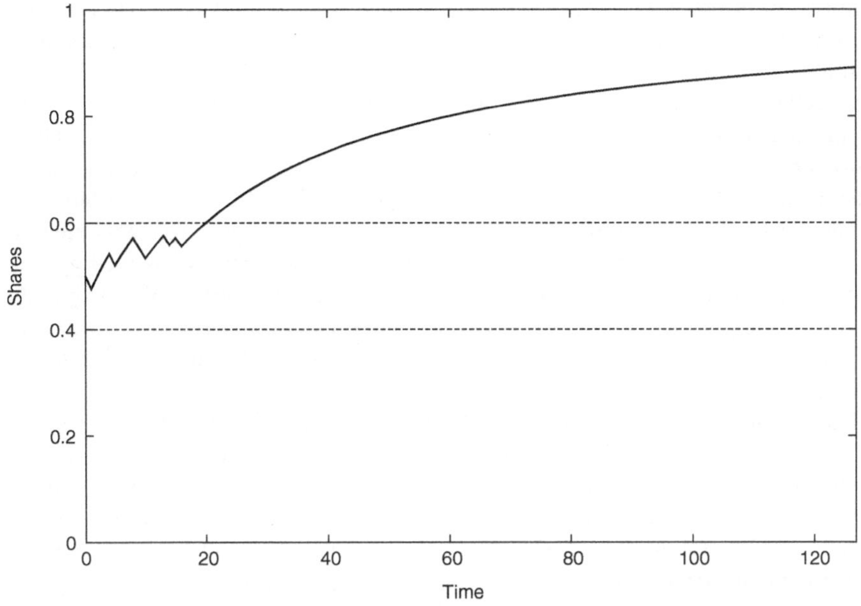

Figure 4.5 Choice process with network externalities and absorbing barriers; generated by computer simulation. The source code is available from the author on request.

Network externalities, bandwagon effects and technological change

Farrell and Saloner (1985, 1986) studied bandwagon effects from a slightly different aspect. Their focus was on how standards are replaced by newer and better standards. This can only happen if at least some agents are prepared to switch just for the utility inherent in the newer standard, even at the cost of losing the network utility.[26] The sequence of decisions is not determined a priori. Rather, at any point in time, any agent may decide to switch or to continue using the incumbent standard, say A. As the focus is the transition to newer standards, $r_B - r_A \equiv \Delta$ is always positive. The decision problem at the time the new technology (standard) becomes available is therefore ($\kappa \in N$ denoting once again a specific actor with specific intrinsic preferences $r_{i,\kappa}$)

$$v(N) - v(1) \leq r_{B,\kappa} - r_{A,\kappa} \equiv \Delta_\kappa. \tag{4.13}$$

If true, κ leaves the incumbent standard's network to become the first adopter of B (or may at least do so if both sides of the inequality are equal and κ is consequently indifferent). Call the number of agents that have adopted the new standard at time t ch_t, the decision problem becomes therefore

$$v(N - ch_t) - v(ch_t + 1) \leq r_{B,\kappa} - r_{A,\kappa} \equiv \Delta_\kappa. \tag{4.14}$$

As v is strictly increasing, the left side of the inequality decreases in ch_t and thus continually as long as ch_t increases, which renders actors with stronger valuation of network utility more prone to join the first movers in adopting the better standard B.

It is obvious from the problem, that the bandwagon effect for the choice of the new technology may cease at some point – specifically if for none of the $N - ch_t$ agents that are still using the older standard the inequality (4.14) holds true. This, however, is still the consequence of a bandwagon effect – specifically that of the incumbent standard – as the only reason not to switch to the new technology is the network effect of the installed base of the old standard, in essence a network effect with related bandwagon potential that had merely been weakened by the appearance of the new technology. The game theoretic properties of this tipping *bandwagon equilibrium*, the sharing of the population between the incumbent and the newcomer technology, have been analyzed in detail by Farrell and Saloner (1985, 1986).

Farrell and Saloner also introduced the issue of the efficiency or inefficiency of switching at a particular time. In the case considered above, it is welfare efficient to switch to the newer standard B since everyone has intrinsic preferences for B and the network utility is the same for both standards (if the eintire population switches). Consequently, the new standard is the social optimum (and also the new Nash equilibrium). If the bandwagon effect is only large enough with a great number of agents adopting the better technology, agents will tend not to switch too early, or, if all agents behave in the same risk averse way, maybe never at all. The outcome is therefore not efficient, a situation termed 'excess inertia' by Farrell and Saloner.

Network externalities, technological change and the public good problem

One further extension is the inclusion of beliefs about the bandwagon effects triggered by one's own decision. Since all agents are indentical except for their intrinsic preferences r, they will – subject of course to the time of switching – all have exactly the same network effect on the rest of the network. It is, however, by no means certain that they are informed about this fact, even more so taking into account that their beliefs about other agent's beliefs also play a major role in this consideration. Call the number of agents a particular agent κ believes to be able to convince to follow her in joining the newcomer B's network $g(\kappa) \leq N - ch_t$. Inequalities (4.13) and (4.14) become[27]

$$v(N) - v(g(\kappa) + 1) \leq r_{B,\kappa} - r_{A,\kappa} \equiv \Delta_\kappa \tag{4.15}$$

and

$$v(N - ch_t) - v(ch_t + g(\kappa) + 1) \leq r_{B,\kappa} - r_{A,\kappa} \equiv \Delta_\kappa . \tag{4.16}$$

Note that this renders obsolete the somewhat heroic assumption that at least the first agent must have such large intrinsic preferences for the new technology that she discards the installed base of the incumbent technology altogether.[28] Given a strong conviction about the imminent takeoff of the new technology and a willingness to take the risk of possibly being wrong, hence a large enough $g(\kappa)$, κ would leave the incumbent network without large intrinsic benefits and fully aware of the risk she is taking.

As will be shown, this situation is conveniently described as a public good problem, the public good being technological progress or specifically the implementation of the innovation of the better technology B. The commitment to be made for the production of the public good in turn is the risk the early adopters have to take. Let it be assumed, as an easy and discrete formalization of this situation, that there are two mutually exclusive network technologies (once again A and B) and let the utility for each agent from using these technologies be $u_A = 1$ (for A) and (for B)

$$u_B = \begin{cases} 2 \text{ if the network } B \text{is large enough} \\ 0 \text{ otherwise.} \end{cases}$$

For simplicity, the payoff functions are assumed to br discrete and not to depend on the exact size of the network in a continuous manner like in the models above. This is, in effect, the most simple case of s-shaped network externality functions that behave very similarly. It is reasonable to assume an s-shaped function to account for both the limited value of a small network and the limited effect of the addition of some more agents to a generally accepted network for the network itself. In between a critical size for the takeoff of the technology may be assumed. Figure 4.6 contains a graphical visualization of the implied class of network externality functions; see section 4.4 for a broader elaboration. Let A be the incumbent technology with all N agents in the system as current adopters, while for B to generate positive utility, at least Ξ adopters are needed. Agents may adopt B at any time; they are well-informed about the game and the payoffs but hold different beliefs concerning the composition of the population with respect to risk aversion and willingness to make commitments in order to achieve general adoption of the better standard. In other words, the situation is subject to strong strategic uncertainty.

It is obvious that agents will either make the commitment and take the risk or not; if we further suppose that agents do not adjust their beliefs without hard evidence (hard evidence being the takeoff of the network of technology B after gaining the support of Ξ agents), those who are inclined to make the commitment make it immediately. Those who do not, however, are able to share the benefits – thus it is conveniently classified as a public good with non-rivalry and non-excludability – and will, as soon as they learn that the network of technology[29] B took off, adopt technology B. Hence, the game comes down to a two-period game and, considering that the second period has a dominant path of action (either non-adopters of the first period switch to B if the network is large enough, or first period adopters switch back to A if it is not) and is perfectly predictable with backward induction,[30] the interesting part of the game is essentially a one-period game.

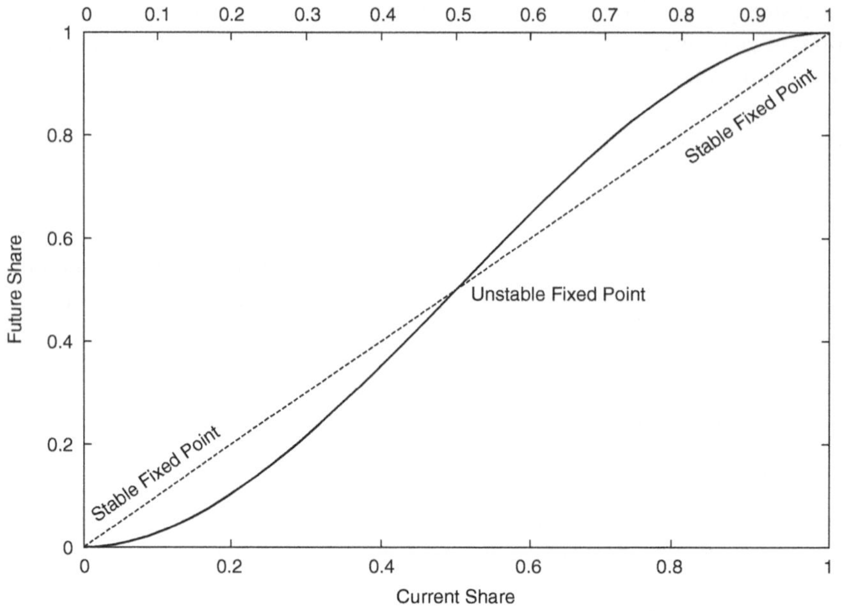

Figure 4.6 Effect of a positive network externality on the development of the shares *s* in the network. Technologies (or standards or firms) with already large shares are in a much better position.

For the trivial two-person case ($N = 2$ and $\Xi = 2$) for which the problem is identical to a simple two-person normal form game, which is furthermore symmetric with two strategies (switch immediately thus making the commitment or wait and see). In this case it takes the form of a simple *stag hunt* game, a subtype of the coordination game. While only the immediate switching of both agents (upper left in Table 4.2) is a Pareto optimum, both coordinated strategy configurations, both committing (upper left) or both staying with technology *A* (lower right) are Nash equilibria.

For an *n*-person case for which only the whole population can lead network *B* to take off, thus $N = \Xi$, the game retains its stag hunt type (Table 4.2).

The game theory solution, both for the two-person and the *n*-person case, is that there is no strictly dominant strategy for any of the players, collective switching is a Nash equilibrium, not switching, however, is as well. Both Nash equilibria are also trembling hand perfect and proper. Nevertheless, these solution concepts have only limited value for the present question since they can only predict that all agents will with certainty play a strategy that leads to a Nash equilibrium (any strategy in the present case) and would if repeated eventually reach one of the Nash equilibria. However, this is not a repeated game.[31] As it is not, there is no way for the players to decide on one of the equilibria. The two equilibria differ in their risk properties, switching is risky, waiting is not. Hence it is clear that

Table 4.2 Technology choice with network externalities and inferior incumbent technology as a normal-form game

		Player 2: Switch to new technology	Player 2: Wait and see
Player 1	Switch	2, 2	0, 1
	Wait and see	1, 0	1, 1

in the absence of proper stochastic predictions of the opponent's choice, any risk averse player chooses not to switch. Doing so would also be self-enforcing – in a repeated game – as the costs, the social loss inflicted by different choices, are in this subtype of the coordination game borne completely by the proponents of switching.

The situation changes further with $N > \Xi$, that is, Ξ adopters of the better technology are sufficient for it to establish a network with large enough revenues (game Γ_{N2}). In this case, agents may safely choose to wait and only join the better network as soon as its success becomes apparent – a rather safe strategy. The only risk is that the actor herself is the last one missing to ensure the establishment of the new technology, a probability that decreases the larger the gap between N and Ξ is.[32]

The nature of this decision problem becomes more apparent when considering two-person normal-form games in a population of $N > 2$ agents with $\Xi < N$ (but $\Xi > 2$) necessary to secure the success of the better technology. The decision of the $N - 2$ agents who are not part of the game is to be seen as the confidence or trust of the two active players in the willingness of the rest of the population to take the risk. In the following game (Table 4.3) this is represented in the anticipated (uniform) probability of other agents to join the network of the new technology φ.

These payoffs in game 4.3 are specifically (retaining the payoffs of Γ_{N1}) 2 for participation in the technologically better network, 1 for participating in the old network, and 0 otherwise, that is for a failed attempt to establish the better technology. The expected payoff in the modified Γ_{N2} are as follows

Table 4.3 Game Γ_{N2}: technology choice as a normal form game in a larger population. Payoff values are computed as in equation (4.21).

		Player 2: Switch to new technology	Player 2: Wait and switch when success is certain
Player 1	Switch	Π_{11}, Π_{11}	Π_{12}, Π_{21}
	Wait	Π_{21}, Π_{12}	Π_{22}, Π_{22}

($j = \{0, 1, 2\}$ denotes how many of the two actively playing agents switch to the better technology):[33]

$$\Pi_{\Gamma,N2} = \sum_{i=\Xi-j}^{N-2} \left(\binom{N-2}{i} \varphi^i (1-\varphi)^{N-2-i} \right) 2 + \sum_{i=0}^{\Xi-j-1} \left(\binom{N-2}{i} \varphi^i (1-\varphi)^{N-2-i} \right) 0 = \sum_{i=\Xi-2}^{N-2} \left(\binom{N-2}{i} \varphi^i (1-\varphi)^{N-2-i} \right) 2 \qquad \text{for switching agents}$$

(4.17)

$$\Pi_{\Gamma,N2} = \sum_{i=\Xi-j}^{N-2} \left(\binom{N-2}{i} \varphi^i (1-\varphi)^{N-2-i} \right) 2 + \sum_{i=0}^{\Xi-sw-1} \left(\binom{N-2}{i} \varphi^i (1-\varphi)^{N-2-i} \right) 1 \qquad \text{for waiting agents.}$$

This equation is to be understood as the stochastic (combinatorial) number of possible combinations of choices S_{CN} of the remaining agents that would ensure the new network functions, hence the number of the specific sequences[34]

$$S_{CN} = \sum_{i=\Xi-j}^{N-2} \binom{N-2}{i} \qquad (4.18)$$

however, each sequence is to be weighted with the likelihood of one specific sequence (of i adopters among $N-2$ agents) to occur

$$p_{i,N-2-i} = \varphi^i (1-\varphi)^{N-2-i}. \qquad (4.19)$$

thus

$$S_{CN,weighted} = \sum_{i=\Xi-j}^{N-2} \left(\binom{N-2}{i} \varphi^i (1-\varphi)^{N-2-i} \right) \qquad (4.20)$$

in turn yielding equation (4.18) above.

Table 4.4 Game Γ_{N2}: technology choice as a normal form game in a larger population for specific values $N = 8$, $\Xi = 5$ and $\varphi = 0.5$.

		Player 2	
		Switch to new technology	Wait and switch when success is certain
Player 1	Switch	84 84	86 44
	Wait	44 86	71 71

The specific values of the payoffs in Γ_{N2} as shown in Table 4.3 are therefore

$$
\begin{aligned}
\Pi_{11} &= \sum_{i=\Xi-2}^{N-2} \left(\binom{N-2}{i} \varphi^i (1-\varphi)^{N-2-i} \right) 2 \\
\Pi_{12} &= \sum_{i=\Xi-1}^{N-2} \left(\binom{N-2}{i} \varphi^i (1-\varphi)^{N-2-i} \right) 2 \\
\Pi_{21} &= \sum_{i=\Xi-1}^{N-2} \left(\binom{N-2}{i} \varphi^i (1-\varphi)^{N-2-i} \right) 2 + \sum_{i=0}^{\Xi-2} \left(\binom{N-2}{i} \varphi^i (1-\varphi)^{N-2-i} \right) 1 \\
\Pi_{22} &= \sum_{i=\Xi}^{N-2} \left(\binom{N-2}{i} \varphi^i (1-\varphi)^{N-2-i} \right) 2 + \sum_{i=0}^{\Xi-1} \left(\binom{N-2}{i} \varphi^i (1-\varphi)^{N-2-i} \right) 1
\end{aligned}
\tag{4.21}
$$

Considering a particular game $\Gamma_{\widetilde{N2}} \in \Gamma_N 2$ with $N = 8$, $\Xi = 5$, and $\varphi = 0.5$ as an example, the matrix form shown in Table 4.4 results, which is a typical prisoner's dilemma with $\Pi_{21} > \Pi_{11} > \Pi_{22} > \Pi_{12}$.

It is thus possible to analyze the technology switching problem with positive network externalities as a public good problem with certain properties of the prisoner's dilemma. The situation is prone to free riding: if agents are sufficiently risk averse, none of them will be willing to take the risk to move first nor will they have sufficient faith in the willingness of other agents to follow.

Evolutionary perspectives on dilemma-prone technological change

An extensive body of literature deals with the possibilities of letting a cooperative solution emerge in modified prisoners' dilemmas (Axelrod, 1984; Lindgren, 1997; Fehr and Gächter, 2000; Fehr *et al.*, 2002; Fehr and Gächter, 2002; Fehr and Fischbacher, 2004; Pacheco *et al.*, 2009; Elsner and Heinrich, 2009; Helbing and Johansson, 2010; Waldeck, 2010; Elsner and Heinrich, 2011). The approaches range from trigger strategies in the repeated game (Axelrod, 1984; Lindgren, 1997) to punishment (essentially also a trigger strategy) which could be shown in public good experiments (Fehr and Gächter, 2000) to specific consequences of the network structure and other environmental conditions in evolutionary and network settings. Specifically, if agents are able to exercise some influence on the network structure (i.e. refusing to engage any other agent than cooperators, 'partner selection'), cooperators tend to perform better and (with a replicator mechanism) take over the population even if information is imperfect and, in particular, limited to the agent's neighbourhood and may only be stored for a certain time (Elsner and Heinrich, 2009, 2011). Also, it has been found that a non-uniform network structure (that is, the agents are not arranged in a regular grid system), particularly a power law structure,[35] has a positive effect on the emergence of cooperation (Pacheco *et al.*, 2009).

Generally, and regardless of the particular mechanism, in order for cooperation to evolve (with rational agents), the expected payoffs of a cooperative strategy, say Π_C, have to exceed the expected payoffs of non-cooperation (defection), say Π_D. In the simple game, this is impossible; the corresponding evolutionary setting shows that universal defection is the only stable equilibrium (see Figure 2.3a). This is certainly a discouraging result; it may also hint that the successful resolution of free-rider situations in relation to technological change is unlikely. That is, while there is no further incentive, agents will most likely not take chances without an appropriate risk premium (which is 0 and thus too low in the free-rider situations if the agent is not a risk-seeker[36]) and will consequently not switch over to the newer technology. However, as mentioned above, there are theoretical mechanisms that facilitate the emergence of cooperation in dilemma prone situations, especially in context of the emergence of informal institutions. This has been confirmed in case studies by Ostrom (2000). Public good experiments were also able to establish that humans do tend to contribute (i.e. not to free-ride) and also to punish other agents that refuse to contribute (Fehr and Gächter, 2000).

On a theoretical level, the emergence of cooperation becomes possible with repeated games. An indefinitely repeated game with a trigger strategy, punishing a defector with a low payoff Π_P (with $\Pi_P < \Pi_C < \Pi_D$), in the future cooperation is the rational choice if future payoffs are discounted with a factor not lower than p^*_{rep}. Note that this factor may also be interpreted as the probability for the game of unknown duration to be continued for at least one further interaction. This critical value is computed straightforwardly from the payoffs and the condition that the expected payoff of cooperation must exceed that of defection

$$E(\Pi_{Cooperation}(\Pi_C, p_{rep})) > E(\Pi_{Defection}(\Pi_D, \Pi_P, p_{rep})) \tag{4.22}$$

Applying the infinite sequences of payoffs for both strategies, defection and the cooperative trigger strategy always yields

$$\Pi_C + p_{rep}\Pi_C + p_{rep}^2\Pi_C + \ldots > \Pi_D + p_{rep}\Pi_P + p_{rep}^2\Pi_P + p_{rep}^3\Pi_P + \ldots$$

$$\frac{\Pi_P}{1-p_{rep}} + \Pi_D - \Pi_P < \frac{\Pi_C}{1-p_{rep}}$$

$$p_{rep} > \frac{\Pi_D - \Pi_C}{\Pi_D - \Pi_P}. \tag{4.23}$$

Axelrod (1984) and Elsner and Heinrich (2009, 2011) go into more detail on this solution. An equivalent solution is attained by allowing direct punishment as a second stage of the game and similar approaches. As mentioned above, a number of further contributing factors have been identified.

4.2.3 Sponsored networks and standard wars

A particularly interesting question that arises from what has been discussed so far concerning network externalities is whether agency on the part of the technologies developers or third parties, such as independent organizations or official regulatory boards, fundamentally changes the picture. This question has again been discussed from the earliest days of research on network externalities. The more traditional scholars like Katz and Shapiro (1985, 1986), Farrell and Saloner (1985, 1986), Choi (1994) as well as Farrell and Klemperer (2007) have addressed the question of state intervention as one of the central themes of their research. Their findings are usually in themselves of mixed nature, sometimes state intervention is useful (for example, to ease what Katz and Shapiro termed 'excess inertia' (Katz and Shapiro, 1986), see above), sometimes it is harmful (deepening the welfare losses compared to the social optimum). Research with a particular interest in the development of the network structure followed over the next decades, including contributions by again Farrell and Saloner (1992), Katz and Shapiro (1994), Economides and White (1994), Shy (1996, 2001) as well as Manenti and Somma (2002).

An interesting related issue is the possibility of standard sponsoring entities extracting some quantity of the additional utility induced by network externalities. In this case, these entities are, of course, vendors of commercial technologies. This is implied in commercial sponsoring of standards, as the enterprises in question must have some incentive to engage in standard sponsoring and inflicting the related costs of heavy initial investments on themselves. This incentive can be targeted at true profits or just the amortization of development cost or, usually, a mixture of both. Consequently, these aspects are also present in the literature tradition mentioned above, for instance in Katz and Shapiro (1985, 1986, 1994), Choi (1994) and Farrell and Saloner (1986). However, there is a second branch of related literature explicitly addressing the resulting industry dynamics including side effects such as gateways between incompatible standards (Farrell and Saloner, 1992; Manenti and Somma, 2002; Dolfsma and Leydesdorff, 2009), tying of standards in interdependent industries (Choi, 2004), software 'piracy' (Shy, 2001) as well as case studies on the matter (David, 1985; David and Bunn, 1988; Arthur, 1990b; David and Steinmueller, 1994; Katz and Shapiro, 1994; Shy, 2002; Baptista, 1998; Swann, 1998, 2000; Stango and Greenstein, 2008; Stango, 2004).

Core aspects of these questions will be analyzed in the following, starting with official or semi-official sponsoring for the purpose of ensuring faster technological progress and socially bettter results. Subsequently, private sponsoring, strategical provision of gateways, standard tying and 'piracy' will be discussed.

Official sponsoring

Official and semi-official bodies[37] have the capacity to actively or passively support new or established standards. The central goal of such a policy is to attain a first-best social optimum, i.e. the standard that ensures the highest possible value of commercial revenues, network utilities and combined net intrinsic utilities. If this is not possible, ensuring a social utility better than the 'market solution' without regulation, as discussed above, may still be attempted. Active support denotes any activity targeted at facilitating the adoption decision of users by either reducing the (adoption) cost c of the favoured technology or increasing the expected network utility. This can be most conveniently accomplished by subventioning the development of the technology itself or establishing appropriate backward compatibility of this technology to the current incumbent standard. In the above notation, the net intrinsic utility terms r_i include both positive and negative private utility of the adopters from using technology i, and hence both actual utility and adoption costs. Thus, active sponsoring (whether by an official or a private entity) increases its net intrinsic utility by reducing the costs, thereby altering the right hand side of an abstract decision condition to discard an incumbent standard A in favour of an alternative B (equation (4.14))

$$v(N-ch_t)-v(ch_t+1)\leq r_{B,\kappa}-r_{A,\kappa}\equiv\Delta_\kappa$$

or, with backward compatibility of the sponsored standard but no forward compatibility of the incumbent, decreasing the left hand side

$$v(N-ch_t)-v(N)\leq r_{B,\kappa}-r_{A,\kappa}\equiv\Delta_\kappa \qquad (4.24)$$

As $r_{B,\kappa}$ rises, the condition is more likely to be true given the environment, especially the relative size of the emerging network ch_t compared to the size of the incumbent network $N-ch_t$. This increases the expectations of potential adopters about how large the emerging network of the new technology will grow. The effect is similar to the one discussed as the adopters project how many other agents may follow that decision in case of switching $g(\kappa)$, with the difference that in this case the agent κ's action is not the cause of their switching, rather, the cause is the official endorsement of the innovative technology and the subsequent wave of adoptions that the agent κ takes part in. Therefore, denoting this similar effect $\tilde{g}$, condition (4.16) becomes

$$v(N-ch_t-\tilde{g})-v(ch_t+\tilde{g}+1)\leq r_{B,\kappa}-r_{A,\kappa}\equiv\Delta_\kappa.$$

Hence, passive sponsoring works as a signal to the community of potential adopters that tends to increase the initial network size and consequently the initial network effects of the sponsored technology (again *B*).[38] Active and passive official sponsoring are often combined, especially since active sponsoring itself bears a signaling power. R&D subsidization is a common example of such a policy. A formally equivalent way for official bodies to accomplish the same effect would be to employ an appropriate fiscal policy, such as a tax reduction or tax exemption or restrictions of market competition for products using the respective technology. Well-known examples of this policy include the promotion of ecological technologies such as feed-in tariffs for renewable energy sources in Germany or car taxation according to fuel efficiency in Europe.[39]

Another, sometimes more subtle but not necessarily less efficient means of official and semi-official influence is passive sponsoring without directly financing one of the competitors. Large entities declaring their support for one standard exercises a signaling effect on private adopters. The (less subtle) definition of a de jure standard protected by law works in the same way.

Note that the case discussed in the current section refers to a setting in which an official body tries to replace an incumbent standard with a new – and supposedly superior – technology. If the installed base of the incumbent is large, however, and backward compatibility is (e.g. for technical reasons) unfeasible, even official bodies may find this prohibitively expensive. In fact, it has been suggested that official bodies have to act early (during a narrow period of opportunity in ensuing standard races that occur when a new industry sector develops (David, 2005)). That case is very similar to the case discussed for private sponsoring in the section below.

The criticism of any efforts at official sponsoring is apparent: there is no way to ensure that the right standard with respect to a first-best social optimum is endorsed. Although governments and official standard setting bodies are likely less biased and better informed than private actors – even more so as this is an industry with increasing returns to scale – they are by no means able to predict the technological progress either of a specific standard or of the industry at large. Neoclassical scholars therefore tend to call for an end to regulation even in the presence of network externalities (Liebowitz and Margolis, 1990, 1994, 1995), while others argue the ambivalent nature of regulation (Katz and Shapiro, 1985; Choi, 1994; Farrell and Klemperer, 2007), and more heterodox researchers imply that a certain degree of policy intervention might serve to guide technological development with better results than the unregulated case can possibly achieve (David, 1985; David and Bunn, 1988; Arthur, 1988, 1989; Swann, 2000).

Private sponsoring

For private entities like firms, as they are not concerned with the social optimum (according to whatever specification), active sponsoring of a technological standard is only an option when the firm can ensure some kind of revenue by engaging in such activities. This revenue may take the form of amortizing past or future R&D costs, infrastructure investments necessary for building the network, other similar types of sunk costs related to the standard, as well as true profits for private consumption or reinvestment by the firm and its shareholders. Considering real-world industries with network externalities, especially newly emerging industries the standards of which may not yet be entirely settled, it seems not unlikely that one or more private firms will have this kind of incentive to engage in sponsoring a candidate technology to become an industry wide standard. The primary reasons for this are the network externalities combined with a general technological tacitness. It takes some time for other firms to gather the required knowledge and to acquire the necessary human capital (by hiring or training) to imitate a technology. The initial developer of this technology will therefore have a certain period of time in which to enjoy considerable monopoly rents. Ex ante, this justifies massive R&D expenditures as well as infrastructure investments. If another firm with a competing technology standard is faster or just more proficient in exploiting the network externalities for their product, this may not only be unfortunate for the less lucky competitor, it might well drive her out of business. Formally, assume an emerging industry with two competing technologies A and B generating heavy network externalities. Potential customers κ face the adoption decision (inequality (4.11)) to choose A if[40]

$$v(N_{ch(\kappa),A}+1)-v(N_{ch(\kappa),B}+1)\geq r_{B,\kappa}-r_{A,\kappa}\equiv \Delta_\kappa .$$

The number of potential adopters is N. Further, let the two technologies be developed by two private actors who are able to dictate a price $z_{i,\kappa} i=\{A,B\}$ for their respective technologies which they may change at any timestep, with any further customer (κ) joining their network. Thus the adopter's decision problem now with prices/costs z explicitly formalized and distinct from r becomes

$$v(N_{ch(\kappa),A}+1)-v(N_{ch(\kappa),B}+1)\geq r_{B,\kappa}-z_{B,\kappa}-r_{A,\kappa}+z_{A,\kappa}\equiv \Delta_\kappa - z_{B,\kappa}+z_{A,\kappa} \tag{4.25}$$

Unable to effect the net intrinsic preferences Δ_κ, and the respective opponent's technology's properties (the size of the network and the price) also being beyond their direct control, they will each try to raise their competitive advantage

$$v(N_{ch(\kappa),i}+1)-z_{i,\kappa}$$

as high as possible, specifically higher than that of the opponent. This includes 'charging' negative costs, i.e. paying the first adopters for 'testing' the technology. In early periods the industry will see a – possibly grotesquely exaggerated – version of Bertrand's price competition. As usual in Bertrand competition, both try to undercut the competitor's price until the price equals the marginal costs (which include the change in future profit expectations). Since future profit expectations rise with every sold product, while the production costs remain constant and roughly zero, the marginal costs of this type may be negative.[41] Thus, the first adopters may be given the technology for free or even at a negative price. If the differences in the net intrinsic preferences are sufficiently small, customers might easily be won over for a standard that they do not prefer intrinsically. If the gap of the network utilities of the two products $v(N_{ch(\kappa),A}+1)-v(N_{ch(\kappa),B}+1)$ is large, the sponsor of the trailing technology will try to invest high sums in paying customers (that is investing in negative $z_{i,\kappa}$) until the resulting spending exceeds the financial scope of that firm and the firm gives up. The victorious competitor can than raise the costs to the tipping point of inequality (4.25) and by doing so appropriate the total social gains resulting from the network utilities as monopoly rents.

If the net intrinsic preferences differ more widely from one customer to another and are sufficiently large compared to the network utility (at the current size of the networks), the picture changes and coexistence of several standards becomes possible in spite of Bertrand-type competition in 'standard wars'. What is different from classical Bertrand competition is that the competitors have – mediated through network effects – good chances to drive any opponent out of the industry. There is thus not price competition for an industry share but rather price competition for a later monopoly status in an industry with increasing returns Demsetz (1968); formal aspects of this are discussed in the following section. It is immediately apparent that in any standard race, first movers have the advantage; an installed base that renders the network utility of their products higher than that of any potential entrant. As a result, private firms may be expected to undergo extreme difficulties to be first movers in any industry they deem profitable, and also to protect their technological knowledge relevant to the standard race from any competitors. In fact, (technological) knowledge is in this case no longer a public good, as its possession is rivalrous with respect to the profits expected from the industry the knowledge refers to. Specific phenomena and strategies of private firms in industries with network externalities will be discussed in the following sections.

Profits in monopolized network industries

The current model allows computation of the appropriable monopoly rents that become available once a competitor manages to gain control of an industry standard after driving out any competing technologies. This value may serve as a benchmark value for other settings, and may possibly even be indicative for more realistic considerations dropping the current perfect information and and perfect rationality requirements. Also, note that this section is rather close to neoclassical considerations of social welfare and producer and consumer rents.

Assume, all alternatives have been driven out of the sector by a victorious standard A. The adoption decision for an adopter κ becomes

$$v(N_{ch(\kappa),A}+1)+r_{A,\kappa}-z_{A,\kappa}\geq 0. \tag{4.26}$$

Let the adopters κ be sorted according to their intrinsic preferences $r_{A,\kappa}$ in descending order. The net intrinsic preferences of the population may now be approximated[42] by a strictly decreasing, monotonic, and if necessary smooth and differentiable function of the number of committed adopters (which is $N_{ch(\kappa),A}$) $r_{A,\kappa}(N_{ch(\kappa),A})$. The function is defined for $0\leq N_{ch(\kappa),A}\leq N$. Note that the intrinsic preferences may be negative. It is therefore possible that the first adopters would not adopt without being paid (in this case $v(1) + r_{A,\kappa}(0) - 0 \leq 0$). However, since the industry is assumed established to the point that A has superseded any competing technologies, early adopters do not have to be considered here. Suppose that for larger values of $n = N_{ch(\kappa),A}$ the adoption decision condition is fulfilled without additional payment[43] from a break even point $N^*_{ch(\kappa),A}$ with $v(N^*_{ch(\kappa),A} + 1) + r_{A,\kappa}(N^*_{ch(\kappa),A}) - 0 = 0$. There might also be an upper boundary $N^{max}_{ch(\kappa),A}$ such that the last few customers suffer intrinsically from the technology A so much (hence having an intrinsic preference $r_{A,\kappa}(N^{max}_{ch(\kappa),A})$ far below zero) that they would have to be paid to adopt it in spite of the large network utility. For simplicity, let it be assumed that $v(n+1)+r_{A,\kappa}(n)-0 > 0$ for $N^*_{ch(\kappa),A} \leq n \leq N^{max}_{ch(\kappa),A}$. Ex post every adopter would be willing to pay her entire positive utility including the ex post network utility $v(N^{max}_{ch(\kappa),A})$ that is received by all adopters. The monopolist can, however, not extract this amount ex ante if the adopters lack faith in the future growth of the network towards $v(N^{max}_{ch(\kappa),A})$. Nevertheless, she will be able to appropriate the ex ante utility that is limited to the intrinsic utility and the network utility resulting from the current network size $v(n)$, and she will extract this entire amount since she controls the number of products sold at a time, thereby being capable to exercise perfect price discrimination. Thus, the monopoly rents are[44]

$$\Pi_{Monopolist} = \int_0^{N^{max}_{ch(\kappa),A}} z_{A,\kappa} - c_{prod,var} \quad dn$$

$$\Pi_{Monopolist} = \int_{N^*_{ch(\kappa),A}}^{N^{max}_{ch(\kappa),A}} v(n+1) + r_{A,\kappa}(n) - c_{prod,var} dn$$

$$- - \int_0^{N^*_{ch(\kappa),A}} v(n+1) + r_{A,\kappa}(n) - c_{prod,var} dn$$

where the second integral is the investment necessary for the network to take off, $I_{network}$[45] and fixed costs and prior development costs could also be included in $I_{network}$, hence making the equation more general

$$\Pi_{Monopolist} = \int_{N^*_{ch(\kappa),A}}^{N^{max}_{ch(\kappa),A}} v(n+1) + r_{A,\kappa}(n) - c_{prod,var} \quad dn - I_{network} \tag{4.27}$$

The situation is shown in Figure 4.7.

Segmented populations

If differences in net instrinsic preferences are large enough, it may be futile for competitors to engage in Bertrand competition in order to take over other technology's networks. The adopters' decision to commit to a specific technology A follows condition (4.25)

$$v(N_{ch(\kappa),A}+1) - v(N_{ch(\kappa),B}+1) \geq r_{B,\kappa} - z_{B,\kappa} - r_{A,\kappa} + z_{A,\kappa} \equiv \Delta_\kappa - z_{B,\kappa} + z_{A,\kappa}$$

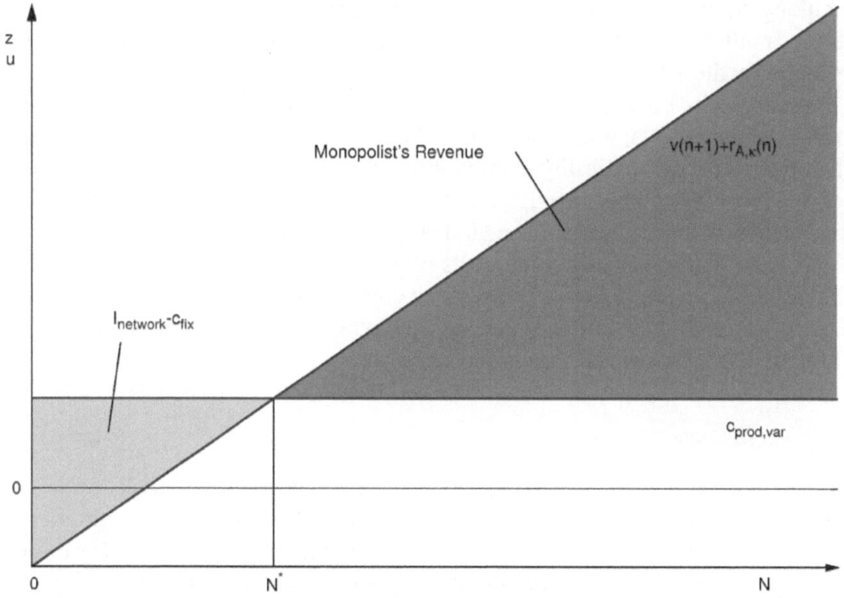

Figure 4.7 Profits in monopolized network industries

as shown above. It was also shown that private standard sponsors, specifically the vendors of the respective standards, will at early stages try to tip the balance of the competition by charging negative prices in order to win over as many adopters as possible, and even agents with intrinsic preferences for other competing standards. If Δ_κ is large enough and different (with different signs) for different actors,[46] this will, however, be overly expensive and also destroys any possible profit opportunities for the time being. Consequently, the strategic decision of one of the standard's vendors to wage such a price war may be rather unlikely. Such a situation may even be deliberately constructed by one of the competitors by adjusting the standard to the needs of a specific subgroup of the group of potential users, which increases the net intrinsic utility of individuals belonging to that group but decreases it for others (niche construction).

What follows is that standards coexist as quasi-monopolies in different segments of a population that differ merely in their preferences. With this coexistence comes the possibility of extracting monopoly rents, though limited by the presence of the competitor who may reconsider her chances to wage a standard war.

Note that with the growth of the networks of the coexisting standards the network utility term of the decision condition $v(N_{ch(\kappa),A} + 1) - v(N_{ch(\kappa),B} + 1)$ changes. If the difference of the network utilities generated by two networks grows larger with the changing size of the two networks, the chances of the vendor of the technology with a larger network and a larger network utility winning a standard war increase considerably. This is even more the case if already committed adopters are allowed to reconsider their decision and switch from the defeated technology to the prevailing one. A possible counter strategy for the vendor with the smaller network is to increase the segmentation if possible. In turn, the imbalance in the network utility terms would be compensated for by a larger net intrinsic utility difference. In a complex network of interrelated technologies, standards and products, this strategy may be quite applicable, considering issues such as backward compatibility, compatibility with related technologies, learning costs, or even style and peer group identity. Literally, the strategy means to break a niche industry off a larger established industry.[47] Concurring with this result, Chen (2010a) finds that lower switching costs may lead to monopolization of a sector while larger switching costs may preserve competition. Though Chen does not discuss niche segments, and in particular deliberate strategic niche construction, his argument is an aspect of the same effect.

Strategic gateways

If waging a true price war is an overly expensive and prohibitively risky strategy, there is another much more cunning option. Given the technological requirements are met, an appropriate device may be used to link the networks of two (or more) technologies together, in turn making the technologies compatible. Such devices are known by several names in different industries. A major step in building modern electricity distribution systems was the introduction of *converters* capable of converting alternating current into direct current and vice versa. For modern electronic hardware, a device that serves to link equipment aligned to different hardware standards is known as an *adapter*. In the software branch, servers that translate from one communication protocol to another are usually named *gateways*. The scope to which these devices are able to influence network effects, network utility and by extension the structure of whole industries was recognized soon after the vivid scientific debate on network externalities commenced by David and Brunn (David and Bunn, 1988; David and Steinmueller, 1994) with later technical analyses by (among others) Farrell and Saloner (1992) and Manenti and Somma (2002). Directly related to this discussion, and very similar in its findings, is the literature tradition on the competition between compatible and incompatible standards (Klemperer, 1987; Katz and Shapiro, 1994; Economides and White, 1994; Choi, 1994; Shy, 2001).

Applying a gateway, the consumers, adoption decision (similar to inequality (4.25)), the condition for adopters deciding for A and not B, becomes

$$\begin{aligned} &v(N_{ch(\kappa),A} + N_{ch(\kappa),B} + 1) - v(N_{ch(\kappa),A} + N_{ch(\kappa),B} + 1) = \\ &0 \geq r_{B,\kappa} - z_{B,\kappa} - r_{A,\kappa} + z_{A,\kappa} \equiv \Delta_\kappa - z_{B,\kappa} + z_{A,\kappa} \end{aligned} \quad (4.28)$$

Thus, the effect of the network utility on the competition between the two now compatible standards is eliminated, as both standards grant access to the now much larger common network.[48] This opens up a variety of strategic options for different kinds of competitors. First, there is the direct effect on the adoption decision of rational adopters. Standard sponsors with strategic disadvantages resulting from a small network or installed base (or, extending the interpretation of the formalism, an inefficient network) will be able to compensate for this disadvantage. It is reasonable to assume that with respect to this effect minor competitors will generally favour compatibility, while major technologies ('market-leaders' or industry leaders) might even work to deliberately introduce incompatibility. However, with local network effects (see below) there is a counter effect which may also give major competitors incentives to support compatibility. If their standard embodies higher intrinsic value to the adopters – which may be due to greater possibilities of developing the technology to be more failsafe and feature-rich – compatibility might work in their favour.[49] Note that this effect is not covered by the model above, as local network effects are not included. The idea is that if customers do specifically value the compatibility to their direct peer group, the groups committed to minor technologies can only be taken over either as the whole peer group at once (which involves a coordination problem of the type analyzed by Cooper and John (1988)), or one by one with additional backward compatibility to enable members that have already changed to the large vendor's technology to be compatible with those that have not yet switched. In this case, it might also be of advantage to the large vendor to repeatedly introduce and remove compatibility[50] to first win over a part of the peer groups, then force the rest to switch.

Another option for competing technology standard vendors is to create compatibility using strategic gateways that work only in one direction. This possibility has also been discussed in the literature on competition with network externalities (Farrell and Saloner, 1992; Shy, 2001; Manenti and Somma, 2002). Without going into details, it shall be noted that this would provide the compatible standard with a much larger network while the network of the other standard remains unchanged. Hence (B being compatible to A but not vice versa) agents choose A if

$$\begin{aligned} &v(N_{ch(\kappa),A}+N_{ch(\kappa),B}+1)-v(N_{ch(\kappa),B}+1)\geq \\ &\qquad r_{B,\kappa}-z_{B,\kappa}-r_{A,\kappa}+z_{A,\kappa}\equiv \Delta_{\kappa}-z_{B,\kappa}+z_{A,\kappa}. \end{aligned} \tag{4.29}$$

This clearly works to the advantage of the incompatible standard – the left hand side of the inequality being positive. Also note, however, that one-way compatibility is not always possible, especially not for communication technologies,[51] and that one-way compatibility is probably not purely one-way even for standards other than those for communication technologies.[52]

Planned obsolescence

The model considered so far is limited to consumers choosing exactly one product, thus adopting exactly one technology exactly once. As argued above, it might realistically be extended to letting the less lucky adopters switch to the industry standard once it becomes clear that their favoured choice will not prevail.

This does not change the dynamics nor the outcome of the model. However, this also limits the maximum revenue a private sponsor would be able to take to the price of the technology (or the product providing the benefit of using the technology) times the number of adopters. Though the monopolist is likely able to exercise price discrimination (charging different prices to different users) the real possibilities of a standard sponsor may be considerably larger. In most cases, it is possible to make use of planned obsolescence. Many goods decay in any case; sometimes it is possible to design the products to depreciate rather fast. This ensures that customers have to renew their equipment periodically and thus continue to buy products and provide the monopolist with further revenue. For immaterial products – say software licenses – this does not work; in other cases, customers may be aware of the quality and use their bargaining power to force the producer to refrain from selling substandard products. This may be accomplished by the threat of establishing a competitor product if the quality available does not satisfy certain expectations. Especially large customers, e.g. firms buying communication technology infrastructure (like glass-fibre wire and complementary equipment), are probably able to exercise such pressure on suppliers.

Nevertheless, with network technologies, the monopolist has a much more convenient way of making use of a very similar strategy. The gradual improvement of a network technology introducing frequent updates with no or limited backward compatibility works in a similar way to planned obsolescence if the producer charges the customers willing to apply the update an update fee or even the full cost of the product again. This phenomenon has been studied in detail as a mixed game theory and neoclassical market model by Choi (1994). Choi finds that this may lead to different regimes depending on consumer preferences and marginal production costs. The monopolist will, for example, choose to design a new version compatible with an older one if established customers would not be willing to pay for the update, and the increased network utility is the only way to attract new customers.[53] Furthermore, Choi argues that the monopolist can also decide to delay the introduction of a product to prevent later competition to the networks of an established consumer base and of new adopters of the improved version. Choi adds some considerations about welfare optimality. Note that with the neoclassical interpretation of social welfare (as the sum of realized utility measured in monetary units) the planned obsolescence as such is not welfare reducing, since the payments made by the demand side are simply transferred as profits to the monopolist. This section applies a simplified version of the effect studied by Choi to the model used above.

For the sake of simplicity, assume first that the demand does not grow. A monopolist has supplied the whole consumer base with a technology (or product) A; now she develops an improved version of the same technology B. This means that the net intrinsic preferences of the adopters are strictly higher for B, thus $r_{B,\kappa} > r_{A,\kappa} \quad \forall \kappa$. With incompatibility, the adoption decision of the customers is consequently

$$v(N_{ch(\kappa),B}+1) - v(N_{ch(\kappa),A}) \geq r_{A,\kappa} - r_{B,\kappa} + z_{B,\kappa} \equiv \Delta_\kappa + z_{B,\kappa}$$

The monopolist may therefore pay a certain number of adopters to switch to the new technology by setting a negative $z_{B,\kappa}$ and thus tipping the network utility part $v(N_{ch(\kappa),B}+1)-v(N_{ch(\kappa),A})$ a little more in favour of B with each adopter. When the condition is finally fulfilled for the remaining agents, the monopolist (again able to exercise price discrimination by setting the number of products sold at a certain price) subsequently appropriates the entire utility. Setting the time steps between incompatible updates too narrowly would lead to losses, as a large number of customers would have to be enticed with a negative price for the update while the revenue acquired as a result of the operation would be rather slim. In fact, there is a trade off between more frequent updates with lower revenues and less frequent updates with higher revenues which are, however, realized at a later time. Offering a backward compatible update, on the other hand, yields a condition that is immediately fulfilled (as long as the monopolist is moderate in setting the price $z_{B,\kappa}$) since $r_{B,\kappa} > r_{A,\kappa}$

$$v(N)-v(N)=0 \le r_{A,\kappa}-r_{B,\kappa}+z_{B,\kappa} \equiv \Delta_{\kappa}+z_{B,\kappa}$$

Hence, and quite intuitively, most agents would be willing to install backward compatible updates while being more selective about incompatible new versions.

This may have a more profound effect in settings with a growing consumer base. The monopolist may foresee the complication that the adoption decision for new adopters might not be in favour of adopting because of a too small network. Therefore, she will consider postponing the introduction of the product in order to have a larger consumer base at the later time. When seen as a continuous process, the monopolist would simply apply updates less frequently – which is, in effect, a variant of the consideration discussed above. It should also be noted that many technologies are complex and the products or infrastructure items allowing the employment of a technology are even more so. This leads to compatibility being much more scaleable than the model considered so far allows. For example, software products are often capable of opening (reading) files produced by older versions (in case of file editing software) or to work with the configuration files from older versions, but not vice versa. With this, the monopolist is able to produce a certain incentive for customers to upgrade to a new version (even paying for it) while largely preserving the integrity of the network and hence the scope of the network utility, especially for new adopters (who will buy the newest version enjoying at least partially the impression of compatibility of all previous versions).

Standard tying

Choi, who developed the model of planned obsolescence and network effects (Choi, 1994), continued to work on industry competition with network externalities. A particularly important issue taken up by Choi (2004) was that standards do not exist as isolated patterns; they are part of a complex structure of technologies mutually relying on one another, and competing with others on different scales.[54]

As a consequence, the success of a technology does not only depend on its own popularity and performance compared to its direct competitors, it is also influenced by its compatibility with a number of other standards in 'adjacent' industries and their development. A personal computer relies on the attuned operation of a large number of hardware interfaces, devices and components all designed according to their respective standards as well as a variety of corresponding software, such as the operating system, the device drivers, office software, networking software, scientific computational or statistical programs, etc. In fact, the specification of the standards is crucial to allow the system to function and its parts to be designed appropriately. For a commercial standard sponsor, it would be beneficial to be active in more than one of these 'adjacent' industries, and even more so to gain a monopoly position in one of those. Control of a standard may be employed in a strategic way in order to give other related standards owned by the same sponsor an advantage. As a result, large systems of different products all developed by one company and all well-attuned to one another may be observed in the industry sectors with especially large network effects – the variety of software packages developed by Microsoft and Apple are just two examples. The fruitful cooperation of Intel, IBM and Microsoft combining their hardware and software standards seems to be a result of the same effect. (For a more detailed account of the history of the development of computer hardware and software standards, see Chapter 6, Glowniak (1998), Bresnahan and Greenstein (1999), West (with co-authors Dedrick and Gallagher) (West and Dedrick, 2001; West, 2003, 2005; Gallagher and West, 2009) or Tanenbaum (2001).)

Choi (2004) calls this effect 'standard tying'. Again, he analyzes the possible, optimal and less optimal strategies of competitors, their effect on the industry dynamics as well as the welfare implications in a mixed neoclassical market and game theory approach. However, in this case, the additional effects of network externalities are not included in his model. This section will apply the model above to this phenomenon.

Consider two tied ('adjacent') industries A and B with two technologies in each competing to become the respective industry standard. Call the technologies A_1 and A_2 for industry A and B_1 and B_2 for industry B; assume further that technologies A_1 and B_1 on the one hand and A_2 and B_2 on the other hand are perfectly tied. Consider the network utility functions $v'(\cdot)$ to be therefore similar to $v(\cdot)$ above, except for being dependent on the installed base of users of not just the technology under consideration but also the technology tied to it. The adoption decision of agents in favour of technology A_1 in industry A will thus be

$$\begin{aligned} & v'(N_{ch(\kappa),A_1}+1, N_{ch(\kappa),B_1}) - v'(N_{ch(\kappa),A_2}+1, N_{ch(\kappa),B_2}) \\ & \geq r_{A_2,\kappa} - z_{A_2,\kappa} - r_{A_1,\kappa} + z_{A_1,\kappa} \equiv \Delta_\kappa + z_{A_1,\kappa} - z_{A_2,\kappa}. \end{aligned} \tag{4.30}$$

The condition for the other industry B is equivalent. Thus, there is another term contributing to network utility that basically increases the strategic maneuverability of the stronger one of the two standard pairings ((A_1B_1) and (A_2B_2)) thus working in favour of a more easy and possibly faster monopolization of the industries. (See also Chapter 5.)

A more generalized formulation would be to allow all four possible pairings $((A_1B_1), (A_1B_2), (A_2B_1)$ and $(A_2B_2))$ with the degree of their tiedness weighted by parameters $\chi_{11}, \chi_{12}, \chi_{21}$ and χ_{22}

$$v'(N_{ch(\kappa),A_1}) = \chi_{11} v'(N_{ch(\kappa),A_1} + 1, N_{ch(\kappa),B_1}) + \chi_{12} v'(N_{ch(\kappa),A_1} + 1, N_{ch(\kappa),B_2})$$
$$v'(N_{ch(\kappa),A_2}) = \chi_{21} v'(N_{ch(\kappa),A_2} + 1, N_{ch(\kappa),B_1}) + \chi_{22} v'(N_{ch(\kappa),A_2} + 1, N_{ch(\kappa),B_2}) \quad (4.31)$$

For simplicity, assuming the network utility function $v'(\cdot)$ to be the simple multiplication of the two installed bases – which is, as required, a monotonically increasing convex function (leading to increasing returns) – allows us to write the network utilities conveniently as a matrix multiplication (σ denoting the unit vectors $\begin{pmatrix}1\\0\end{pmatrix}$ for A_1 and $\begin{pmatrix}0\\1\end{pmatrix}$ for A_2[55])

$$v'(N_{ch(\kappa),TEC}) = (N_{ch(\kappa),TEC} + 1)\sigma_{TEC} \begin{pmatrix}\chi_{11} & \chi_{12}\\ \chi_{21} & \chi_{22}\end{pmatrix} \begin{pmatrix}N_{ch(\kappa),B_1}\\ N_{ch(\kappa),B_2}\end{pmatrix} \quad TEC = A_1, A_2 \quad (4.32)$$

This makes the model suitable to be modelled using replicator dynamics, which in turn allows investigation of its equilibrium and stability properties. This, however, is a method other than simple game theory (or integrated neoclassical style markets and game theory) as discussed in the current section; hence, this point will be advanced further in the context of the complex systems approach to network externalities in section 4.3.5.

Piracy

Up until now, technology standards or the products providing their benefits were treated as if they were generally similar to any other product in a monopolized sector: the monopolist can choose the price as well as the quantity produced at any time. However, as mentioned in the remarks on the neoclassical theory of monopolies and oligopolies in section 4.1.2, this may effectively not be the case with increasing returns to the number of users. Traditional neoclassics has no reason to deal with product 'piracy' as a major effect: production inflicts production costs that are more or less the same for all competitors, therefore it does not only not make sense to imitate specific products of other producers, the vendor of the pirated product would also not even care. Monopolies in sectors with network externalities, however, will need to use price discrimination by controlling the output quantity in order to collect the full amount of profits appropriable to them. This is because with every further customer the product gains value, thus the reservation prices of future adopters rise continuously and the monopolist would try to increase the price as well. Any imitator (neglecting for the moment the effects of law enforcement) could destroy this regime by imitating the product and selling it in large quantities at low prices.

Call the true variable unit production cost $c_{prod,var}$ (and assume it to be equal for the monopolist and any imitator). Note that these costs of production are charged to the customers, being included in the net intrinsic preferences r_κ. Any further financial modifications made by the monopolist are instead part of the term z, which may serve the monopoly to win the first adopters (set negative in this case) and (otherwise) contains the monopoly rents appropriated, hence $\Pi_{Monopolist}$. The monopolist's profit function (as defined above in the section on profits in monopolized network industries, equation (4.27))

$$\Pi_{Monopolist} = \int_{N^*_{ch(\kappa)}}^{N^{max}_{ch(\kappa)}} v(n+1) + r_\kappa(n) - c_{prod,var} \quad dn - I_{network}$$

is reduced by the imitator to the returns of an effective price equal to the price $z_{Im} + c_{prod,var}$ the imitator charges. Say, the adopters reservation price surpasses $z_{Im} + c_{prod,var}$ at $n = N^{Im}_{ch(\kappa)}$, and the imitator succeeds in selling to N_{Im} customers. The monopoly profits become[56]

$$\begin{aligned}\Pi_{Monopolist} = & \int_{N^*_{ch(\kappa)}}^{N^{Im}_{ch(\kappa)}} v(n+1) + r_\kappa(n) - c_{prod,var} dn + \int_{N^{Im}_{ch(\kappa)}}^{N^{max}_{ch(\kappa)}} z_{Im} - c_{prod,var} dn \\ & - N_{Im}(z_{Im} + c_{prod,var}) - I_{network} \\ = & \int_{N^*_{ch(\kappa)}}^{N^{Im}_{ch(\kappa)}} v(n+1) + r_\kappa(n) dn \\ & + (N^{max}_{ch(\kappa)} - N^{im}_{ch(\kappa)} - N_{Im})(z_{Im} - c_{prod,var}) - I_{network}\end{aligned} \quad (4.33)$$

The imitator on the other hand acquires a profit of $\Pi_{Im} = z_{Im} N_{Im}$. The monopolist can try to force the imitator out of the industry with Bertrand type competition, as discussed above in the case of competition between standards. However, it is reasonable to believe that she will not. In the above case of competition between standards, the immediate reward for the victorious competitor is the appropriation of monopoly profits, while the less lucky vendor of the failed standard loses any profit-making opportunity and retains the costs (which, considering development costs and the costs of the standard war, are probably considerable). The imitator faces just the variable costs of production and will sell only if the monopolist's price is above the variable production costs and therefore there is a possible imitator price $z_{Im} + c_{prod,var} < z_{Monopolist} + c_{prod,var}$ such that $z_{Im} > 0$, which guarantees her immediate positive revenues. She will also be flexible enough to cease the production as soon as the monopolist sets the price below variable costs (which for the imitator are also the marginal costs) and reestablish her hold in the industry as soon as the price rises again. The reason for this is that the imitators do not have

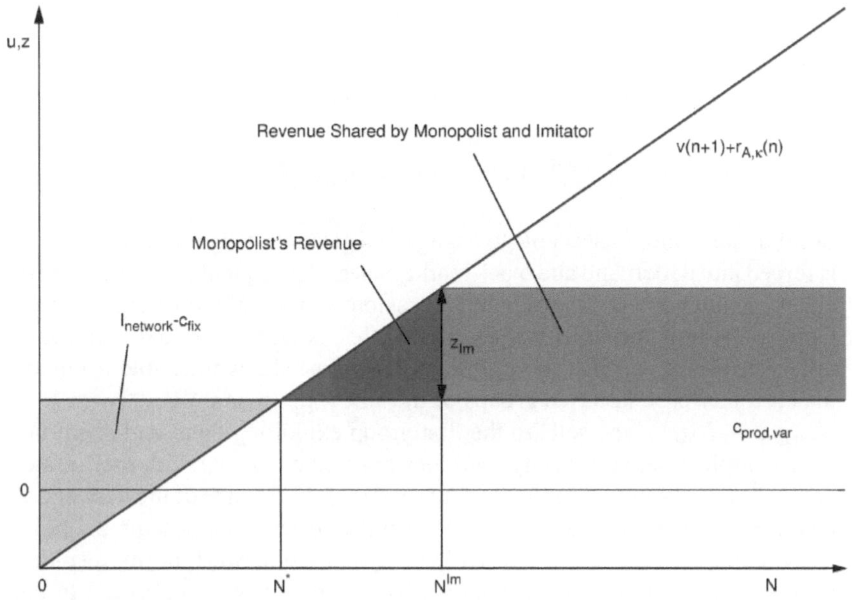

Figure 4.8 Profits in monopolized network industries with product pirates

to build and maintain a network technology of their own, they just participate in exploiting the network utility of the monopolist's network. The resulting situation is illustrated in Figure 4.8.

Of course, the monopolist's position is nevertheless rather strong. She faces four major options that are more favourable than an unwinnable standard war.

- In most jurisdictions, product 'piracy' is illegal. The imitator will therefore face a certain risk of being captured and fined or (in case of persons, not firms, juridically 'natural persons') tried which drives down her expected profits. In terms of the market model, additionally to the production cost the imitator has to add an appropriate risk premium c_{risk} to the unit price she can offer.[57] The monopolist, having to offer a price at least as low as the imitator's price $z_{Im} + c_{prod,var} + c_{risk}$, but facing lower costs than the imitator $c_{prod,var}$, will in turn be able to acquire a profit equal to c_{risk} per unit sold. This raises her expected profits to

$$\Pi_{Monopolist} = \int_{N^*_{ch(\kappa)}}^{N^{Im}_{ch(\kappa)}} v(n+1) + r_\kappa(n) \quad dn$$
$$+ (N^{max}_{ch(\kappa)} - N^{im}_{ch(\kappa)} - N_{Im})(z_{Im} - c_{prod,var} + c_{risk}) - I_{network}$$

and, as the price is such that the imitator does not gain, $z_{Im} = 0$, at which point the imitator ceases to produce, hence also $N_{Im} = 0$

$$\Pi_{Monopolist} = \int_{N^*_{ch(\kappa)}}^{N^{Im}_{ch(\kappa)}} v(n+1) + r_\kappa(n) \quad dn$$
$$+ (N^{max}_{ch(\kappa)} - N^{im}_{ch(\kappa)})(z_{Im} - c_{prod,var} + c_{risk}) - I_{network}$$

- Second, the monopolist is able to engage in legal marketing while the imitator is forced into underhand and black market sales. Consequently, it is justified to assume a much larger base of potential customers for the standard monopolist than for the imitator, at least if the monopolist exhausts the legal framework to take measures against the imitator's business. She is thus able to divide the customer base into two groups of the size $N_{legalist}$ and N_{pirate} (with $N = N_{legalist} + N_{pirate}$); she sells to the first group exploiting them and acquiring their complete network utility and does not bother to deal with the 'pirate' group.[58] This allows rather high profits and shrinks the size of the network of potential adopters compared to the case of the perfect unopposed monopoly. The members of the first group will even yield the network utility resulting from the network size being enlarged by the 'pirate' group (N_{pirate}) to the monopolist. Under the assumption that preferences and utilities of the legalist and pirate groups are identically distributed,[59] the monopolists expected profit rises to

$$\Pi_{Monopolist} = \int_{N*}^{N^{Im}} v(n+1) + r_\kappa(n)dn + \int_{N^{Im}}^{N^{max}} \frac{N_{legalist}}{N_{legalist} + N_{pirate}}$$
$$v(n_{legalist} + n_{pirate} + 1) + r_\kappa(n)dn - I_{network} \qquad (4.34)$$

For a graphical illustration, see Figure 4.9. Note that this effect also exists when the monopolist is incapable of perfect price discrimination. The monopolist would in effect give the product (or technology) away for free to the 'pirate' group (by means of the pirate) and subsequently with an installed base of adopters in place, charge the legalist group high prices. This strategy will (both with and without perfect price discrimination) also work when competing for the industry with another standard. In fact, as a part of a standard war, the standard that is pirated more often will succeed in establishing a larger network and will eventually win the race to become a commonly accepted industry standard. The monopolist would prefer perfect price discrimination if she had the choice (which she does not); however, tolerating 'piracy' (called 'profitable piracy' in the literature; for details, see Le Texier and Gordah (2011)) combined with this strategy saves a part of the monopoly profits.[60] Thus, 'piracy' may reduce the monopoly profits but may still be favourable to the monopolist. Shy provided a similar formal analysis of 'piracy' as a part of his extensive discussion of a number of phenomena of network externalities (Shy, 2001) and in several earlier papers (Navon *et al.*,

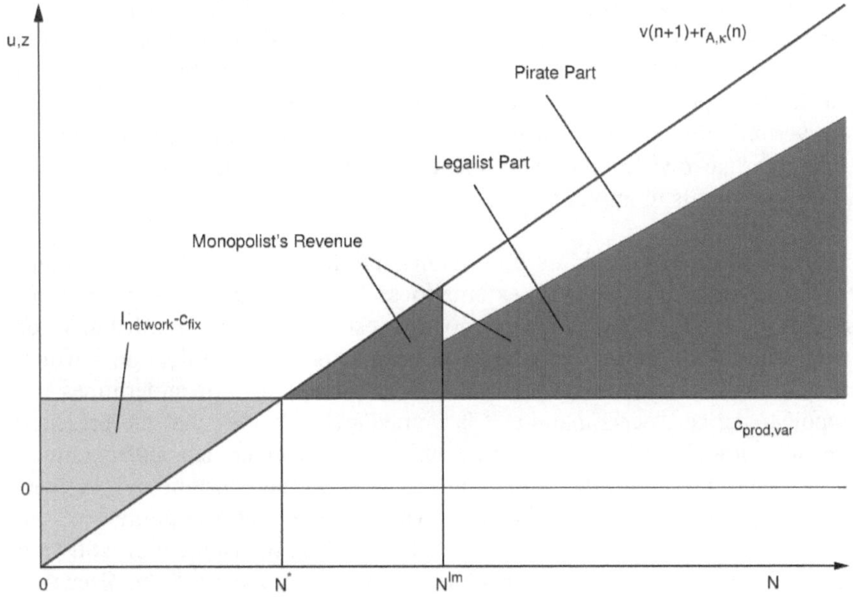

Figure 4.9 Profits in monopolized network industries with piracy and clearly separated pirate and legalist consumer groups

1995; Shy and Thisse, 1999). Prasad and Mahjan (2003) discuss possible corporate strategies along similar lines, going as far as to recommend that monopolies always start with a strategy of tolerating 'profitable piracy' in order to establish a sufficiently large network.[61] For similar discussions see King and Lampe (2003), Peitz and Waelbroeck (2006), Le Texier and Gordah (2011) and Darmon *et al.* (2009). Peitz and Waelbroeck (2006) also provide an extensive overview of the literature on positive and negative effects of piracy. This effect is hard to show in empirical evidence; however, there is anecdotal evidence. Several officials of large software companies have been quoted as stating they are aware of this effect and aggressively incorporate it into corporate strategy.[62] There is, however, also a group of scholars trying to reject the idea of this effect.[63]

- As the monopolist has technical control of the standard, she can acquire an advantage over the imitator by changing the specifications as often as she likes. This advantage exists until the imitator eventually catches up, and comes at the cost of additional unnecessary incompatibility introduced into the system, thus decreasing the network utility.
- Finally, in an economic system of complex interdependent technologies, many of them with technology standards and network externalities, the monopolists in several sectors are likely working together (even if not belonging to the same company) in order to attune their technologies and to assemble the end products. The imitator would be unable to sell her product if

> she is not accepted into this system. However, this is rather unlikely, given that the monopolist would strongly oppose cooperating with the imitator. For the standard sponsors in related sectors, cooperation with the monopolist as the developer of the standard would be much more valuable than cooperation with some product pirate. Hence, notwithstanding the resourcefulness of clever imitators, in such a heavily integrated system their perspectives might be rather narrow, leaving the monopolist with most of her ability to extract monopoly rents unimpaired.

As the considerations in this section show, product 'piracy' is an important effect in industries with network externalities. The extraction of revenues from these industries depends on high previous investments in the establishment of the network and the subsequent ability to exercise price discrimination – which makes imitation profitable. Imitation (product piracy) in turn undermines the monopolist's price discrimination. It is immediately obvious that the presence of product 'piracy' works directly in favour of the consumer side, shifting much of what would have been the monopoly rent back to the consumers. However, the long-term effect may be different – which is the central argument of the advocates of strong intellectual property rights and firm punishment of 'software piracy'.[64] Notwithstanding some specifics[65] of the IT economy (to be discussed in Chapter 6), software 'piracy' is of course just the latest widely known and widely discussed example of the phenomenon. The profit incentives the developer of technology standards expects for the time the technology is introduced, the standard generally accepted and the network established is crucial for her initial investment decisions. Thus, with decreasing expected revenues as a result of product 'piracy', the willingness to innovate and engage in R&D efforts fades as well, which may result in an underprovision of new technologies and hence possibly also stronger monopolies and lock-ins for once-established standards.

However, considering that the 'pirated' industries are not isolated but are subject to competition between technologies to become the industry-wide standard, the long term effects of 'piracy' may not be that simple. For example, as supported by a wide range of scholars (discussed in more detail above) and confirmed by evidence from case studies, 'piracy' serves to establish the network of adopters of a technology that enables the sponsor of that technology to drive competitors out of the industry and act as a monopolist later on. This strategy does not work against imitators as they engage in another kind of competition, not challenging the monopoly of the sponsored standard but waging a price war with the monopolists very own product designs.

4.2.4 Adverse selection of industries

Broadening the view of the previous sections on sponsoring of network technologies, it may justly be stated that adopters of network technologies face more uncertainty with respect to this adoption decision; the more promising technologies are competing to become the industry-wide standard. A 'promising' technology in turn refers to a technology with a large enough network[66] already in place, or a powerful sponsor such that adequately high investments in building such a network ('negative costs', 'paying' adopters or test-users, $-z_{i,\kappa}$) are possible and sufficiently likely to succeed in creating an installed base for this technology. It is reasonable to assume that potential adopters evaluate the risk, cost and investment opportunity of making an adoption decision by taking into consideration (among other facts and figures) the uncertainty about which of the competing technologies will prevail. Illustratively, if the utility an adopter expects in case the adoption decision was correct is U, then the expected utility of investing in one of two competing technologies with currently equal chances to become the industry standard for industry A is $\frac{U_A}{2}$, while the expected value of adopting a technology without noteworthy competitors in industry B is U_B. Further, for industry 1, the variance of the expected utility is rather high, namely $\frac{U_A^2}{4}$, while for the second adoption decision the expected value is realized with certainty (variance is 0). Choosing between equally attractive industries A and B (thus $U_A = U_B$) to employ technologies or products as a part of the enterprise infrastructure, rational adopters would choose the higher expected utility $U_B > \frac{U_A}{2}$. Even with the first infrastructure investment (i.e. industry A) being much more useful, $U_B = \frac{U_A}{2}$, most real-world adopters (humans who are mostly risk averse though heterogeneous, see Smith (1989)) would choose the smaller variance $0 < \frac{U_A^2}{4}$.

Consider a very simple model of infrastructure investment opportunity in an emerging or rapidly growing industry sector. Let $i = A, B, \ldots n$ be industries, for all of which there are a number $s_i > 0$ (but for some industries just 1) of competing sponsored technology standards available. Hence, all industries with more than one possible standard start with a price war. Further, a number of potential adopters choose some of the industries (according to their budget) to buy infrastructure from. All expected utilities U_i are positive but depend on the number of adopters, and thus unknown ex ante. Furthermore, the adopters expect adoption costs to be insignificant compared to utilities of 'correct' adoption decisions (i.e. adopting technologies that indeed become industry standards).

As the adopters are unable to evaluate the relative expected utilities of the industries in advance, their only means of distinguishing the investment opportunities are the number of competing standards in the respective industries. In fact, neglecting the only available information, they would be forced to resort to using the Laplace criterion $E(U_A) = E(U_B) = \ldots E(U_i)$ (everything may occur with equal probability), leading to the expected utilities from adopting technologies that are unopposed in their industries $(i|s_i = 1)$ being multiples of the expected utility of every other adoption decision.

$$E(U_i|s_i = \tilde{s}) = \frac{U_i}{\tilde{s}}$$

Any alternative assumption regarding rational decision making of the adopters that takes this consideration into account instead of using the Laplace criterion results in even more strengthening this effect.

Consequently, adopters will choose to invest first in industries with established industry wide standards. From the viewpoint of increasing returns and network externalities this is the optimal solution. However, the result still holds if the technologies' sponsors extract monopoly rents, in fact rational adopters are more willing to invest in more monopolized sectors. This may strip them of their utility as shown above, as well as hinder further technological advances, as alternative technologies are not being explored and the monopolist has no strong incentives to engage in further R&D.[67]

Changing the setting slightly, assume every industry starts with at least two competitors and give consumers the financial scope to invest in exactly one technology (representing the dominant or an out of date standard of one of the industries) and the possibility to retard the time of their investment. It is obvious that risk averse agents, again facing a public good problem regarding commitment to one of the technologies and contributing to its network, will choose not to invest at all. If the control they have over their action is, however, imperfect; that is, if they nevertheless invest with a small probability, representing their 'trembling hand', some adoptions will occur resulting more or less in the scenario considered above. Some industries are first dominated by one technology (attracting further investment) then monopolized and rendered into very successful standards with huge networks of adopters and fantastic monopoly rents. The faster the industry is monopolized the quicker this happens, leaving other industries with fruitful technological competition far behind. A process of selection governs this inter-industry development. Monopolized industries are favoured, heterogeneous industries are largely discarded (because of their lack of widely accepted standards) and may eventually become extinct. If heterogeneity can be seen as hinting at great potential and a promising future for this technology, this mechanism is not only selection, it is adverse selection as described by Akerlof in his example of the 'markets for lemons' (Akerlof, 1970): selection works in favour of bad specimens; therefore a neoclassical market subject to adverse selection in its pure form will continuously shrink and eventually cease to exist. In this case, it is of course not a neoclassical market but rather a dynamic system of industrial development with increasing returns taking the form of network externalities. It can, however, safely be stated that it works to reduce the diversity of innovations as well as of the paths of future developments in technology. This may lead to a significant slow-down, in if not a general stagnation of, technological progress, depending on what effects of monopolized industries and reduced diversity in technologies on technological change are assumed.

4.3 Non-ergodicity and path dependence in models of network externalities: the complex systems perspective

Contemporary to Katz and Shapiro's development of their market theory of network externalities, another group of scholars took a very different approach to the same phenomenon. They were not so interested in finding the concrete strategies perfectly rational agents with perfect information would take, nor in assessing the effects of these interactions and possible state interventions on the general welfare. Their focus was rather on the development of the industry as such. As a result, they emphasized different aspects of the problem, such as the path dependence of the technological and organizational development and technological lock-ins, aspects which likely show profound 'welfare' effects of their own if investigated in a neoclassical market context. Without the neoclassical assumption of a consistent system of prices and values, the concept of the aggregated welfare does of course not make sense.[68] Nevertheless, and notwithstanding the conceptual differences, the degree to which the findings point in the same direction is remarkable.

This different approach was first taken by Arthur *et al.* (1982), with subsequent contributions by Arthur *et al.* (1987), Arthur (1988, 1989, 1990b), Arthur and Lane (1993), Dosi *et al.* (1994), Dosi and Kaniovski (1994), Kirman (1997) and Winter *et al.* (2003). The basic idea is to model technology choice not as a game theory model focusing on the rationality of the agents (as Katz and Shapiro as well as Farrell and Saloner and others following the mixed game theory and neoclassical market approach did) but as a stochastic process. The development of the idea starts with a simple Eggenberger–Pólya urn scheme[69] as developed by Eggenberger and Pólya (1923) at the beginning of the twentieth century. The number of coloured balls of (two) different colours in an urn evolves according to a stochastic sequence, in the standard Pólya process using a rule such that the likelihood for a share of a colour to increase is equal to its current share of the total number of balls in the urn. This process tends to conserve the current shares. However, other rules for similar schemes are possible giving rise to a class of models, the *generalized urn schemes* (Arthur *et al.*, 1982; Dosi *et al.*, 1994; Dosi and Kaniovski, 1994).

4.3.1 The simple Eggenberger–Pólya process

Before considering generalized urn schemes and the possibility of modelling the development of technology standards using generalized urn schemes, this section presents the simple case of the Eggenberger–Pólya process, as also considered by Eggenberger and Pólya in their original paper (Eggenberger and Pólya, 1923). Let there be an urn containing a number of balls with two different colours A and B (say red and black,), n_A balls of colour A (red) and n_B balls of colour B (black) with $n = n_A + n_B$. One ball is drawn from the urn with subsequent replacement, that is, the ball is put back into the urn. Additionally, one more ball of the same colour as the one that was drawn is placed into the urn. Consequently, with each draw (each timestep in time t), one ball is added to the urn and the stocks n, n_A and n_B change in time and will therefore be denoted as n_t, $n_{A,t}$ and $n_{B,t}$.

The Eggenberger–Pólya process $\frac{n_{A,t}}{n_t}$ develops according to

$$\begin{aligned}\frac{n_{A,t+1}}{n_{t+1}} &= \frac{n_{A,t}}{n_t} - \frac{\frac{n_{A,t}}{n_t}}{t+2+1} + \frac{\zeta_{t+1}\left(\frac{n_{A,t}}{n_t}\right)}{t+2+1} \\ &= \frac{n_{A,t}}{n_t} - \frac{\frac{n_{A,t}}{n_t} - \zeta_{t+1}\left(\frac{n_{A,t}}{n_t}\right)}{t+3} \qquad (4.35)\end{aligned}$$

where $\zeta_{t+1}\left(\frac{n_{A,t}}{n_t}\right)$ is a sequence of (Bernoulli distributed) random numbers such that

$$\zeta_{t+1}\left(\frac{n_{A,t}}{n_t}\right)=\begin{cases}1 \text{ with probability } & \frac{n_{A,t}}{n_t}\\ 0 \text{ with probability } & 1-\frac{n_{A,t}}{n_t}=\frac{n_{B,t}}{n_t}\end{cases}. \tag{4.36}$$

The interpretation of this is quite straightforward. Consider the upper line in equation (4.35). The first term represents the current share of balls with colour A. The second term represents the reduction of this share caused by the increase of the number of balls in the urn by 1 (neglecting the probability that the added ball might also be of colour A). The last term is the possible increase in the share of type-A balls in the urn if the additional ball is of colour A, which happens with probability $\frac{n_{A,t}}{n_t}$. The expected value of the sequence $\zeta_{t+1}\left(\frac{n_{A,t}}{n_t}\right)$ is

$$E\left(\zeta_{t+1}\left(\frac{n_{A,t}}{n_t}\right)\right)=1\cdot\frac{n_{A,t}}{n_t}+0\cdot\left(1-\frac{n_{A,t}}{n_t}\right)=\frac{n_{A,t}}{n_t}$$

yielding an expected value of the share of type-A balls $\frac{n_{A,t+1}}{n_{t+1}}$ of

$$\begin{aligned}E\left(\frac{n_{A,t+1}}{n_{t+1}}\right)&=\frac{t+2}{t+3}\frac{n_{A,t}}{n_t}+\frac{\zeta_{t+1}\left(\frac{n_{A,t}}{n_t}\right)}{t+3}\\&=\frac{t+2}{t+3}\frac{n_{A,t}}{n_t}+\frac{\frac{n_{A,t}}{n_t}}{t+3}=\frac{n_{A,t}}{n_t}\end{aligned} \tag{4.37}$$

Equation (4.37) shows that every combination of balls is stable in this one-dimensional Eggenberger–Pólya process. More specifically, the expected value in equation (4.37), since it returns exactly the initial share, holds not just for one timestep but for arbitrarily many steps, including $t\to\infty$.[70]

$$\lim_{T\to\infty}E\left(\frac{n_{A,t+T}}{n_{t+T}}\right)=\frac{n_{A,t}}{n_t} \tag{4.38}$$

There are infinitely many possible distributions of shares between 0 and 1 and any possible distribution of shares is a stable limit.

The case considered so far is the one-dimensional Eggenberger–Pólya process; it is one dimensional because there is only one independent state variable, e.g. $\frac{n_{A,t}}{n_t}$, while the other one ($\frac{n_{B,t}}{n_t}=1-\frac{n_{A,t}}{n_t}$) is completely determined by the first one. However, the properties of the Eggenberger–Pólya process are general and also hold for a d-dimensional system, the shares now being a $d+1$-dimensional vector $X_t=\begin{pmatrix}x_{1,t}\\x_{2,t}\\\dots\\x_{d+1,t}\end{pmatrix}$. Equations (4.35) and (4.36) become in this case (with

$i = 1, ..., d+1$)

$$x_{i,t+1} = x_{i,t} - \frac{x_{i,t} - \zeta_{i,t+1}(x_{i,t})}{t+3} \qquad \forall i f \qquad \text{(Arthur \textit{et al.}, 1982, 8)}$$

$$\zeta_{i,t+1}(x_{i,t}) = \begin{cases} 1 \text{ with probability} & x_{i,t} \\ 0 \text{ with probability} & 1 - x_{i,t} = \sum_{j \neq i} x_{j,t} \end{cases} \qquad \forall i$$

(Arthur *et al.*, 1982, 5)

and (4.38) becomes

$$\lim_{T \to \infty} E(x_{i,t+T}) = x_{i,t} \qquad \forall i \tag{4.39}$$

The stability property of the one-dimensional case is retained; at (any) time t the expected division of the 'population' of balls in the urn at any future time $t+T$ (including $T \to \infty$), X_{t+T} equals the current shares of the $d+1$ types X_t.

4.3.2 Generalized urn schemes

There is an infinite number of different rules according to which balls of different colours could be added to an urn using a random draw from the urn as an input for the rule. The simple Eggenberger–Pólya process implies the specific rule, namely exactly one ball is drawn and exactly one ball matching the colour of that which was drawn is added to the urn. To mention some examples for other possible rules, the first three of which were suggested by Arthur *et al.* in their seminal paper (Arthur *et al.*, 1982):

1. Draw ι balls, determine the type (colour) that the most of these ι balls belong to and add one ball of this colour, if two or more types are present equally often in the sample, draw again. ($\iota = 1$ with two types, i.e. a one-dimensional system is identical to the Eggenberger–Pólya process) (Arthur *et al.*, 1982: R1, p. 61)
2. Draw ι balls, determine the type (colour) that the *fewest* of these ι balls belong to (or, if, applicable, that is not represented in the sample at all) and add one ball of this colour. If undecidable by with this draw, draw again (Arthur *et al.*, 1982: R3, p. 61).
3. Draw a sequence of balls until r balls of one type (colour) have been drawn; add a ball of this colour (Arthur *et al.*, 1982: R2, p. 61).
4. Continuous rule (as opposed to discrete): determine the current shares $x_{i,t}$ and add a ball of (any) type i corresponding to an increasing probability function $p(x_{i,t})$, e.g.

$$p(x_{i,t}) = \frac{3x_i^2 - 2x_i^3}{\sum_{j=1}^{d+1} 3x_j^2 - 2x_j^3}.$$

(The probability functions add up to 1: $\sum_{i=1}^{d} p(x_{i,t}) = 1$.)

5. Continuous rule: determine the current shares $x_{i,t}$ and add a ball of (any) type i corresponding to a decreasing probability function $p(x_{i,t})$, e.g.

$$p(x_{i,t}) = \frac{1 - 3x_i^2 + 2x_i^3}{\sum_{j=1}^{d+1} 1 - 3x_j^2 + 2x_j^3}.$$

The continuous rules are more easy to fit into the development equations of the process (though not quite as realistic as rules for the necessarily discrete drawing and replacement of balls in urns). In fact, the discrete rules will have to be transformed into continuous probabilities in order to analyze the stability properties of the urn scheme in question. This can be done by computing the probability for each type to gain another ball given the current X_t.[71] Generalizing further that a number n_0 of balls is in the urn in the beginning ($t = 0$), the generalized urn scheme is defined as a process

$$x_{i,t+1} = x_{i,t} - \frac{x_{i,t} - \zeta_{i,t+1}(x_{i,t})}{n_0 + t + 1} \qquad \forall i \qquad \text{(Arthur } et\ al.\text{, 1982, 8)}$$

with

$$\zeta_{i,t+1}(x_{i,t}) = \begin{cases} 1 \text{ with probability} & x_{i,t} \\ 0 \text{ with probability} & 1 - x_{i,t} = \sum_{j \neq i} x_{j,t} \end{cases} \qquad \forall i$$

Consequently,

$$\zeta_{i,t+1}(x_{i,t}) = p(x_{i,t}) + \beta_{i,t+1} \qquad \text{(Arthur } et\ al.\text{, 1982, 5)}$$

where $\beta_{i,t+1}$ is the term for deviations of the binary function $\zeta_{i,t} \in \{0, 1\}$ from its mean $p(x_{i,t})$. Thus, $\beta_{i,t+1}$ is independently identically distributed with mean 0 and standard error $p(x_{i,t})(1 - p(x_{i,t}))$.

As the X_t are $d + 1$ dimensional vectors (one of the elements, however, is determined by the requirement $\sum_{i=1}^{d+1} x_{i,t} = 1$), the process defines a d-dimensional system of recurrence equations (the equations being d equations out of the $d + 1$ defined by equation (Arthur *et al.*, 1982, 8)[72]) which in turn has the structure of a replicator model. (The elements of) X_t are the state variable, the above recurrence equation (equation (Arthur *et al.*, 1982, 8)) defines the development of the system (development equation) and the probability functions $p(x_{i,t})$ are the relative fitnesses of the $d + 1$ types of 'agents' that form a part of the replicator model.[73] The probabilities are Borel functions (measurable functions) mapping $p(\cdot) : [0, 1]_{d+1} \to [0, 1]_{d+1}$.[74] In fact, they must also sum up to 1 for all agents

$$\sum_{i=1}^{d+1} p(x_{i,t}) = 1$$

as is true for relative fitnesses. This is achieved by using a vector of absolute fitnesses $f_t = (f_{i,t})_{\forall i}$ from which the definition of a relative fitness p_t that sums up to 1 is straightforward

$$p_{i,t} = \frac{f_{i,t}}{\sum_{j=0}^{d+1} f_{j,t}}$$

Note that rules 4 and 5 do exactly that, the absolute fitnesses being $f_{i,t} = 3x_i^2 - 2x_i^3$ and $f_{i,t} = 1 - 3x_i^2 + 2x_i^3$ respectively in rules 4 and 5.

The above rules may stand for different regimes of technology adoption, assuming that the agents are homogeneous except for the stochastic sequence of deviations $\beta_{i,t}$ from the mean of the distributions, $p(x_{i,t})$. Any general adoption rule for infinite streams of agents (infinite time) can be defined in such terms. Consequently, using such a system to model real-world systems of technology competition, it is possible to gain some understanding of the process, which may be very different from the processes of technology choice as considered in the mixed game theory and market models in the sections above. Furthermore, the models used in the mixed game theory and market approach can then be analyzed from another, broader, point of view. From there, it may even be possible to make predictions and implicit or explicit policy recommendations as are usual in neoclassical papers.

For this, the most crucial questions are those concerning the existence and stability properties of fixed points (equilibria) in the system. The given system of stochastic recurrence equations yields the following expected value for X_{t+1}

$$E(x_{i,t+1}) = x_{i,t} - \frac{x_{i,t}}{n_0+t+1} + \frac{p(x_{i,t})}{n_0+t+1} + \frac{E(\beta_{i,t+1})}{n_0+t+1} \tag{4.40}$$

and, since $E(\beta_{i,t+1}) = 0$

$$E(x_{i,t+1}) = x_{i,t} - \frac{x_{i,t}}{n_0+t+1} + \frac{p(x_{i,t})}{n_0+t+1} \tag{4.41}$$

Equilibria are obviously those points for which $E(x_{i,t+1}) = x_{i,t}$, hence, equation (4.41) yields the equilibrium condition

$$x_{i,t} = x_{i,t} - \frac{x_{i,t}}{n_0+t+1} + \frac{p(x_{i,t})}{n_0+t+1} \quad \forall i$$

$$\frac{p(x_{i,t})}{n_0+t+1} = \frac{x_{i,t}}{n_0+t+1} \quad \forall i$$

$$p(x_{i,t}^*) = x_{i,t}^* \quad \forall i \tag{4.42}$$

Acknowledging that this is independent of time t (since the system is time invariant), therefore using the notation X instead of X_t (and x_i instead of $x_{i,t}$)

and writing $p(\cdot)$ as a vector function of X, this becomes

$$p(X^*) = X^* \tag{4.43}$$

The system exists in the d-dimensional phase space X which is constrained between 0 and 1 in each dimension $X = [0, 1]_d$. The set of equilibria is a subset of this phase space. Now, define a subset $\mathcal{B}$ which comprises the stable equilibria (also called stable fixed points, or attractors), that is, independently from t (equation without number (Arthur *et al.*, 1982: p. 62))

$$\mathcal{B} = \{X: \quad p(x_i^*) = x_i^* \quad \forall i\} \tag{4.44}$$

such that for an arbitrarily small environment ε of $\mathcal{B}$ (equation 2, part 1 (Arthur *et al.*, 1982))

$$p(x_i) > x_i \quad \text{for } x_i \in [x_i^* - \varepsilon, x_i^*] \quad \forall i \tag{4.45}$$

and (equation 2, part 2 (Arthur *et al.*, 1982))

$$p(x_i) < x_i \quad \text{for } x_i \in [x_i^*, x_i^* + \varepsilon] \quad \forall i. \tag{4.46}$$

This means that the system moves dynamically towards the equilibrium in its local (arbitrarily small) environment. With the random vector sequence $\beta_{i,t}$, the system is still stochastic but with the stochastic process being independently identically distributed with mean zero. There is consequently a corresponding deterministic system (that is equivalent to the expected values of the stochastic system, hence simply discarding β from equation (4.40)). The definition of the set of stable equilibria $\mathcal{B}$ in the stochastic system in equations (4.44) through (4.46) above is equivalent to the set of equilibria in the deterministic system. The conditions (4.45) and (4.46) correspond to asymptotic stability of the corresponding deterministic system – which may be shown by a number of methods. For continuous functions (and $p(x)$ is continuous) the most convenient is probably to establish that the modulus of the dominant eigenvalue is smaller than one. Alternatively, it could be shown that the motion of the Lyapunov functions is negative outside $\mathcal{B}$ (the way chosen by Arthur (1988) and Kirman (1997)), or that the Lyapunov exponents are negative.[75]

The central piece of Arthur *et al.*'s seminal paper (Arthur *et al.*, 1982) is their theorem that states that if the set $\mathcal{B}$ is not empty, the system converges towards $\mathcal{B}$ with certainty (with probability 1). The full proof requires a lengthy consideration of stochastic approximation; therefore, it shall suffice for the current study to give a short outline. The local stability of the components of $\mathcal{B}$ follows from the definition of $\mathcal{B}$; the system will almost certainly remain in the neighbourhood of a stable fixed point $X^* \in \mathcal{B}$ once it has reached this part of the phase space. All other parts of the system are, however, disequilibrium states. The dynamic moves about the phase space until the system gets trapped in the close neighbourhood of one of the stable fixed points.

From this, it is also obvious that generalized urn schemes are not ergodic if the set $\mathcal{B}$ is not empty (in other words if the system has one or more stable fixed points): parts of the phase space are unreachable from other parts of the phase space. In fact, the likely[76] flow of the trajectories is also directed in parts of the phase space; trajectories approach the attractors but (almost certainly) do not leave them. This, in turn, implies that the development of the system is *path dependent* and not reversible. If $\mathcal{B}$ contains two unconnected components, that is two distinct attractors, the path dependence furthermore does not just apply to the development of the system towards the attractor, but also to the result. Once the system has reached the neighbourhood of one of the attractors, it will (almost certainly) not leave this area of the phase space to approach another one. Since the attractors are, however, just local attractors,[77] there is no way to predict the outcome of the dynamic process without knowledge of both the sequence of random values $\beta_{i,t}$ and the initial values X_0.

Returning to the above examples for urn scheme rules alternative to the original Eggenberger–Pólya urn scheme, the dynamics of simple one-dimensional ($d = 1$) cases shall be discussed in the following to demonstrate the scope of the model and the nature of its findings in a more illustrative way. The cases therefore comprise two types of balls (or adoption decisions for technologies), the shares are conveniently written as x_t and $1 - x_t$. For the two continuous rules (R4 and R5), the equilibrium and stability properties can be assessed much more easily than in case of the discrete rules, which lead to rather complicated transition probability functions. The continuous transition probability for R4 and R5 is given as a perfectly continuous and differentiable function. The equilibria can therefore easily be computed by setting

$$p(x) = x$$

which applied to the respective transition probability functions yields the three equilibria for R4[78]

$$x^{R4,1} = 0$$

$$x^{R4,2} = 0.5$$

$$x^{R4,3} = 1$$

For R5, the equation is still a third degree equation[79] and yields three equilibria as well; however, only one of the three lies within valid parameters $0 \leq x \leq 1$

$$x^{R5,1} = 0.5$$

$$(x^{R5,notvalid} = 0.5 \pm \frac{\sqrt{5}}{2})$$

The stability is conveniently analyzed using the eigenvalues of the replicator dynamic system, a single recurrence equation. Hence, the dominant eigenvalue (in this case the single eigenvalue of the system) must have a modulus smaller than 1 in order for the system to be stable. The eigenvalue is computed from the Jacobian matrix $\mathcal{J}$ of the system such that ($\mathcal{I}$ being the unit matrix)[80]

$$det(\mathcal{J} - \lambda\mathcal{I}) = 0 \tag{4.47}$$

As the systems are one-dimensional, the Jacobian matrices have only one element $\mathcal{J} = \left(\frac{\partial x_{t+1}}{\partial x_t}\right)$, which applied to equation (4.47) yields the same single element of the Jacobian matrix as the system's single eigenvalue $\lambda = \left(\frac{\partial x_{t+1}}{\partial x_t}\right)$. However, the eigenvalue is not static since the recurrence equation is of higher than first order; it is a function of x. Therefore, the stability property varies between different points in the phase space, and, in essence, the eigenvalue assumes different values for the different fixed points. For R4, the eigenvalue computes as

$$\lambda = \frac{\partial x_{t+1}}{\partial x_t} = \frac{t+n_0}{t+n_0+1} + \frac{6x_t - 6x_t^2}{t+n_0+1} \tag{4.48}$$

and for the equilibria assumes the values (exclusively real and positive, and therefore identical to their modulus)

$$\lambda(x^{R4,1}) = \lambda(0) = \frac{t+n_0}{t+n_0+1} < 1$$

$$\lambda(x^{R4,2}) = \lambda(0.5) = \frac{t+n_0+1.5}{t+n_0+1} > 1$$

$$\lambda(x^{R4,3}) = \lambda(1) = \frac{t+n_0}{t+n_0+1} < 1$$

In turn, the interior equilibrium $x = 0.5$, for which the population would be equally shared between the two types (or technologies), is unstable; the complete standardization equilibria are stable. This system is depicted in Figure 4.10a. In this illustration, the crossings of the 45°-line denote equilibria; those that occur from below (Arthur and Lane (1993) term them 'upcrossings') are unstable, those from above are attractors. The eigenvalue following from R5 is

$$\lambda = \frac{\partial x_{t+1}}{\partial x_t} = \frac{t+n_0}{t+n_0+1} - \frac{6x_t - 6x_t^2}{t+n_0+1}. \tag{4.49}$$

For the sole fixed point resulting from R5 that falls within valid parameters, the (real and positive) eigenvalue becomes

$$\lambda(x^{R5,1}) = \lambda(0.5) = \frac{t+n_0-1.5}{t+n_0+1} < 1$$

which means that this equilibrium is a stable fixed point. The system is illustrated in Figure 4.10b; it is designed as the exact reverse of R4, favouring weak

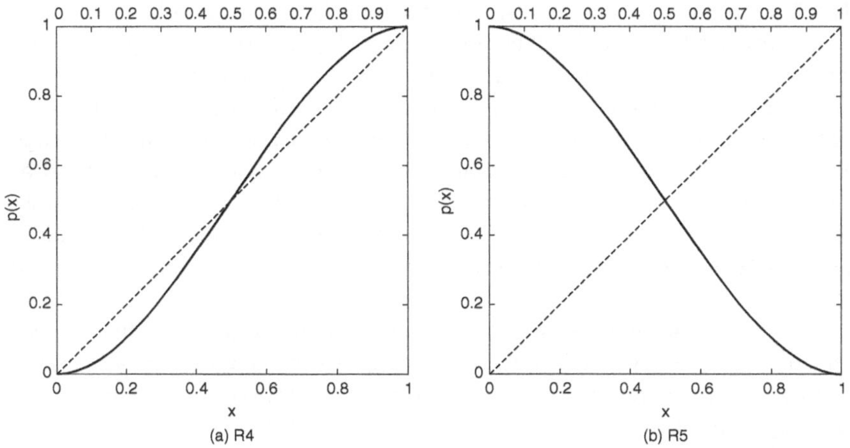

Figure 4.10 Pólya urn systems according to rules r4 and r5 with equilibria and stability properties. Crossings of the 45°-line from above are stable equilibria.

types over strong ones; for technology choice, this would of course imply an unrealistically high preference for technological diversity. Nevertheless, as the following sections show, situations with interior (industry sharing, not perfectly standardizing) attractors are conceivable and not at all unrealistic.

4.3.3 Modelling technology diffusion with urn schemes

The seminal paper on generalized urn schemes by Arthur *et al.* in 1982 does not detail applications of the model; however, it is mentioned that it could be used to model technology choice patterns as well as a number of other real-world phenomena. However, Arthur applied the urn scheme approach to technology choice shortly thereafter (Arthur, 1983) developing a model more closely related to those discussed in the network externality section (4.2.1) above. Drawing on empirical and theoretical work by several other authors (including Paul David and Richard Nelson) he develops the argument of increasing returns to adoption (the term 'network externality' was introduced later) as a result of knowledge effects. Once a technology is available, people learn how to use it, can teach others, etc. Note that this argument is similar to Arrow's (1962) learning by doing argument in his growth theory (see section 3.3.3). He adds the higher probability of a widely used technology to receive greater funds (and more thought) for further development as an alternative interpretation. In later papers (Arthur *et al.*, 1987; Arthur, 1988, 1990b), Arthur and his co-authors illustrated the mechanism extensively with regional industry organization dynamics, citing empirical work on the role of spin-offs for regional economic organization in the USA. As knowledge about a technology becomes available in a region, the willingness to innovate using this technology rises, as does the willingness to invest. Caused by the tendential immobility of human capital – humans want and need a certain regional social environment which they are usually unwilling to give up – this leads to a regional buildup of knowledge corresponding especially to technologies that are traditionally strong in the region. Arthur and Lane (1993) finally extended (and formalized) Arthur's earlier argument of network externalities caused by the path dependence of the development of knowledge and research. They stress that any industrial product requires a certain level of skills to operate, and widespread information about the existence (and possibly the good quality) of the product helps the success of the product. The required information has to be accumulated by early adopters and will diffuse by word of mouth.

What is interesting about Arthur and his collaborator's argument is that they do not focus on and barely mention true network goods. It is easy to apply the same reasoning to technologies that explicitly rely on networks, such as communication technology and hard- and software standards, the type of products that were analyzed by the contributors to the mixed game theory and market approach discussed in the sections above, the type of products which the term 'network externality' was originally meant to fit (Katz and Shapiro, 1985). This example also suggests of course that network externalities are a rare effect that applies to a particular group of goods, and thereby disturbs the reasoning of the standard market approach and requires special treatment with game theory methods. (The proponents of this line of research may or may not have seen their investigations in this way.) However, the group of scholars around Arthur and Kaniovski had something very different in mind. Their examples are designed to apply to a very much broader area of microeconomics: if virtually any product and any technology requires standards, and relies on slowly diffusing tacit knowledge and specifics of the 'topology' (in the case of the regional clustering argument), increasing returns are no special case but rather a central issue that has been insufficiently dealt with in the neoclassical theory.[81]

The approach to this problem that Arthur, Kaniovski and their co-authors were proposing was a general framework going beyond diminishing and constant returns, and analyzing microeconomics from a complex systems perspective that emphasizes among other things the effects of path dependence and positive feedbacks. Later contributions by both Arthur and Dosi and Kaniovski are consequently more confident in announcing a new general class of models (Dosi and Kaniovski, 1994; Dosi *et al.*, 1994) and also more extensive in their analysis of the implications of their findings for economics in general (Arthur, 1988, 1989).

In particular, the relevant findings of this line of research for the economics of innovation and technological change came down to a number of important aspects that shall be discussed in the remainder of this section.

Positive feedback Positive feedback denotes effects that are able to act as a catalyst for regenerating that very same effect. Naturally, this mechanism works simply to put systems out of balance. Either the effect occurs as a rate or velocity high enough to accelerate the effect (until, at least in more complex, real-world systems some other restriction halts the effect), or it does not, in which case the effect will eventually break down. The limit case at which the progress of the effect remains constant is usually unstable. To underline the connection of positive feedbacks and technology diffusion, consider for instance – as in the urn scheme according to rule R4[82] above – the rate at which a particular technology is chosen. If this rate is high enough at the start, the rate continues to increase; otherwise this technology will likely be entirely cornered within a rather short time.

Positive feedbacks are the central mechanism of what may be called evolutionary systems. Even very small differences in systems generating sufficiently strong positive feedback are amplified by this feedback mechanism to exercise a profound effect on the system as a whole. In effect, for an entity in an evolutionary system – in this case a technology standard candidate – to be successful, it must gain a small advantage over its competitors. In case of patterns like technology standards that rely heavily on network effects, this advantage is a larger or more suitable[83] network depending on the specific shape of the choice process, in this model the urn scheme transition probability function. Arthur in one of his early studies honoured this insight by mentioning the significance of 'historically small events' in the paper's title (Arthur, 1983). In fact, the profound importance of positive feedbacks was anticipated or at least suspected earlier by scholars like Veblen and Kapp – without of course being able to explain their full scope. They termed this 'circular cumulative causation'; for an overview, see Berger and Elsner (2007) and Berger (2009a).

Path dependence Positive feedback implies path dependence. The system has several attractors, thus several possible outcomes. Which of those is realized depends on the initial values and possibly on stochastic influences. In one-dimensional systems with continuous dynamics, this means simply that there is an unstable fixed point (a repeller) and the trajectories move towards different fixed points on either side of the repeller. The path thus depends on the relative position of the initial state to the repeller. As an illustrative example, see the analysis of the dynamics of the urn scheme governed by rule R4 above (Figure 4.10): it was found that the system has two attractors, the equilibria $x^{R4,1} = 0$ and $x^{R4,3} = 1$ with a repeller, $x^{R4,2} = 0.5$ in between. In these systems, the 'trajectories' are nothing more than the direction of movement along the coordinate of the single variable (x). In higher dimensional systems, trajectories may be almost parallel in the beginning, then diverge gradually and finally approach the different attractors. In a stochastically perturbed system like the urn schemes currently discussed, the perturbations are likely to shift the system in the first (parallel) phase between trajectories (of the corresponding deterministic system[84]) approaching different attractors. Consider as an example a system of technology choice between two alternative technologies and an irrelevant third option, e.g. either a third technology with considerably weaker network effects or the option not to choose a technology. The system develops as illustrated in Figure 4.11, the irrelevant option being at the bottom left. Note that there is a limit path for exactly equal shares of the two technologies, a single limit trajectory that does not approach one of the attractors but rather a saddle point with a share of 50% for both technologies (and 0% for

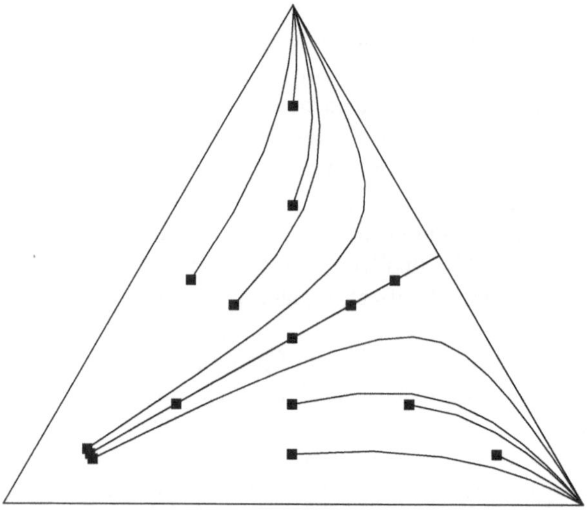

Figure 4.11 Simplex of the development of the shares of three technologies, two of them (upper and lower right) with network effects, one without (lower left). The process is path dependent and sensitive to initial conditions. Simulated (source code available from the author on request) for a number of values, squares indicate the starting points, lines the trajectories.

the irrelevant option). For perturbed systems, the probability for this point to be the final outcome for $t \to \infty$ is zero.[85]

Non-ergodicity (and 'broken ergodicity') An ergodic system is a dynamic system the behaviour of which remains equal for the entire phase space. Dynamic systems with fixed points (whether attractors or repellers) or other structures that absorb trajectories or alter their flow are never ergodic, and hence non-ergodic. Non-ergodicity is equivalent to path dependence: if the properties of the dynamic flow change over the phase space, the starting point (the initial condition) becomes important, as do small perturbations.

The phase space of a globally non-ergodic system may be partly ergodic, in which case these parts of the phase space may be called ergodic components. The transition of the system from an ergodic component to non-ergodic parts leads to the breaking of the ergodicity and, in a way the emergence of a complex structure. This is justified, since ergodic systems or components can by definition not be complex; they in themselves are completely homogeneous, possibly stochastic but not complex, non-ergodic systems develop a higher (though not necessarily high[86]) degree of complexity. As an application to technological progress, consider humanity (or its predecessors) before the development of intelligent thought. The development of technology in this system is completely ergodic, a pure side effect of other aspects such as instincts and fortune. Proto-technologies come into being and get lost – probably at a rather constant rate. The development of higher cognitive abilities, communication and knowledge (in turn technology facilitating the development of technology) established positive feedback loops and turned the technological progress of humanity into something completely different.[87] Positive feedback prevents this different system from collapsing; ensuring that it remains in non-ergodic areas of its phase space.

Predictability The urn scheme governed by rule R5 as considered above is equivalent to diminishing returns: variety is valued so much that the technology with the smaller network is always chosen with the higher probability. In turn, the process is non-ergodic but predictable. Its dynamics approaches the single fixed point, an attractor, rapidly. The same is true for a system with constant returns. A rule R6 could be defined with a transition probability function that does not depend on the current shares and assigns equal probabilities to all $d+1$ competing technologies, thus $p(x) = \frac{1}{d+1}$. This system has a single attractor as well, the components of the attractor (the equilibrium vector of the shares X^*) equivalent to the transition probability vector $X^* = (p(x))_{\forall i=1,\ldots,d+1}$.

A non-ergodic system with multiple attractors is, however, not generally predictable. Multiple attractors imply increasing returns to adoption. In this system, it is predictable that the system will converge to some point in the set of attractors B for $t \rightarrow \infty$. As the attractors are very different, in fact they stand for different technologies becoming the industry wide standard, the more interesting prediction (both for adopters, vendors and other involved agents) would be which point in B will be approached. This, however, is only possible with knowledge of the initial path of the trajectory until it moves into the force field of an attractor.

The unpredictability of the outcome in generalized urn schemes with more than one attractor translates into the consequence that no equilibrium dominates another and none of the involved technologies is inferior or superior ex ante. This is a classical example for true uncertainty; strategic behaviour is difficult in this situation. Possible predictions are at best probabilistic. However, small differences or 'historically small events' (Arthur, 1983) may translate into different paths, and in complex real life systems agents such as innovators are never aware of any small effect that might have an impact on the technological development. Considering this, the necessary conclusion is that it is not only a random number sequence that is unknown and might be approximated by a probabilistic model. Rather, the probability distribution itself is not known.

Lock-ins It is not enough that technology choice systems with network externalities, which cause positive feedback, are unpredictable, they are also not necessarily path-efficient. In a system with more than one attractor, the general or social desirability of one of the attractors (one of the technologies) might be greater than that of the others. The system, however, remains unpredictable and the other attractors are unabatedly possible outcomes.

What is more, once the system has moved into the force field of an attractor representing an inferior technology, there is no way (within the so-defined system) to leave it and switch to the path-efficient technology. Arthur and David established the term 'lock-in' to describe this effect (Arthur, 1983; David, 1985).

Specific policy recommendations were added by David (2005). The time for a regulator (or a large enough organized and coordinated group of potential adopters) to act is when they are able to exercise influence. This is before any of the technology standard candidates can become the industry standard. Suppose they knew exactly which candidate was going to be the superior choice, they would simply have to organize sufficient endorsement within a narrow time 'window of opportunity' and the problem would be settled. Unfortunately, it is usually difficult enough to understand the current economic effects of current technology – and anticipating future effects of new technologies is certainly a messy business. In David's metaphorical language, it means that the 'blind giant' (the entity with the power to make the technology choice process path-efficient but without 'sight', without knowledge of the process) has just to step through the narrow 'window of opportunity' (David, 2005). Note that it may sometimes be better to endorse some technology to generate certainty for the adopters, thus preventing lots of agents from choosing the wrong standard ('angry orphans' (David, 2005)) and possibly also adverse selection of industries as discussed in section 4.2.4 above. This may, however, also make things worse, as it deliberately destroys a certain variety of standards and for good or bad delivers the industry into the hands of the developers of the chosen standard.

Arthur demonstratively discusses this question of predictability, ergodicity and path dependence in the context of the regimes of diminishing, constant and increasing returns in his 1989 paper (Arthur, 1989). Constant and diminishing returns are the regimes of technology choice as they are traditionally considered (with some notable exceptions) to be more or less based on neoclassical assumptions; the positive feedback mechanisms, however, lead to increasing returns. This implies an entirely different mode of technological progress. Predictability is limited and path efficiency is not guaranteed. The urn scheme approach does not dismiss the traditional neoclassical analysis with constant or diminishing returns, it rather places it into context, clarifying assumptions, pointing out consequences, and providing a framework which is able to cover different regimes. Increasing returns are represented, for example, by rules R1, R2 and R4 above, while R3 and R5 are equivalent to diminishing returns; the case of constant returns is simply a constant transition probability function.

4.3.4 Rational technology choice of two groups of homogeneous agents and its dynamic properties

The mixed game theory and market approach as discussed in section 4.2 can be rewritten as a generalized urn scheme. In fact, Arthur, Ermoliev and Kaniovski studied a game theory technology choice model in several variations in a number of papers (Arthur, 1983; Arthur *et al.*, 1987; Arthur, 1989) to do exactly that.

This model is rather close to those discussed by other scholars committed to the mixed game theory and market approach, particularly Katz and Shapiro (1985, 1986).

Starting with the model discussed in section 4.2.1, the group of agents has to be redefined to be less perfectly informed in such a way that they cannot take into account the agents that have yet to adopt a technology. Their information is thus limited to the net intrinsic preferences and the current strength of the network effects of both technologies. The adoption decision for agent κ (to choose technology A, not B) is

$$v(N_{ch(\kappa),A}+1)-v(N_{ch(\kappa),B}+1)\geq r_B-r_A\equiv\Delta.$$

Again, $v(\cdot)$ denotes the network utility generated by the network of the customers already committed to the respective technology plus one (for the agent κ who is currently choosing), r stands for the net intrinsic preferences of agent κ for the two technologies. If the inequality holds true, κ chooses A, otherwise she does not and will revert to adopting B. Let the net intrinsic preferences be homogeneous in two subgroups of the population. One prefers A and consequently has a negative Δ, the other prefers B and has a positive Δ.

The population thus initially comes down to a binary random number sequence, half of the agents joining the network of technology A and half of them joining technology B. This is true as long as the network effects (the left side of the inequality) are weak enough not to interfere with the impact of the net intrinsic preferences. Consequently, there are also two specific values for the size difference of the two networks $v(N_{ch(\kappa),A}+1)-v(N_{ch(\kappa),B}+1)$ such that both groups will choose the stronger one, the network of technology A, if $v(N_{ch(\kappa),A}+1)-v(N_{ch(\kappa),B}+1)$ is large enough, or the network of technology B if the (then negative) value of $v(N_{ch(\kappa),A}+1)-v(N_{ch(\kappa),B}+1)$ is small enough. Arthur (1983) calls these critical values the *absorbing barriers* of this technology choice process.

The dynamics of this process (Figure 4.12 provides an idealized illustration), thus the corresponding urn scheme is not continuous. It has two jump discontinuities at the critical values of the strength of the network externality. The middle segment has a stable fixed point which is, however, subject to stochastic disturbances. For the other two segments, all agents will always choose just one technology, namely the one with the larger network. Hence, complete dominance of the sector by each one of the two technologies are two further stable fixed points. Figure 4.12 assumes that the critical values of the network externality, the absorbing barriers, correspond to a specific sector share not to an absolute utility value (which is also possible but has a transition probability function which changes in time).

Three things are apparent from this:

- First, Arthur *et al.*'s (1982) theorem on the convergence towards attractors in generalized urn schemes may be applied: for systems with absorbing barriers

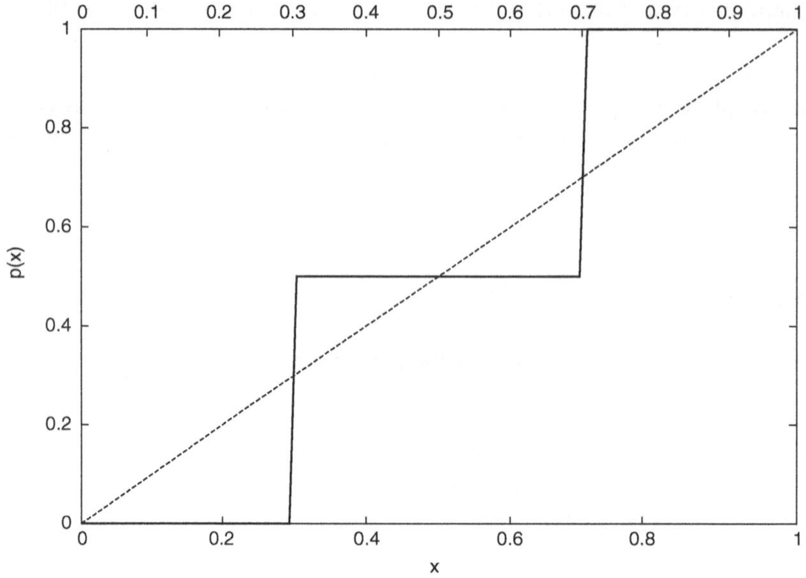

Figure 4.12 Absorbing barriers process with equilibria and stability properties. Crossings of the 45°-line from above are stable equilibria.

given in absolute utility values, the probability of the system being in the vicinity of the sector sharing attractor in the middle segment approaches zero as $t \to \infty$. The process in the middle segment is basically a random walk and the probability for a random walk to never leave a certain interval in infinite time approaches zero. (For systems with absorbing barriers given in sector shares, that probability remains positive and significant.) This effect is less striking than in the urn schemes considered above, it may take a longer time. Nevertheless, the system generally predicts a lock-in effect with ex ante unpredictable outcome and uncontrollable path-efficiency. The results of the mixed game theory and market model and the model of positive feedback in a complex system are remarkably similar.

- Second, compared to the mixed game theory and market models, generalized urn schemes are the less narrow approach of the two, allowing for a more global view on the nature of the process of innovation, technology choice (and diffusion) and technological progress.
- Third, depending on the exact model, both approaches easily allow for both a stable sector sharing attractor and a solution of global standardization as well as combinations of those two (where in turn the outcome depends on initial values and random influence). The quality of the models possible in this wide variety have to be judged from the resemblance to real-world systems they are able to bring about.

4.3.5 Technology tying reconsidered

The effect of 'standard tying' (Choi, 2004) was discussed above in the context of a mixed game theory and market approach to network externalities. It has also already been mentioned that the derived model is particularly well-suited to being studied using replicator dynamics, which in this case may be interpreted as an urn scheme.

The theoretical setting is two specialized industries for each of which the place of the industry standard technology is contested by two (or more) candidates. There is cross-industry standard tying: compatibility is required (it generates greater network utility) and the sponsors may be active in both industries. The basic question is how this system would develop and what influence the link between the technologies has on the 'standard war'. It is immediately obvious that a commercial sponsor of the industry standard in one industry would be able to corner the competitor in the other industry. The more general case is of course that of a number of technologies i (the shares being $p_i(t)$ at time t) in one industry A and a number of technologies j (shares $q_i(t)$ at time t) in the other industry B, with a compatibility term $\chi_{ij} \geq 0$ for each possible combination contributing to the utility of potential customers and thereby the fitness and future evolutionary performance of both industries. Figure 4.13 provides a stylized illustration of a transition probability function for such a pattern: both transition probabilities in both industries depend on the current sector shares in each of the industries.

Consider a replicator model for both industries that uses the expected payoffs $\Pi_{A,i}$ and $\Pi_{B,j}$ with $i,j = 1,2$ as the fitness terms in the following way

$$\frac{\partial p_i(t)}{\partial t} = (\Pi_{A,i}(t) - \phi_A(t))p_i(t)\forall i \tag{4.50}$$

and equivalently for the other industry

$$\frac{\partial q_j(t)}{\partial t} = (\Pi_{B,j}(t) - \phi_B(t))p_j(t)\forall j, \tag{4.51}$$

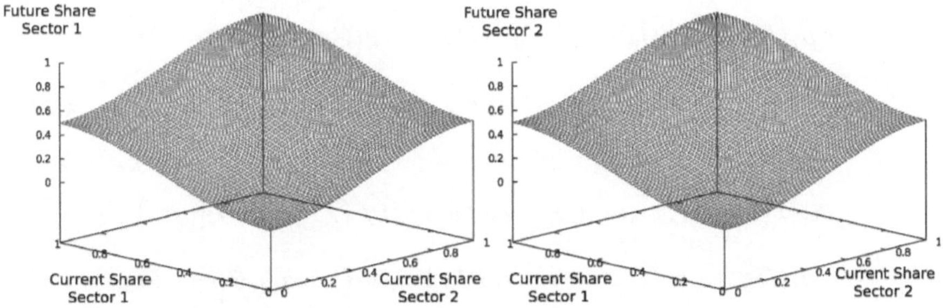

Figure 4.13 Transition probabilities in two tied sectors.

where $\phi(t)$ are the average fitness terms

$$\phi_A(t) = \sum_i p_i(t)\Pi_{A,i}(t), \qquad \phi_B(t) = \sum_j p_j(t)\Pi_{B,j}(t), \tag{4.52}$$

a common replicator model with differential equations (in fact, following the model in equation (2.20)). The model hence gives the development of the sector shares in time as depending on the payoffs of the technology in question and that of its competitor. For above average payoffs the sector share increases, for below average payoffs it decreases. For convenience, the explicit notations t for each term shall be dropped for the following considerations. Let the unit vectors be denoted by $\sigma_1 = \begin{pmatrix} 1 & 0 \end{pmatrix}$ and $\sigma_2 = \begin{pmatrix} 0 & 1 \end{pmatrix}$. The expected payoffs compute, following equation (4.32), as

$$\Pi_{A,i} = p_i\sigma_i \begin{pmatrix} \chi_{11} & \chi_{12} \\ \chi_{21} & \chi_{22} \end{pmatrix} \begin{pmatrix} q_1 \\ q_2 \end{pmatrix}. \tag{4.53}$$

and

$$\Pi_{B,i} = q_i\sigma_i \begin{pmatrix} \chi_{11} & \chi_{21} \\ \chi_{12} & \chi_{22} \end{pmatrix} \begin{pmatrix} p_1 \\ p_2 \end{pmatrix}. \tag{4.53'}$$

Using $p_2 = 1 - p_1$ and $q_2 = 1 - q_1$, this becomes

$$\begin{aligned}
\Pi_{A,1} &= p_1(q_1(\chi_{11} - \chi_{12}) + \chi_{12}) \\
\Pi_{A,2} &= (1 - p_1)(q_1(\chi_{21} - \chi_{22}) + \chi_{22}) \\
\Pi_{B,1} &= q_1(p_1(\chi_{11} - \chi_{21}) + \chi_{21}) \\
\Pi_{B,2} &= (1 - q_1)(p_1(\chi_{12} - \chi_{22}) + \chi_{22})
\end{aligned} \tag{4.54}$$

The industries are sufficiently described just considering one of the two competing standards, standard 1, i.e. shares p_1 and q_1 since the difference of the shares from 1 gives the respective other competitors share $p_2 = 1 - p_1$ and $q_2 = 1 - q_1$. Applying the payoffs (i.e. fitness) (4.54) to the replicator equation (4.51) yields

$$\begin{aligned}
\frac{\partial p_1}{\partial t} &= p_1(\Pi_{A,1} - p_1\Pi_{A,1} - (1 - p_1)\Pi_{A,2}) = p_1(1 - p_1)(\Pi_{A,1} - \Pi_{A,2}) \\
&= p_1(1 - p_1)(p_1(q_1(\chi_{11} - \chi_{12}) + \chi_{12}) - (1 - p_1)(q_1(\chi_{21} - \chi_{22}) + \chi_{22})) \\
&= -p_1^3(q_1(\chi_{11} - \chi_{12}) + \chi_{12} + q_1(\chi_{21} - \chi_{22}) + \chi_{22}) + p_1^2(q_1(\chi_{11} - \chi_{12}) \\
&\quad + \chi_{12} + 2q_1(\chi_{21} - \chi_{22}) + 2\chi_{22}) - p_1(q_1(\chi_{21} - \chi_{22}) + \chi_{22}) \\
\frac{\partial q_1}{\partial t} &= q_1(\Pi_{B,1} - q_1\Pi_{B,1} - (1 - q_1)\Pi_{B,2}) = q_1(1 - q_1)(\Pi_{B,1} - \Pi_{B,2}) \\
&= q_1(1 - q_1)(q_1(p_1(\chi_{11} - \chi_{21}) + \chi_{21}) - (1 - q_1)(p_1(\chi_{12} - \chi_{22}) + \chi_{22}))
\end{aligned}$$

$$= -q_1^3(p_1(\chi_{11} - \chi_{21}) + \chi_{21} + p_1(\chi_{12} - \chi_{22}) + \chi_{22}) + q_1^2(p_1(\chi_{11} - \chi_{21}) + \chi_{21} + 2p_1(\chi_{12} - \chi_{22}) + 2\chi_{22}) - q_1(p_1(\chi_{11} - \chi_{22}) + \chi_{22}) \quad (4.55)$$

There are consequently three equilibria for each industry

$$p^*_{1,1} = 0 \qquad q^*_{1,1} = 0$$
$$p^*_{1,2} = 1 \qquad q^*_{1,2} = 1$$
$$p^*_{1,2} = \frac{q_1(\chi_{11} - \chi_{12}) + \chi_{12}}{q_1(\chi_{11} - \chi_{12}) + \chi_{12} + q_1(\chi_{21} - \chi_{22}) + \chi_{22}} \quad q^*_{1,2} = \frac{p_1(\chi_{11} - \chi_{21}) + \chi_{21}}{p_1(\chi_{11} - \chi_{21}) + \chi_{21} + p_1(\chi_{12} - \chi_{22}) + \chi_{22}}.$$

Each combination of any of the p^*_1 and any of the q^*_1 is a viable fixed point. The first two cases (for both sectors, i.e. $p^*_{1,1}, p^*_{1,2}, q^*_{1,1}, q^*_{1,2}$) are the corner solutions where the respective industry is completely monopolized. The third solution for both industries ($p^*_{1,3}$ and $q^*_{1,3}$) in turn may, depending on the compatibility terms χ, be anywhere in the phase space.

To assess the stability, consider the Jacobian matrix

$$\mathcal{J} = \begin{pmatrix} \frac{\partial^2 p_1}{\partial t \partial p_1} & \frac{\partial^2 p_1}{\partial t \partial q_1} \\ \frac{\partial^2 q_1}{\partial t \partial p_1} & \frac{\partial^2 q_1}{\partial t \partial q_1} \end{pmatrix} \quad (4.56)$$

where the elements are each long polynomials.[88] For the corner cases, the Jacobian turns out to have negative elements in the main diagonal and otherwise 0

$$\mathcal{J}(0,0) = \begin{pmatrix} -\chi_{22} & 0 \\ 0 & -\chi_{22} \end{pmatrix} \qquad \mathcal{J}(1,1) = \begin{pmatrix} -\chi_{11} & 0 \\ 0 & -\chi_{11} \end{pmatrix}$$
$$\mathcal{J}(1,0) = \begin{pmatrix} -\chi_{12} & 0 \\ 0 & -\chi_{12} \end{pmatrix} \qquad \mathcal{J}(0,1) = \begin{pmatrix} -\chi_{21} & 0 \\ 0 & -\chi_{21} \end{pmatrix}.$$

In turn, for each of these cases, the two eigenvalues are negative and identical

$$\lambda(p^*_{1,1}, q^*_{1,1})_{1/2} = -\chi_{22}$$
$$\lambda(p^*_{1,2}, q^*_{1,2})_{1/2} = -\chi_{11}$$
$$\lambda(p^*_{1,2}, q^*_{1,1})_{1/2} = -\chi_{12}$$
$$\lambda(p^*_{1,1}, q^*_{1,2})_{1/2} = -\chi_{21}. \quad (4.57)$$

For the interior solutions involving $p^*_{1,3}$ or $q^*_{1,3}$, the stability may turn out very differently, depending on the compatibility parameters χ_{11}, χ_{12}, χ_{21} and χ_{22}. Consider, for instance, the following three cases:

- Perfect compatibility of any standard with any of the two standards in the other industry, $\chi_{11} = \chi_{12} = \chi_{21} = \chi_{22} = 1$. This is also the case in which the industries are not tied since there is no asymmetric incompatibility. $p^*_{1,3}$ and $q^*_{1,3}$ compute in this case as $p^*_{1,3} = q^*_{1,3} = 0.5$. The eigenvalues are (identical) $\lambda(p^*_{1,3}, q^*_{1,3})_{1/2} = 0.5$, hence the fixed point is a perfect repeller.

- Pairwise compatibility between $A1$ and $B1$ on the one hand and $A2$ and $B2$ on the other hand (the first standards in both industries and the second standards in both industries) $\chi_{11} = \chi_{22} = 1$, $\chi_{12} = \chi_{21} = 0$. $p^*_{1,3}$ and $q^*_{1,3}$ compute in this case as a relation $p^*_{1,3} = q^*_{1,3}$ (and hence the shares of the other two standards $1 - p^*_{1,3} = 1 - q^*_{1,3}$). The eigenvalues differ depending on the values $p^*_{1,3}$ and $q^*_{1,3}$ assume. They are negative (stable) for very low values and very high values (i.e. close to the corner cases $p^*_{1,1}, q^*_{1,1}$ and $p^*_{1,2}, q^*_{1,2}$ while the other two corner fixed points are neutral fixed points with $\lambda_{1/2} = 0$ since $\chi_{12} = \chi_{21} = 0$); at $p^*_{1,3} = q^*_{1,3} = \frac{1}{2} \pm \frac{1}{2\sqrt{3}}$, a phase transition occurs, the dominant (largest real part) eigenvalue becomes positive and the fixed point becomes unstable (a saddle point). The situation is shown in Figure 4.14.
- Pairwise compatibility between $A1$ and $B2$ on the one hand and $A2$ and $B1$ on the other hand $\chi_{11} = \chi_{22} = 0$, $\chi_{12} = \chi_{21} = 1$. This situation mirrors the second case above with $p^*_{1,3} = 1 - q^*_{1,3}$, hence the corner cases $p^*_{1,1}, q^*_{1,2}$ and $p^*_{1,2}, q^*_{1,1}$ are stable while the other two corner fixed points are neutral fixed points. The system undergoes the same phase transition as in the second case.

To sum up, if all compatibility parameters are positive, all border cases are stable while internal cases may or may not be stable but tend to be stable for asymmetric

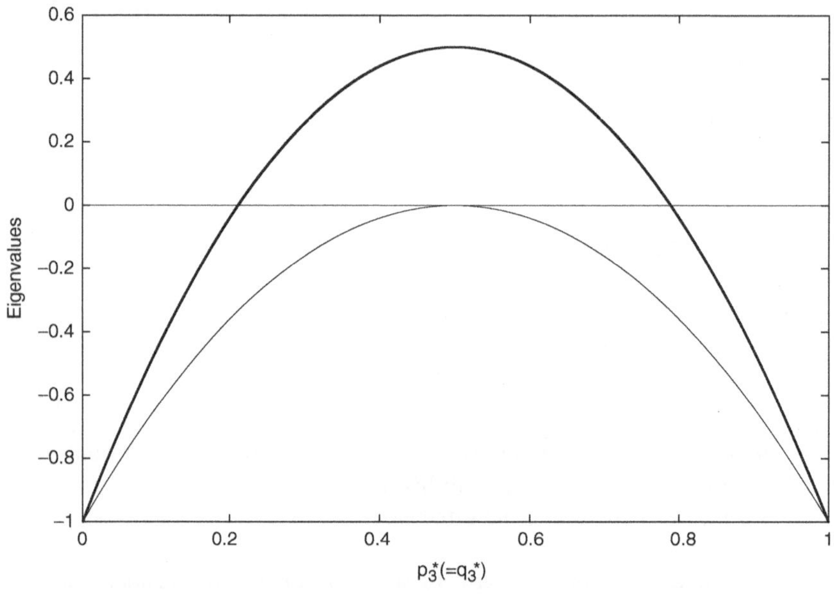

Figure 4.14 Eigenvalues resulting for the standard usage shares $p^*_{1,3} = q^*_{1,3}$ in a system with two completely tied industries with one standard in one industry being perfectly compatible with only (and exactly) one standard of the other industry (compatibility parameters $\chi_{11} = \chi_{22} = 1$, $\chi_{12} = \chi_{21} = 0$).

fixed points and unstable for symmetric ones, though depending on the parameters. There may also be more complex attractor structures.

4.3.6 Local network externalities

Following the rapid development of graph theory and the theory of games on networks (for an overview, see Goyal (2007) or Easley and Kleinberg (2010)), starting in 2007, several scholars came up with the concept of *local network externalities* (Sundararajan, 2008; Lee *et al.*, 2006; Delre *et al.*, 2007; Uchida and Shirayama, 2008; Pegoretti *et al.*, 2009). Their research seems to be unrelated and develops along different lines, drawing on both the mixed game theory and market approach and the complex systems approach. Local network externalities take two different shapes:

- First, the goods or technologies under investigation may be network technologies in the strict sense. In this case, the adopters do not usually care which kind of technology predominates in some very distant part of the network. The relevant issue is which technology is used by their own peer group, if any. This points out an inaccuracy the literature on network goods (many of the contributions following the mixed game theory and market approach as discussed above) has dragged along since its beginnings. Sundararajan (2008) in turn applied local network effects to this approach. Though Sundararajan used a slightly different formulation, the resulting model is pretty straightforward. Depending on the set of her direct neighbours N_κ, an agent κ will adopt a technology A if and only if (let $n_{A,\kappa}$ denote the number of adopters of technology A among κ's direct neighbours)

  $$v(n_{A,\kappa}) + r_{A,\kappa} \geq 0 \qquad (4.58)$$

  Where $v(\cdot)$ denotes the monotonically increasing network utility and $r_{A,\kappa}$ is the net intrinsic preference of κ for A. For the local network effects to become important, $r_{A,\kappa}$ has to be negative for some (or, more interesting, all but some) agents κ; otherwise anyone would immediately adopt it since there is until now no competing technology.
- On the other hand, the common knowledge of all available technologies, the true preferences for these technologies and possibly even the intrinsic preferences of other agents are rather heroic assumptions to make. Even taking into account that advertisements in certain media may have a global effect, it is utterly unrealistic to assume that agents generally pay attention to advertisements and are able to extract the relevant and true information about the respective goods or technologies. The logical step to take in order to construct a more realistic model is to apply 'epidemic' dynamics, the spread of the information step by step along the network. The line of research concerned with this effect (Lee *et al.*, 2006; Delre *et al.*, 2007; Uchida and Shirayama, 2008; Pegoretti *et al.*, 2009) usually specifies that the direct neighbourhood

of an adopter instantly or with a time lag acquires the relevant information about a technology.

The fundamental effect that arises from local network externalities is that the process of convergence to one of the fixed points is slowed down considerably. Intuitively, just a few early (intrinsically motivated) adopters are scattered about the network. Most of the remaining agents have yet to come in contact with them, and among those directly neighbouring the early adopters only a fraction will be immediately convinced to join the technology adopters. Thus, the technology spreads rather slowly. In fact, if there is more than one technology in the system, the two technologies may even conquer different parts of the network, hence not resulting in a full standardization even though the network externality dynamics itself works strongly in favour of monopolization. As Lee *et al.* put it '[a] local bias can persist if customers interact only with well-known others .... In this case, the winner-take-all hypothesis may not be valid' (Lee *et al.*, 2006: p. 1846).

So far, a network has been assumed abstractly and ad hoc with n nodes $\kappa = 1, 2, ..., n$ each of which has a set of n_κ direct neighbours (n_κ is also called κ's *degree*) N_κ. The justification for this is immediately provided by the simple observation of real-world agents that direct interaction matters – especially in the case of network goods or network technologies and network externalities caused by information, (tacit) knowledge – and that the direct interaction does not occur between any two of the n agents (except for very small groups) but rather only between agents who happen to 'be connected' or, more intuitively, to know each other. Different network topologies may, however, result in very different dynamics. It is obvious that the complete network where any agent is any other's direct neighbour $\{n_\kappa = n \backslash \kappa\}_{\forall\kappa}$ (*total connectivity*) is identical to the above situation without a network. There are no local network effects or, more accurately, the scope of the local network effects is global such that global and local network effects cannot be distinguished. Consequently, all properties discussed above for populations without a network topology are valid for the total connectivity case. A regular lattice network on the other hand with a rather small degree (direct neighbourhood) of the nodes compared to the total size of the network is radically different. Consider a simple cellular automaton like chess board topology using the Moore neighbourhood (the eight adjacent squares being direct neighbours), hence all agents are of degree $\{n_\kappa = 8\}_{\forall\kappa}$. Starting with a few early adopters of two rival technologies randomly scattered on the topology, and proceeding to let each agent with two or more neighbours who have adopted the same technology follow in choosing this technology, this will almost always result in a sharing of the sector (see Figure 4.15).

Indeed, this kind of grid lattice and the Moore or the von Neumann neighbourhood structure (the latter being comprised of just the four orthogonally and horizontally adjacent squares) has historically been used for most models studying local effects in economics, including Schelling's famous segregation simulations (Schelling, 1971). Lee *et al.*'s (2006) argument clearly holds most significantly for local lattice neighbourhoods. However, those researching local

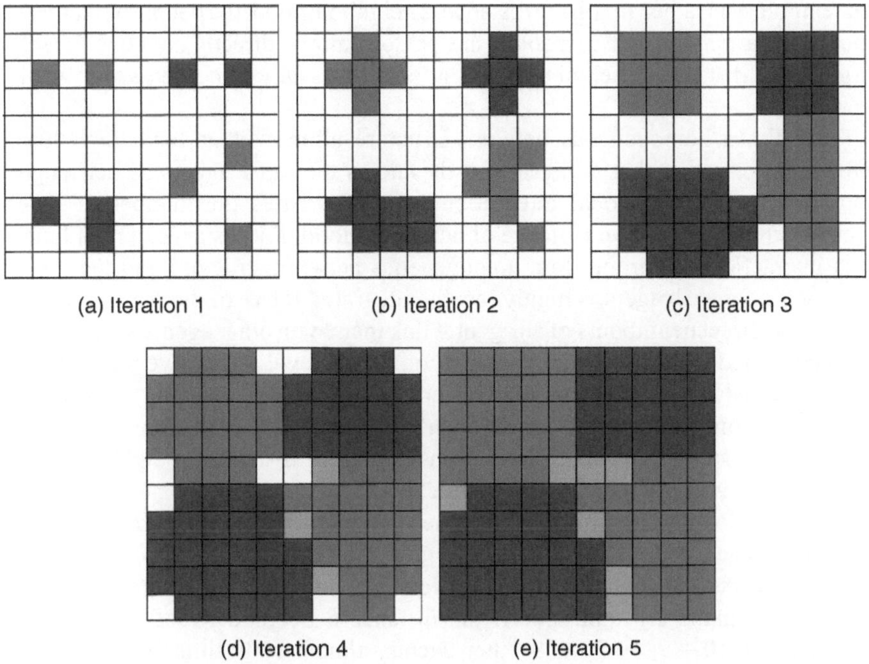

Figure 4.15 Local network effects in a cellular automaton. Every square with at least two adjacent squares of one colour (dark grey or grey) takes that colour in the next iteration. Squares with more than two adjacent squares of each colour stay 'neutral' (light grey). The situation does not result in a global lock-in.

network externalities (including Lee *et al.*) have been aware that this particular case does not resemble some of the most important properties of real-world systems (except possibly some geographic, area, or literal neighbourhood effects). Especially, distances between two random agents in the network (as measured by the steps through the network from one agent to another the shortest possible path requires) are excessively long in lattice networks. The following subsections will go into detail on network topologies that meet this requirement better, as well as other requirements such as strong local clustering, scale-freeness of the distribution (power law distributions), and the positive correlation of direct neighbour's degrees (Boccaletti *et al.*, 2006; Newman, 2005).

Local network effects in Watts–Strogatz random graphs

One of the doubtlessly most striking findings in microsociology and microeconomics is Watts' and Strogatz's (Watts and Strogatz, 1998; Watts, 1999) 'small worlds' phenomenon. Any person on earth, as Watts and Strogatz's illustrative story has it, is no more than six steps away from any other person's social neighbourhood. That means the direct social environment of the average contemporary is rather small, the distance-2 neighbourhood, the people not directly known by a person, but to people the person knows directly, is considerably larger. The distance-6 neighbourhood is said to usually encompass the whole world.

Basically, a homogeneous network structure allows either for a high local integration ensuring that an agent mostly knows the same agents as her neighbours, or for a high global integration which increases the number of other agents included in a larger distance of neighbourhood. For instance, a population with interactions according to a homogeneous network structure and ten direct neighbours for each agent is highly locally integrated if five of the ten contacts of each of the direct neighbours of an agent *A* link them with other agents in A's direct neighbourhood (*grid network*). That way, on a local level almost everyone knows each other (which is a realistic feature for modelling the real-world), but on the global scale, only about[89] $\frac{10\times5}{5} = 10$ agents are added to the neighbourhood with each distance step of neighbourhood. Considering the distance-6 neighbourhood, any agent knows not more than 60 people. On the other hand, the network could be strongly globally integrated if just one (namely the agent herself) of the ten contacts of each of my neighbours belong to the direct peer group of an agent *hierarchical network*. That way the network has no strong local integration but globally the paths are quite short: the distance-6 neighbourhood of an agent encompasses $10 + 9^5 = 59{,}059$ other agents, almost 1000 times more than in the other setting.

Note that while grid networks are not effective for transmission of information, diffusion of knowledge, etc., the hierarchical network is inherently unstable. The breakdown of one node (corresponding, for social networks, to an agent leaving the network) may split the network into two or several smaller networks – for instance, in the above case with ten direct neighbours, the loss of one agent may split the network in up into ten networks of much smaller size.

Watts' and Strogatz's famous idea was to combine short distances and strong local clustering by starting with a regular grid and rewiring a portion of the links in the local neighbourhood to connect to random other agents. This leaves the strong local integration intact while providing shortcuts to otherwise more distant parts of the network. Moreover, the combination of clustering and short paths can be carefully controlled using the probability of rewiring the individual edges $\omega \in [0, 1]$. $\omega = 0$ means no rewiring, and thus the unmodified regular lattice topology, while $\omega = 1$ corresponds to a completely random network. The degrees of the nodes (the agents) in this network remain of course constant and identical for all nodes.[90]

Using Watts–Strogatz random graphs is also the most common approach for studying local network effects with 'epidemic' models in the literature (Lee *et al.*, 2006; Delre *et al.*, 2007; Uchida and Shirayama, 2008; Pegoretti *et al.*, 2009). Lee *et al.* (2006) and Pegoretti *et al.* (2009) report consistently that different technologies share the sector for low values of ω. A critical value for which the outcome changes suddenly and dramatically occurs at about $\omega \approx 0.2$; for larger values of ω, the winner-take-all solution holds. The sharpness of this critical value (how fast the probability of the winner-take-all solution approaches 1) as well as its exact value depends probably on the uniform degree of the nodes in the underlying Watts–Strogatz random graph. Lee *et al.*'s (2006) finding that the winner-take-all solution does not hold for local externalities in certain network topologies refers to low values of ω. Lee *et al.* argue, however, that this also depends on the presence of complementary goods (hence other industries tied to the current one as detailed in the previous section).[91] Both Lee *et al.* (2006) and Pegoretti *et al.* (2009) conduct their simulations using the uniform node degree $n_\kappa = 10$. Uchida and Shirayama (2008) run different simulations for n_κ between eight and fourteen. Delre *et al.* (2007) work with an even more localized network with even fewer direct neighbours per agent, $n_\kappa = 4$.[92] As these numbers are rather small compared to a usual real-world neighbourhood, it is also likely that the winner-take-all result is not that unlikely in real-world networks. Considering the estimated size of human peer groups (or rather the cognitive limit for the size of the peer group), also called Dunbar's number (Dunbar, 1993), which is of the order of 100 to a few hundred,[93] a real-world network would have be expected to exhibit a much higher density. This may be even more true for other types of actors in the socio-economic system, such as firms, since they are not constrained by a cognitive limit in that sense. The higher degree of the nodes of such real-world networks would bring the network closer to a total connectivity network, even if a rather small fraction of the links connect to more distant parts of the network (representing a very small ω).

Moreover, the diffusion speed[94] reaches a maximum for $\omega = 0.1$ (Delre *et al.*, 2007) or slightly higher (Pegoretti *et al.*, 2009) because for smaller values, the diffusion is limited to too few areas of the network in the beginning, while for larger values the network clustering is too low for local groups of adopters to exercise a strong enough influence on more reluctant agents (that is, those agents with a strongly negative intrinsic preference $r_{.,\kappa}$). Pegoretti *et al.* (2009) report further that for imperfect information the picture changes again. They use a slightly more difficult model including local network externalities in the technology choice function (as do Lee *et al.* (2006), Delre *et al.* (2007) and Uchida and Shirayama (2008)), but for an imperfect information case (which is compared to the perfect information benchmark) also adding a layer of local network effects based on the distribution of information or knowledge. For the perfect information setting, they assume that all agents are immediately aware of the available technologies and accordingly their preferences, for imperfect information they are only aware of a (possibly empty) subset of the available technologies. With a small probability, they just learn about a technology (interpreted as advertisement), which is necessary to place the information somewhere in the network in the first place; more commonly, however, they learn from neighbouring adopters. This imperfect information setting leads to a more smooth transition between negligible probability for a winner-take-all outcome for the regular localized lattice grid topology $\omega = 0$, and close to certain probability for Watts–Strogatz random graphs with about $\omega > 0.1$. The probability function assumes a smooth s-shape between $\omega = 0$ and $\omega = 0.1$ reaching about 20% for $\omega = 0.01$. Thus, the most likely prediction obtained by studying Watts–Strogatz random graphs that resemble the real-world most closely (those for a rather small ω around or below 0.1) is that a complete monopolization of the sector is probable though not certain. From the divisibility of the network and the model altogether, it follows that the likelihood of oligopoly solutions with just a few of the competing technologies left rises as well.

Using Watts–Strogatz random graphs and imperfect information instead of perfect information and regular lattice topologies levels out the mechanism reported by Lee *et al.* (2006). Considering local network externalities may have a certain influence, resulting in weaker and non-global lock-in effects; however, this does not hold as strictly as it may seem at first sight. Imperfect information as well as a not strictly localized topology (e.g. a Watts–Strogatz random graph or a scale free network) and a number of other effects (technology tying, sponsoring, etc.) restores the global network externality structure in part or entirely.

Power laws

As mentioned above, Watts–Strogatz random graphs still insufficiently resemble real-world networks in some respects (Boccaletti *et al.*, 2006). Specifically, Watts–Strogatz random graphs are not scale free, that is, the properties in terms of clustering, path length and degree correlation change depending on the level of aggregation that is considered. In effect, instead of 'true' clusters as observed in the real-world, Watts–Strogatz random graphs consist of a continuum of highly localized lattice patterns superseded by a thin structure of shortcuts to distant regions of the same global lattice. Hence, although a short average path length is guaranteed by the shortcuts, the overall topology is governed by the lattice grid and is rather uniform.

This is different for so-called 'scale free networks' (Boccaletti *et al.*, 2006; Kim *et al.*, 2007; Uchida and Shirayama, 2008) where the degrees (the number of direct neighbours) of the nodes are distributed according to a power law[95]

$$p(n_\kappa) \sim \Upsilon n_\kappa^{-\alpha} \tag{4.59}$$

where n_κ are the degrees, $p(n_\kappa)$ are their statistical relative frequencies and α and Υ are positive constants.[96] Υ is the normalization parameter which ensures that the relative frequencies add up to 1

$$\int_{min(n_\kappa)}^{\infty} p(n_\kappa) dn_\kappa = 1. \tag{4.60}$$

This condition determines the value of the normalization parameter.[97] Equation (4.59) is thus the strictly decreasing probability density function of the power law distribution. More illustratively, the double logarithmic (log-log) plot between the degrees and the frequency of nodes of this degree is linear.

This becomes more obvious when writing the probability density function (4.59) in logarithmic form

$$ln(p(n_\kappa)) \sim ln(\Upsilon n_\kappa^{-\alpha}) = -\alpha ln(\Upsilon) \quad - -\alpha ln(n_\kappa) \tag{4.61}$$

A network with node degrees following a power law is consequently scale free in the way that a specific multiplicative increase in node degree, say by factor h, results in always the same multiplicative reduction $H(h)$ of the frequency of nodes of the higher degree compared to the lower degree nodes. For example, considering a hypothetical network where the node degrees are distributed according to a power law with $\alpha = 2$, the relative frequency of degree-1 nodes is $\Upsilon \times 1^{-2} = \Upsilon$. An increase in degree by two ($h = 2$) results in the relative frequency $\Upsilon \times (1 \times h)^{-2} = \Upsilon \times 2^{-2} = \frac{\Upsilon}{4}$, and thus a decrease to $H(2) = \frac{1}{4}$ the frequency mentioned earlier. Increasing the degree again by the same factor (now from 2 to 4) lets the frequency drop by the same rate $\Upsilon \times 4^{-2} = \frac{\Upsilon}{16}$. This relation holds no matter which initial degree we apply the factor h to. More formally, the distribution is said to be scale invariant

$$p(hn_\kappa) = H(h)p(n_\kappa). \tag{4.62}$$

This holds also for subgraphs of the network as long as the selection of the nodes for the subgraphs is sufficiently random, since the probability density does not differ from one node to another node.

Furthermore, it has been found that aggregating closely connected nodes to clusters using a suitable method (Kim *et al.* (2007) use the *random sequencing* method[98]), thus constructing a smaller meta-network where the clusters of the original network are single nodes ('supernodes') leaves this property untouched (Kim *et al.*, 2007). The node degrees in the aggregated network are still power law distributed, which means that power law networks are self-similar and show the same topological properties both in the entire network, in subgraphs of the network and at an aggregated level.

Power law distributions have been found to be present in a number of socio-economic phenomena, including income distribution (Mandelbrot and Hudson, 2004) (specifically the Pareto distribution which is a subtype of power law distributions) price variations (specifically long term development of prices[99]) (Mandelbrot and Hudson, 2004), firm size (Axtell, 2001), industrial agglomerations (Andriani and McKelvey, 2007), social agglomerations (Newman, 2005) and, for the current context especially, social networks including decision making structures in organizations (Andriani and McKelvey, 2007). Other networks that are potentially relevant for information distribution and technology choice patterns, such as the structure of the internet and the geographical allocation of people as well as infrastructure, have also been found to follow power laws (Newman, 2005). Boccaletti *et al.* (2006) found that those networks are further (apart from high clustering and short average path length that are usually present in power law networks) characterized by positive correlations of the degrees of neighbouring nodes (in turn, highly connected nodes tend to connect to other highly connected nodes).

Uchida and Shirayama (2008) found that this also has a profound effect on the technology choice (or in their example, equivalently, product choice) dynamics. They compare the properties of different network topologies with respect to this issue. Apart from regular lattices, purely random and Watts–Strogatz random graphs they consider three different power law topologies, one of which – the *connecting nearest neighbour* topology[100] – has the properties that were observed in real-world networks according to Boccaletti *et al.* (2006). Uchida and Shirayama (2008) use a design for their simulations that is slightly different from those by Lee *et al.* (2006), Delre *et al.* (2007) and Pegoretti *et al.* (2009) as discussed above. They assume that adopters have to pay a rather costly switching fee should they reconsider their options and decide to join another network. As a result, most network topologies including Watts–Strogatz random graphs result in more or less equal sector shares in their simulations. They report that the connecting nearest neighbour power law topology mechanism yields a remarkably stable asymmetric outcome, with one technology (or good) acquiring about 60–80% of the sector sales even for settings with high switching costs. As their simulation setup is probably too restrictive, this finding may hold an interesting explanation for the prominence of monopolistic or strongly oligopolistic industry structures in many different economic sectors, and especially in technology-intensive industries. Assuming that successful establishment of an industry standard technology will likely be followed by enormous revenues and a phase of growth and expansion of the firm's activities, this phenomenon may also be related to Axtell's (2001) result that firm sizes themselves are power law distributed. One way or another, Uchida and Shirayama (2008) have shown that power law networks lead to distinct patterns in technology choice that are different from the outcomes of regular lattice graphs or random graphs with uniform degree, though these differences have not yet been studied comprehensively.

4.4 The shape and form of network externalities

So far, the effects of network externalities on rational decision making, particular games, and corresponding population dynamics have been considered extensively. Little has, however, been said about the *form* a network externality takes. This reflects the very little attention given to this question in the literature on network effects. Of course, the network effects considered so far do have a form and a very well-defined one – the definition has just been more or less implicit, or at least by assumption, without extensive discussion of its shape. That is especially true for the mixed neoclassic and game-theoretic models discussed in section 4.2;[101] the population dynamics models in section 4.3 have explicitly defined a number of different rules resulting in different dynamics; however, still without discussion of why exactly this form should be relevant for models of network effects.

Swann (2002) discusses three types of functional forms of network externalities (which he names after the scholars who first mentioned them), two of which are proper network effects while one (Sarnoff's law) is the simple market share discussed from a network effects perspective.

SARNOFF'S LAW

A broadcasting network follows Sarnoff's Law, i.e. a broadcaster, the vendor, broadcasts to all users while the users are not directly interdependent. The aggregate utility of a network with N agents or customers is directly proportional to N (given that users are homogeneous and thus generate identical utilities). The user's utility is constant in N. Thus, this form of a network effect is not a network effect in the strict sense but rather a simple market share effect. Consequently, N has no effect on the dynamics of competition between two or several networks. If the probability of the network growing, i.e. of the next agent choosing this network, depends on the vendor's revenue (utility), the dynamics follows strictly the 45°-line (see Figure 4.16).

METCALFE'S LAW

A network of simple direct interdependence (i.e. two-way communications); the case usually (more or less implicitly) assumed in most of the literature on network externalities follows Metcalfe's Law. The aggregated utility of a network of size N is proportional to the number of possible different interactions, i.e. $N(N-1)$, while the utility to the individual agent follows the number of interactions possible for that agent, hence $N-1$. The system dynamics, if two such networks with N_1 and N_2 users respectively compete, yields, if the aggregate utility is the decisive factor, a probability for a network (say, the first one) $p_1(N_1, N_2)$ to grow (i.e. the next user chooses the network)

$$p_1(N_1, N_2) = \frac{N_1^2 - N_1}{N_1^2 - N_1 + N_2^2 - N_2} \tag{4.63}$$

which is an s-shaped curve in the current usage share of that network $\frac{N_1}{N_1+N_2}$ (see Figure 4.16).

REED'S LAW

A network that focuses on possible group structure (i.e. how many possible different groups can be formed in that network) follows Reed's Law. Reed's Law yields an aggregated utility of 2^N in a network of size N, while the utility to the individual user (i.e. how many different groups a user can be in on a network of N users) is 2^{N-1}. The competition dynamics of two such networks (again of N_1 and N_2 users) are straightforwardly

$$p_1(N_1, N_2) = \frac{2^{N_1}}{2^{N_1} + 2^{N_2}}. \tag{4.64}$$

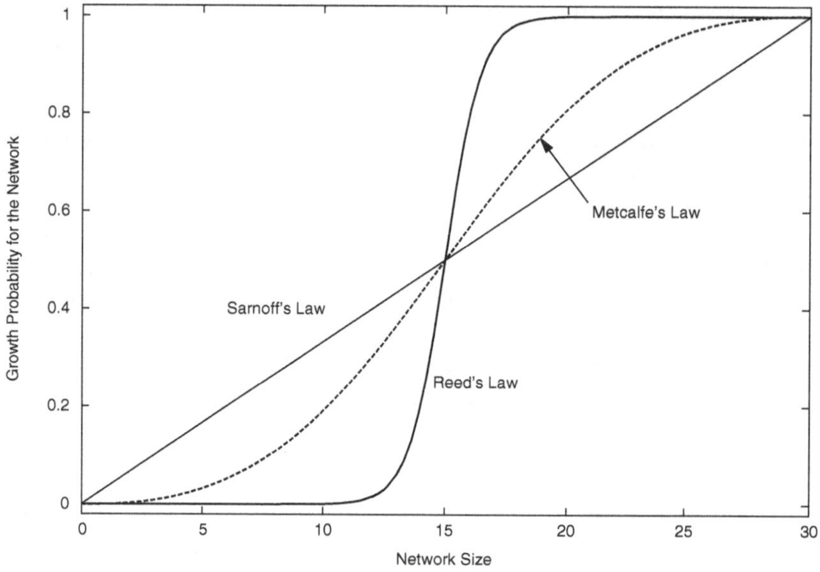

Figure 4.16 Dynamics resulting from three different network externality laws.

This leads also to an s-shaped curve in the current usage share of that network $\frac{N_1}{N_1+N_2}$; however, with a sufficiently large number of users, a much steeper one than in Metcalfe's Law (see Figure 4.16).

As Swann remarks, it is very likely that real-world network effects do not generally follow such laws, but are most likely subject to upper boundaries or at least decreasing utilities beyond a certain number of users or beyond the limits of the respective user's peer group, social group or wider environment (e.g. language group). This would lead to a falling slope of the network effect functions for higher numbers of users or higher usage shares; that is, the upper stable equilibria in Figure 4.16 would not be those of complete monopolization $\frac{N_1}{N_1+N_2} = 1$ but rather internal fixed points with some (asymmetric) sharing of the industry (since the network externality function would cross the 45°-line once more for a value $\frac{N_1}{N_1+N_2} < 1$). This corresponds to the argument concerning local network effects (see section 4.3.6).

The consideration in this section is not a central element of the current study; however, it is an important aspect for the consideration of how to model network effects both in the current chapter and in the model chapter (Chapter 5).

4.5 Some empirical indications of network externalities and lock-ins

The scope of network externalities and their consequences in real-world economics are difficult to assess. Lock-ins as a particular consequence may lead to the highest though indirect 'costs' inflicted by network externalities. Other, direct costs of network externalities, such as the directly diminished utility resulting from negative effects on the utility of users of alternative standards, are very limited (provided these users may just, at a cost, switch their standard for the more successful one) and hence most likely insignificant compared to the former type of costs. Nevertheless, the direct costs are much more easy to quantify by estimating the switching costs (and possibly the lower utility for the time until the switching is complete). For an oligopoly market with perfect rationality and perfect competition, but subject to network externalities, it is justified to assume that the firms set their prices such that the adopters to not have an incentive to leave the network, and hence that the chosen price is lower than the price of the opponent, switching costs and difference of the network size and network utility included. Alternatively, a similar computation could be conducted for the vendor side: a network vendor only has the incentive to wage a 'standard war' with a competitor if the profits resulting from the combined networks (if she succeeds in the standard war) are higher than the profits from her own network and the switching costs of all users she would have to cover to win them over combined. This consideration implies a connection of price, switching costs and network utility. Using this method, Shy (2002) computes the switching costs in Israeli telephone networks as about half the price,[102] for the Finnish banking sector as zero for the largest bank in the number of customers and almost the discounted life-time fees of an account for the major competitors. Shy does, however, recognize that the assumptions are heroic and the phenomenon is more complex; however, his results indicate strongly that switching costs are generally substantial.

As mentioned, switching costs are only a small part of the social and economic costs caused by lock-ins. For the larger structural part, including the costs inflicted by failure to switch to better technologies, failure to use better standards in the first place, monopolization, decreased variety of alternatives and failure to conduct research in alternative technologies, the evidence is, however, anecdotal and implausible.[103] It may suffice to say that costs and utilities in counter-factual arguments are generally difficult to assess; as lock-ins are path-dependent, an attempt to compare the real course of history with an alternative trajectory (what if we were using Dvorak-keyboard layouts?) is counter-factual. The nature of the phenomenon, however, shows (as well as the analytical models by Arthur and others that were discussed in sections 4.2 and 4.3), that the costs are enormous. David (1985, 1992, 2005), Arthur *et al.* (1987), David and Bunn (1988), Arthur and Lane (1993), David and Steinmueller (1994), Stango (2004) and Stango and Greenstein (2008) give a great number of historical examples for lock-ins, including, as mentioned, the QWERTY-keyboard, the competition of computer operating systems during the 1980s, the systems of direct current and alternating current electricity supply in the late nineteenth and early twentieth centuries, videotape types and recorders in the early 1980s, the 'browser wars' between the web browsers Netscape Navigator and Microsoft Internet Explorer in the 1990s, and others. The approach of network externalities and lock-ins as a whole, including the theoretical models by Arthur and Katz and Shapiro (see sections 4.2

and 4.3) has also fiercely been criticized, especially by Liebowitz and Margolis (1990, 1994, 1995) who concede that network effects exist but claim that the neoclassical market is still efficient, i.e. that both network externalities and lock-ins are either a trivial and in ex ante terms[104] perfectly optimal phenomenon, or a fairytale (depending on the exact phenomenon in question). Nevertheless, the debate was not only one between proponents of the neoclassical market and heterodox scholars researching network effects, but exhibits (with sidelines) a more complex structure. One of the most recent contributions to it, for instance (by Hall and Günther (2011)) argues for a broader view of the phenomenon in the context of Veblenian institutionalization.

It may, however, also be noted, that standardization as a phenomenon is linked to the phenomenon of the technological lock-in: a standard is valuable only if it has substantial support – in which case, however, the system is locked-in and alternative standards introduced to the system would stand no chance (if not vastly superior or heavily sponsored). As will be shown in the simulations in Chapter 5, a system with a sequence of technological lock-ins and (once the locked-in technology is sufficiently outdated) break-outs experiences the highest growth rates in the phase of the lock-in (or standardization).

4.6 The scope of network effects: hidden networks of structural change

Arthur (1988), Dosi *et al.* (1994) and Kirman (1997) as well as David (2005) and De Langhe and Greiff (2010) took steps to establish a general theory of economics with inherent network externalities. This introduces a notion of economics fundamentally different from the neoclassical market and general equilibrium approach. There are a number of good reasons that suggest Arthur and his fellow advocates of generalized models of lock-in and network effects were indeed right. David and collaborators (David, 1985; David and Bunn, 1988; David and Steinmueller, 1994) as well as Stango (2004), Greenstein (Stango and Greenstein, 2008) and Arthur himself (Arthur, 1990a, 2007) give a number of historical examples where the establishment of standard designs for new technologies had great significance (though other aspects are important as well). Among others, as discussed above, the development of typewriter keyboard layouts and videotape formats, technologies of electrical power transmission and geographical allocation of industries (and especially innovative startups) as well as all explicitly defined types of measurement units and design or quality standards, have been subject to and sensitive to the historic circumstances under which they have been developed and under which they eventually reached wide acceptance. However, network externalities and resulting lock-in effects also exist in areas of socio-economic life where it is less evident at first sight.

- **Strong technological standardization** helped industrial society to emerge. The modern industrial society came into being with what is called the 'industrial revolution' and occurred first in England in the eighteenth century.

It is clear that the rate of technological progress in this era was fueled by a number of crucial innovations all interacting with one another that gave rise to forms of economic organization. Economists have analyzed the rate of innovations during this period (Freeman and Perez, 1988; Mokyr, 1992; de Groot and Franses, 2009) and have sometimes tried to come up with a general theory explaining this transition from an in effect post-feudal society to the leading industrial state of the world (Mokyr, 1992). Much of this transition remains unclear or unknown since the organization of that society was very different from the one we know and the economic measures used today were unknown in this time. A certain aspect may be involved: the standardization of measurement units and money during the centuries before the industrial revolution and to some extent the standardization of technology that co-evolved with the rise of industry. The extent of this effect is debated. Mokyr, for instance, while acknowledging that network effects play a part in the industrial revolution in England dismisses them as a general explaination based on the fact that they were 'not very important on the markets of consumer goods'[105] (Mokyr, 1993). The question may, however, not be whether or not companies employ a railway of a certain rail-width as a public transportation service or put it to use in an industrial context or mining operations. Rather, it may be crucial that a number of public and private entities actively develop the industrial organization, relying specifically on the utilization of steam power and vehicles on rails for transportation – a fact recognized as important by both Mokyr (1992) and other scholars (Carlsson and Stankiewicz, 1991; Hughes, 2000).

- **General coordination problems** not merely involving two persons for a limited time, but a larger number of entities for possibly the entire assessable future, show the same structure as the network effects discussed above. They involve multiple equilibria (in turn different standards or different solutions) and positive feedback: agents adhering to the same solution experience positive spillovers for one another. While this is formally the same situation as discussed above with respect to network effects, especially in the section on generalized urn schemes (section 4.3), coordination problems are abundant in socio-economic systems throughout history. This ranges from simple questions, such as which side of the road to drive on or which kinds of deals persons of certain social status may or may not strike, to the macroeconomic coordination failures discussed by Cooper and John (1988) and even very complicated patterns such as languages. Modern languages notably gained their integral current form just recently (in historic terms, one to several centuries ago), co-evolving with the spread of books and the technologies of printing as well as the rising nation states. In essence, nation states are rather arbitrary entities; they gain their robustness and historical stability merely through the co-evolving language, culture and literature (which in turn becomes the state's national language, culture and literature) the influence of which extends over the same geographic area and their population (Anderson, 1991 [1983]).

- **General concepts of socio-economic life** such as money (more generally a universally accepted common exchange intermediary) and social entities exhibit the same structures as network externalities, as there are huge numbers of possible candidates for 'money' and even more possibilities for organizing a population into social and economic groups. Notably, for the network effect to exist, it is not necessary that the historic development of these structures is observed. It may be argued that both money and social stratification are historically state sponsored but this probably constitutes nothing other than a case of public sponsoring of one standard leading to an eventual lock-in situation. For both effects mentioned above – the universally accepted common exchange intermediary and social classes – their self-organized emergence has been shown by simulation studies by Howitt and Clower (2000) and Axtell *et al.* (2000), respectively. Bourdieu has discussed social stratification as consisting of different dimensions including personal wealth, personal skill and knowledge as well as the capability to interact in certain social contexts (Bourdieu, 1986). There are probably more, similar examples.
- **Institutions** work in a way resembling network externalities. As discussed by various scholars over the course of the decades since the development of advanced methods of game theory, social dilemma situations are conveniently solved by applying a trigger strategy – a cooperative strategy with the *credible threat* of punishment should the other player(s) not cooperate (Axelrod, 1984; Ostrom, 2000; Fehr and Gächter, 2000, 2002; Fehr *et al.*, 2002; Bowles and Choi, 2002; Bowles and Gintis, 2003; Fehr and Fischbacher, 2004; Elsner and Heinrich, 2009). In effect a social norm which solves the dilemma and increases the overall benefits of the actors must first be established; once it is in place, however, agents are able to coerce other agents into following this norm. In fact, the norm does not even necessarily have to increase the agent's benefits; if agents adhere to the norm and are willing to defend it by punishing defectors, the norm will be sufficiently stable to hold in practice. This in turn resembles the fact that social norms do not always have to be problem solving in that sense; they may simply exist – caused and perpetuated by their evolutionary stability or the ceremonial value human actors may derive from a structured social environment with *ceremonial institutions* (Bourdieu, 1986; Bush, 1987; Elsner, 2012). The easiest way to model social norms is a simple two-person prisoners' dilemma (as the easiest case of this class of problems); see Table 4.5.

  Considering a cooperative but resentful[106] strategy that punishes defection in the indefinitely[107] repeated setting of the same game, such as *tit-for-tat*,[108] this can be modelled as a prisoners' dilemma which is repeated with a certain probability and with a certain probability only. (That probability will be called p_{rep}.) If this probability is high enough, the prisoners' dilemma structure is transformed into a coordination game type; specifically, if in the repeated game the payoffs of tit-for-tat players against themselves are higher than the

Table 4.5 Prisoners' dilemma

		Player 2: Cooperate	Player 2: Defect
Player 1	Cooperate	Π_{11} Π_{11}	Π_{21} Π_{12}
	Defect	Π_{12} Π_{21}	Π_{22} Π_{22}

with $\Pi_{21} > \Pi_{11} > \Pi_{22} > \Pi_{12}$

expected payoffs of defectors against tit-for-tat players,

$$\Pi_{Defect/TFT} < \Pi_{TFT/TFT}.$$

As detailed in section 4.2.2, the sufficient condition for this to hold is[109]

$$p_{rep} > 1 - \frac{\Pi_{11} - \Pi_{22}}{\Pi_{21} - \Pi_{22}} = \frac{\Pi_{21} - \Pi_{11}}{\Pi_{21} - \Pi_{22}}$$

It is straightforward to extend this setting to an n-person situation in which agents play $n-1$ simultaneous two-person prisoners' dilemmas with the $n-1$ other agents. It is also formally equivalent to a second-stage punishment game as proposed by Fehr *et al.* (2002), which allows agents to reduce payoffs of their counterparts at a certain cost after the main stage of the game (within the current repetition) is finished. When cooperating during the main stage, subsequent punishment of a defector reduces the payoff for the latter – being equivalent to the reduction from Π_{21} to $\frac{\Pi_{22}}{1-p_{rep}} + \Pi_{21} - \Pi_{22}$ as experienced by an opponent in tit-for-tat (though the structure differs with respect to Π_{12}, which, however, is not crucial for the outcome of the game as long as $\Pi_{22} > \Pi_{12}$). The situation is prone to network externalities, the social norm gradually gaining strength and value with its wide acceptance. With more than one possible social norm, the development between norms and towards probably only one of them is also path dependent[110] and subject to eventual lock-in.

With a number of arguments presented for network externalities and lock-in effects present not only in several rather narrow environments of network technologies and economic geography but in central aspects of social and economic life over huge timespans, the current study will now proceed to discuss the interrelation of technological change as considered in Chapter 3 and network effects as discussed in the present chapter. The following section will outline some extensions to models of economic growth discussed above that take network effects into account and assess the general nature and scope of the influence of network effects on economic development by means of technological change. The following chapter

(Chapter 5) will then proceed to consider a general model of technological change, network effects and economic growth.

4.7 The nature of technological change with network effects

Lock-in effects, if seen as an integrative feature of economic processes have a profound effect on economic growth. Technological change is one of the standard examples for situations with lock-ins as well as a central aspect of economic growth. It acts as a device inducing network effects into the economics of growth. Nevertheless, this aspect has – with some notable exceptions (León-Ledesma, 2000; Silverberg and Verspagen, 2005, 2007) – rarely been addressed.

The Keynesian (Harrod, 1939; Samuelson, 1939a,b; Kaldor, 1940) and later the Kaldorian traditions of growth literature (Kaldor, 1972, 1975; Dixon and Thirlwall, 1975; Lorenz, 1987; León-Ledesma, 2000, 2006) have pointed to inherent instabilities caused by feedback systems that are unable to adjust to a common equilibrium because of time lags and path sensitive motion. In the 1960s economists further called for serious consideration of demand-pull effects instead of reducing economic growth to a production side phenomenon (Schmookler, 1965, 1966),[111] an effect that emphasized the Keynesian disequilibrium-inclination. Finally, Kaldor was the first to arrive at the conclusion that this might lead to increasing returns and, following those, a positive feedback loop (Kaldor, 1972, 1975). In turn, increasing production would lead to a capital inflow which allows the production of even larger quantities and from this again increased quantities of capital in the production process. Kaldor's idea was formalized shortly thereafter (Dixon and Thirlwall, 1975). This is different from the purely functional notions of increasing returns to capital (or human capital) as they are modelled in the theories of endogenous growth[112] as well as some earlier AK-type growth models,[113] because the increasing returns are recognized to be the consequence of a more complex feedback system. This system transcends both unrealistic growth beyond all limits (Solow's critique (Solow, 1994)) and convergence towards universal optimal values.[114] Internalizing the feedback loops' parameters easily and quite plausibly prevents unstoppable growth. The disequilibrium properties – discussed below in more detail – on the other hand predict the path dependent, potentially (with stochastic effects) not ex ante predictable, and between sectors or regions divergent, development.

So far, this has nothing to do with network externalities. In this Kaldorian model, the increasing returns are caused by demand effects of other sectors or regions. However, the effect is generally identical to what results from network externalities. Consider a simple growth model

$$\begin{aligned} k_t &= y_t \\ y_{t+1} &= F(k_t) \end{aligned} \tag{4.65}$$

where y denotes output, k capital – both changing in time t - and $F(\cdot)$ is a function that will capture network externalities in the invested capital. As the relation of capital and output is trivial, the system comes down to the simple recurrence equation

$$y_{t+1} = F(y_t). \tag{4.66}$$

The network externality shall take the form of a power function of the positive parameter γ

$$y_{t+1} = F(y_t) = \frac{y_t^{\gamma}}{\gamma}. \tag{4.67}$$

To develop the network externality, the input capital used for production ($y_t = k_t$) causes increased output effects, the more capital is involved. This holds as long as $\gamma > 1$; if γ falls below 1 the network externality degenerates and becomes a 'negative network effect'.

The system has two equilibria

$$y_1^* = 0$$
$$y_2^* = \sqrt[\gamma-1]{\gamma} \tag{4.68}$$

as shown in Figure 4.17. Using the single eigenvalue of the recurrence equation

$$\lambda = \frac{\partial y_{t+1}}{\partial y_t} = \frac{\gamma y_t^{\gamma-1}}{\gamma} = y_t^{\gamma-1} \tag{4.69}$$

the stability of the equilibria can be assessed[115]

$$\lambda(y_1^*) = \begin{cases} 0 & \text{for } \gamma > 1 \\ 1 & \text{for } \gamma = 1 \\ \rightarrow \infty & \text{for } \gamma < 1 \end{cases}$$
$$\lambda(y_2^*) = \gamma \tag{4.70}$$

In turn, y_1^*, the trivial equilibrium that stands for no production at all, is stable if $\gamma > 1$. The other equilibrium, y_2^*, is stable for $0 < \gamma < 1$, which is equivalent to diminishing returns (what was called a 'negative' network effect above, or just the neoclassical standard assumption to the property of capital as a productive factor). At $\gamma = 1$, a phase transition occurs, on the other side of which this fixed point becomes unstable, the trajectories lead away from this point. This, in turn, is the crucial property of the positive feedback present in the system. If the initial value of capital is larger than the repeller y_2^*, growth ensues; if it is not, the production collapses. As mentioned above, the unrealistic feature of infinite growth could easily be avoided by letting γ fall slowly as the system moves through the growth area above the repeller y_2^*, which would cause the system to drift to the left until it eventually hits the upper equilibrium at some low value of γ. As y_2^* is stable for

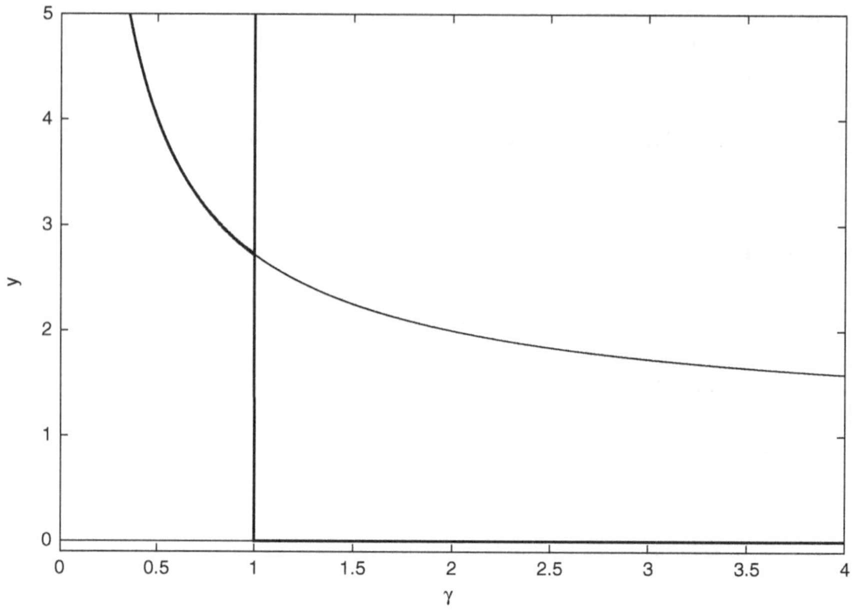

Figure 4.17 Equilibria of the simple path dependent growth model; bold lines denote stable equilibria.

$\gamma < 1$, this would also be an attractor and the system would remain in its vicinity, ceasing to grow at this point.

To see that the model represents a network externality, consider the system for a moment to be composed of two different types of capital that develop identically to the system (4.67)

$$\begin{pmatrix} y_{1,t+1} \\ y_{2,t+1} \end{pmatrix} = \begin{pmatrix} \frac{y_{1,t}^{\gamma}}{\gamma} + 0 \\ 0 + \frac{y_{2,t}^{\gamma}}{\gamma} \end{pmatrix}. \tag{4.71}$$

The two types represent alternative but incompatible networks, capital input may be allocated in the beginning, following which the two parts of the system develop a dynamic motion according to their initial values. For critical initial amounts of capital sufficient for one network to take off but insufficient for both, the success or eventual collapse of the system, and in fact of both systems, depends entirely on the initial allocation of the available capital.[116]

In the spirit of this simple structure – though using other models which are less stylized and closer to the economic interpretation (Dixon and Thirlwall, 1975; León-Ledesma, 2000) – a slowly developing Kaldorian tradition has been evolving since the 1970s. The central prediction, as mentioned, is that the technological development is path dependent and non-ergodic. Either a network established around a technology or a technological paradigm is strong enough to sustain further development, or the development will stagnate or eventually collapse. Lots of small historical events have an impact on whether a technology finds itself on the expanding or the collapsing side of the network externality dynamics. Most of these small impacts are insignificant and as such irrelevant to the larger picture of the technological development of the respective industry, but some of them are not. It is impossible to predict in advance which technologies will be successful and to what extent. It may not even be predictable that a (in the sense of at least one) technology will be successful in a given time period. It might be possible to operate with the principle of large numbers – the relative variance of the sum of independent identically distributed random variables decreases with the number of random values thus making the process more predictable for large numbers of variables or simply at the large scale. This, however, neglects the possibly complicated interrelation of feedback processes between different technology systems or industries that are partly responsible for the network effect in the first place. Details on the structure and possible prediction of network effects on technological development on the large scale and in general will be discussed in the following chapter (Chapter 5).

Several scholars working with Kaldorian (in Lorenz's (1987) and León-Ledesma's (2000) case) or similar disequilibrium models (in Keen's (1995) case) have pointed out that a certain number (three usually being sufficient) of independent subsystems with their independent feedback mechanisms may give rise to deterministic chaos in their interaction in the aggregated system. These may be several components of the economic system as studied by Keen (1995) or also the coupling of several economic sectors (or equivalently geographic regions) as modelled by Lorenz (1987). Though not specifically discussed in the context of network externalities, the edge of chaos is also and particularly an issue of the growth effects of network externalities: network effects are structurally similar to the patterns modelled in the mentioned models, with an additional inclination to occur in complex multi-industry contexts. So far, neither the effects of deterministic chaos driven by network externality structures on economic growth and technological change, nor their feedback on the network externality itself and the potentially resulting lock-in effects, have been analyzed in detail. Their presence has, however, been noticed in the literature (Dolfsma and Leydesdorff, 2009). Their potential importance for the industry structure, particularly as a possible solution or also a possible cause of technological lock-ins has been emphasized. Some further consequences are obvious. The systems gain an exceptionally complex structure; minimal changes may lead not only to significant changes but to almost unpredictable changes in almost unpredictable parts of the system and with unpredictable direction. However, the resulting complex system may also experience extended phases of stability, dissipative patterns and emergent structure that exist in the chaotic surroundings, in spite of, and potentially caused by, the very mechanisms of the network effects. For instance, a dominant technology with the surrounding industry locked-in to this technology does certainly guarantee stability in the technological process.

This dominance lasts until the technology in question will be so utterly inadequate with respect to more modern technologies that eventually a (possibly catastrophic) change of the technological paradigm occurs.

Another tradition of research has made an approach to model technological change as influenced by some kind of network effect. In this case, however, the focus is not the production process but rather the nature of innovation. The basic idea was conceived by Kauffman (1993) as a general evolutionary mechanism and subsequently applied to the economic issue of technological change by Kauffman *et al.* (2000). The model is discussed in more detail in the previous chapter in section 3.4.4. A number of agents engage in a local search for better technologies (innovative research) in an n-dimensional technology landscape. Depending on the shape of the landscape, the search radius and the currently known technologies, the economy-wide innovation speed varies in time. As soon as a new state of the art technology is detected, the agents search the spacial vicinity of this location in the technology space with a good chance of finding slight improvements. After a while, the search radius increases as the immediate vicinity is explored, the likelihood of 'follow-up innovations' declines and so does the innovation speed, until a new radically different better technology in some other region of the technology space is detected. In other words, great innovations spark a boost of 'follow-up innovations', thus providing a research side positive feedback loop. Further, the more agents that immediately follow the cutting edge innovations the faster the 'follow-up innovations' are made. Though this is clearly a network externality structure, Kauffman *et al.* (2000) do not explicitly use this interpretation. A similar model was considered by Silverberg and Verspagen (2005, 2007). They reduced the technology space to a two-dimensional entity, however explicitly including cumulativeness of technological change. Agents may research any point in the technology space but only those directly connected to the base (the 'lower' edge of the bounded two-dimensional search space) are considered 'operational'. In effect, people may be aware of possible future technologies though they currently are far beyond their reach in that they know the problem to be solved and the applying laws of physics and nature. Thus, achieving a scientific breakthrough linking a group of connected technologies to the operational area may instantly lead to a number of innovations putting the now operational technologies to use.

In all, there are two major types of network externalities that may induce significant effects in economic growth. First, those of the research side that make further advances more likely after a significant innovation has been made. These may be responsible for clustering of innovations in time, such that industries undergo phases of slow and rapid technological change as discussed by Kauffman *et al.* (2000). Second, there are those that increase utility with wide adoption of a standard. The reduction of production costs as a result of standardization is a mere variant of this second type. At this point it is of minor interest if efficiency gains occur in the process of production or consumption. The result is simply that the scope of profits for the vendor widens either because the production costs have decreased or because the utility derived by the consumers has risen. The latter effect will, for normal goods, increase the reservation price of the consumers, which normally (for normal elasticities of demand) enables the vendors to either set a higher price or serve a larger number of customers, depending on the assumptions on the industry structure and the type of competition. The resulting higher revenues may or may not be appropriated by the vendors or be transferred to the consumer side via lower prices or – in case of increased utility – remain with the consumers.

The models discussed in the following chapter (Chapter 5) will go more into detail on both the types of network effects and the question of how to model them properly, and the implications of those models for economic growth and technological change.

4.8 Conclusion: from network effects to economic growth

Network effects have been found to lead to technologically inefficient development, and in turn inefficient allocation, since once a network grows, it generates a positive feedback on this very growth process. Network effects further spark bandwagon movements, as the choice of one agent may induce others to follow. Public and private sponsoring may influence the choice process while the critical 'windows in time' (David and Bunn, 1988) are still open. A wide range of strategies including niche construction, gateways, planned obsolescence, standard tying and profitable piracy are also available to standard owners. Path dependence, non-ergodic dynamic systems, and unpredictable outcomes, as well as a strong tendency towards monopolization as a result of the positive feedback are further structural properties of industries with network effects. For local network effects the picture might be different and depend heavily on the network structure. A tendency to a strongly asymmetric industry is nevertheless still present.

There is evidence for network effects in a large variety of industries and other environments all over the socio-economic system, including formal and non-formal institutions, infrastructure and technology standards as well as all kinds of coordination problems.

It may therefore be strongly suggested not to neglect network effects in theories of economic growth and technological change. The results of the current chapter give reason to suppose such models with network effects differ from the mostly purely deterministic growth models discussed in Chapter 3 in several ways:

- They are themselves path dependent. This aspect was detailed in section 4.7.
- They may amplify small events and may therefore not be ex ante predictable. This does not mean that the result is strictly random – on the contrary, there may be very robust and stable complex patterns; the distribution of likely outcomes may also very well be assessable.
- They may also support persistent diversity in terms of different development paths for different industries and different regions.

- If the system has sufficiently many components they may even be chaotic and may lead to the emergence of stable patterns at an aggregated level, a phenomenon frequently reported for complex systems.[117]
- And almost certainly they will lead to technological paradigm changes on a regular (or irregular but recurring) basis. This may contribute to the explanation of growth cycles.

The next chapter will focus on developing possibilities of including network effects in theories of economic growth and technological change, and assessing the possible and established results.

5 A complex model of growth, technological change and network effects

Having given an extensive account of both the phenomenon of network externalities including the resulting path dependent processes (Chapter 4) and the major theories of economic growth and technological change (Chapter 3), this study will take the argument one step further in this chapter. After presenting and analyzing the mostly macroeconomic growth theories, Chapter 3 concluded that while many interesting propositions have been raised and major progress has been made in explaining the data series, the state of the art is still very far from understanding many of the observed phenomena. Especially, the integration of the micro-level into macro-level growth theories is mostly limited to a very stylized approach that essentially assumes the micro-level away by defining agents as homogeneous and not directly interdependent. Thereby macro- and micro-levels become proportional, with the population size acting as the proportionality factor. One notable exception are the models of the evolutionary growth theory, Nelson and Winter's mechanism (Nelson and Winter, 1974, 1982a; Nelson *et al.*, 1976) and its diverse adaptations. (For details, see section 3.4.) What was exceptional about this mechanism was that it combined a micro-level model with a well-behaved aggregated data stream (thus, an endogenous macro-level) generated by the micro-level entities. The theoretical interpretation of the device (innovative and imitative research) fits well with the Schumpeterian approach to innovation, and is also the single central theoretical framework used in virtually all subsequent evolutionary theories of growth[1] (Conlisk, 1989; Silverberg and Lehnert, 1993, 1994).

Nevertheless, to achieve reasonable simplicity and concentrate on the core of the argument, these evolutionary models deal with otherwise conventional economic assumptions. As discussed in Chapter 4 this may not be appropriate. Chapter 4 concluded that network effects do generally cause externalities (network externalities) that lead to lock-ins, the path dependent development of economic and technological dynamics and a number of non-conventional effects that are discussed extensively in the economic literature, but mostly just with respect to the sectors where they are most prevalent – typical network technologies and the information and communication technology sector. The analysis in Chapter 4 does, however, strongly suggest that network effects, network externalities and the resulting phenomena like lock-ins, path dependence, emergent pattern formation, etc., occur all over the economy. (The argument is most obvious from the universal presence of common standards.) To construct a theory of economic growth and technological change with a proper representation of the micro-level, network externalities should be included. This shall be attempted in this chapter. Furthermore, it will be argued that network effects that result in externalities, lock-ins, etc., are not merely a refining or complicating addition to growth theories but may be used as a device that generates similar effects and may be used alternatively (or complementary) to Nelson and Winter's mechanism.[2] The goal of the theory and models developed and discussed in this chapter is to explain:

- Superlinear growth patterns, i.e. the growth rate does not strictly decline as in linear settings[3] nor does it simply oscillate around a linear function.
- Circular patterns in growth rates, i.e. the growth rate alternates smoothly between higher and lower levels. This does not necesarily follow a regular pattern since the estimated cycle lengths differ in time and between regions. (See Figure 3.7.)
- Possible occasional catastrophic declines with sudden negative growth rates, e.g. financial crises. As these may in part be caused by influences that may not strictly be part of the economic system (policy decisions, natural disasters, 'speculative mania'), they may be difficult to model and would possibly require an elaborate representation of the financial system in the model as attempted in Keen's model of financial instability (Keen, 1995). This effect will not play a central role in the models considered in the current chapter as the objective is to model the effect of network externalities and keep other parts of the model as simple as possible. It is difficult to estimate the extent of such phenomena since a large variety of effects interfere with any data series on crises, including random influences and political decisions. Many of the downturns observable in real-world growth data may be – correctly or incorrectly – attributed to such influences as shown in Figure 5.1.
- The co-evolution of the technological system with mostly slow technological change but the occasional development of radically new technologies which may induce peaks in the number of innovations, as reported by de Groot and Franses (2009).
- Oligopolistic industry structures on the micro- or sector-specific-level especially for innovative technologies. For detailed examples, see the case study in Chapter 6. This includes frequent 'standard wars' (West, 2003; Stango, 2004; Bresnahan and Greenstein, 1999) as well as, for some outflanked proprietary standards, the dedicated creation of niche industries by essentially breaking off from the greater industry (Bresnahan and Greenstein, 1999). The oligopolistic structure persists but does usually not lead to swift and irrevocable monopolization.
- Interconnectedness of technologies up to the point that the possibility space for new technologies is shaped by historic predecessors of standards: infrastructure relies on older infrastructure, early cars used the same streets

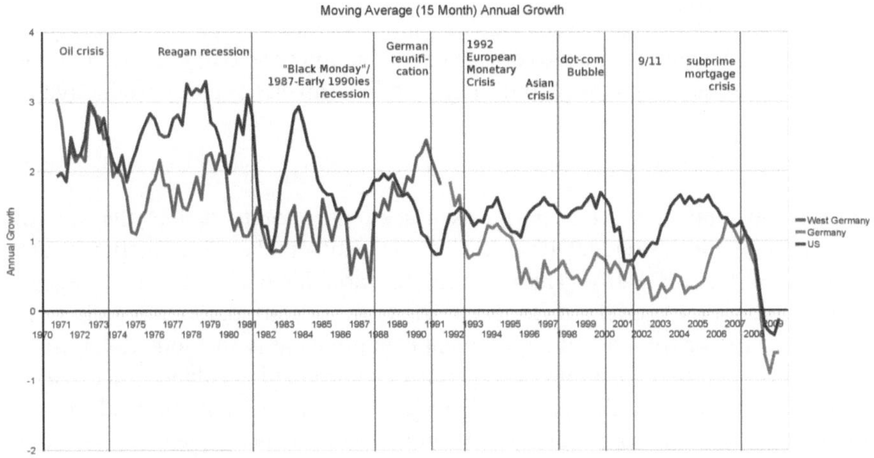

Figure 5.1 Annual GDP per capita growth rates of Germany and the USA (three-month data) with a 15-month moving average filter applied. Historically important economic events are marked and labeled. (Data from Bureau of Economic Analysis (2010) and German Federal Statistical Office (2010).)

designed and used for older means of transportation, the internet technology was developed to work with the telephone network (Glowniak, 1998), etc.

This chapter is organized as follows. The focus will be a model of the technological development of a single industry with network externalities and their economic consequences. This model is introduced in section 5.1. The following section (5.2) discusses the implications of this model for the macro-level and the micro-to-macro transition using several related models and working with both analytical methods and simple computer simulations. Finally, section 5.3 investigates the possibilities of computer simulation, transforms the model introduced in section 5.1 into a simulation model and presents simulation results. It is followed by a short conclusion.

5.1 A stylized formal model of network externalities, economic growth and technological change

There are two different points in the technological–economic system where network externalities can potentially play a major part in the processes of growth and change: first, the production or consumption process may become more efficient as a result of network externalities without involving major changes in the technologies in use; second, the improvement of applicable technologies may increase as a result of network externalities resulting in step-wise increases in production and (or) consumption efficiency.[4]

5.1.1 Production, growth and technological change without network effects

Consider a very simple (stylized) device of production of an output y as a function of capital k, labour l, and the labour–output ratio a (inverted labour productivity).

$$y = k \times \frac{1}{a} \times l \tag{5.1}$$

where the capital k is seen simply as a scaling factor to denote the size of production of the individual producer or the entity under investigation. Let i be the index for this entity. These are micro variables that may – in an agent-based variant of the model – differ from firm to firm, from region to region, from sector to sector, or from any other relevant economic entity to another. Any such agent is assumed to have a number of entities of production (plants) each with a specific labour–output ratio a_j and each designed to produce a normalized amount of output $y_{i,j} = 1 \forall j$ at its command. The number of plants an agent can afford is defined by the available capital k_i.

$$y_i = \sum_{j=1}^{k_i} y_{i,j} = \sum_{j=1}^{k_i} \frac{1}{a_j} \times l_{i,j} \tag{5.2}$$

And as $y_{i,j}$ is normalized,

$$y_i = \sum_{j=1}^{k_i} y_{i,j} = k_i. \tag{5.3}$$

Noting that (5.2) is not only valid as a sum but also item-wise for each j

$$y_{i,j} = \frac{1}{a_j} \times l_{i,j} \tag{5.4}$$

and substituting equation (5.3) into equations (5.2) and (5.4), hence

$$y_i = k_i = \sum_{j=1}^{k_i} \frac{1}{a_j} \times l_{i,j} \tag{5.5}$$

and

$$1 = \frac{1}{a_j} \times l_{i,j}, \tag{5.6}$$

the production equation can be written as the labour demand of agent i as a function of the involved a_j with fixed output (k_i)

$$\sum_{j=1}^{k_i} 1 \times a_j = \sum_{j=1}^{k_i} l_{i,j} = l_i \tag{5.7}$$

Note that the normalization of the output to 1 per plant (i.e. k_i for firm i) simply reduces the complexity of the labour demand dynamics. Below, the output dynamics will be re-internalized by allowing k_i to expand.

Furthermore, the amount of available labour is fixed at[5] L_0 and is (for simplicity) assumed to be allocated efficiently to the most efficient entities of production (plants), in turn those with the lowest a, and consequently the highest labour productivity.[6] That is, labour is the scarce factor of production in the model (that prevents unlimited expansion of production with any current technology), while any other potentially scarce factors of production like human capital, the capital stock or the institutional background of the economy are assumed normalized to unity for simplicity. With this, the production function normalized above (equation (5.5)) to $y_i = k_i$ becomes non-trivial again as only a part (the most efficient part) of the k_i existing plants produce output. The better the technologies in use (in terms of the labour–output ratio) by entity i, the more can be produced, thus the larger the output y_t. Figure 5.2 gives a simple illustration of the system with just one entity, two different labour supply values L_0 (dark and light curves) and rather arbitrary labour–output ratios for the plants. As can be seen, increasing the labour supply leads to increasing output; this relation as such, and notwithstanding other effects introduced below, shows decreasing (or at most constant[7]) returns: an increase in available labour force by more than 60% (from 2.2 to 3.5 units) leads to an output increase of just one third of the previous value from 6 to 8 units of output. Note that with the dynamic construction of new plants (as k_i grows) this becomes a vintage model.

Simple economic growth may be defined by introducing a feedback from production to capital (in turn denoting the number of plants), that is capital as

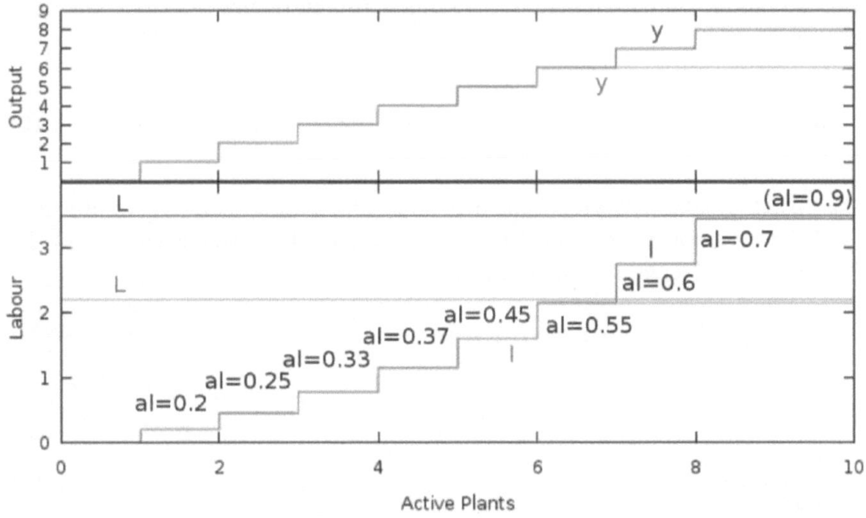

Figure 5.2 Illustration of the system (plants, labour and output) with two different labour supply values (dark curves with $L_0 = 2.2$, light curves with $L_0 = 3.5$).

a function of the output of the previous period

$$k_{t+1} = k(y_t). \tag{5.8}$$

This, however, is not the focus of the current analysis. The more important pattern is technological change. The discovery of a better technology by an entity i is assumed to happen with a probability p each turn, lowering the labour–output ratio of the best technology available to the entity $a_{t,i}$ by a factor $0 < \rho_{inno} < 1$. Hence

$$a_{t+1,i} = \begin{cases} \rho_{inno} \quad a_{t,i} & \text{with probability } p \\ a_{t,i} & \text{otherwise} \end{cases}. \tag{5.9}$$

5.1.2 Network externalities of production efficiency

In order to account for network effects, assume there are a number of technology types. Technologies of the same type (and only those) are compatible, thus generating network effects in the same – larger – network. As technologies improve over time with better labour–output ratios, these types may be seen as developing technology lines (see Figure 5.4). Note that innovation is thus cumulative, a trait observed in real economic systems Silverberg *et al.* (1988). Let s_{TEC} denote the shares of the respective kinds of technology (TEC) of the total production where each plant j of each agent i employs exactly one technology. Several plants of the same or different agents may use the same technology type. The number of active plants in the population using technology (TEC) divided by the total number of active plants in turn equals s_{TEC}.

The first type of network externalities works by reducing the labour–output ratio; the labour demand function (5.7) transforms to reduce the labour requirements in s_{TEC}, e.g.

$$\sum_{j=1}^{k_i} \frac{a_{j,TEC}}{s_{TEC}} = l_i. \tag{5.10}$$

See also Figure 5.3. As a consequence, for any technology (TEC) currently used in active production by agent i (that is, also served by the available labour L_0), the labour required decreases in its share s_{TEC}

$$\frac{\partial l_i}{\partial s_{TEC}} = \frac{-a_{j,TEC}}{s_{TEC}^2} \quad \forall j: \quad a_{j,TEC}$$

thus

$$\frac{\partial l_i}{\partial s_{TEC}} < 0. \tag{5.11}$$

The efficiency rises with this type of network externalities. It would be appropriate to allocate labour among the plants not only using (minimizing) the standalone

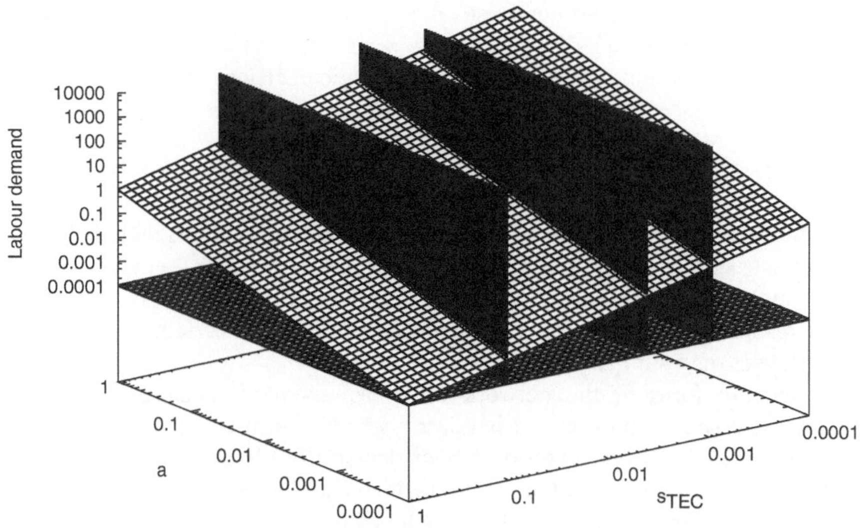

Figure 5.3 Labour demand function of a single plant; all axes logarithmic. The vertical functions mark three technology levels, $a = 0.1$, $a = 0.04$ and $a = 0.0009$.

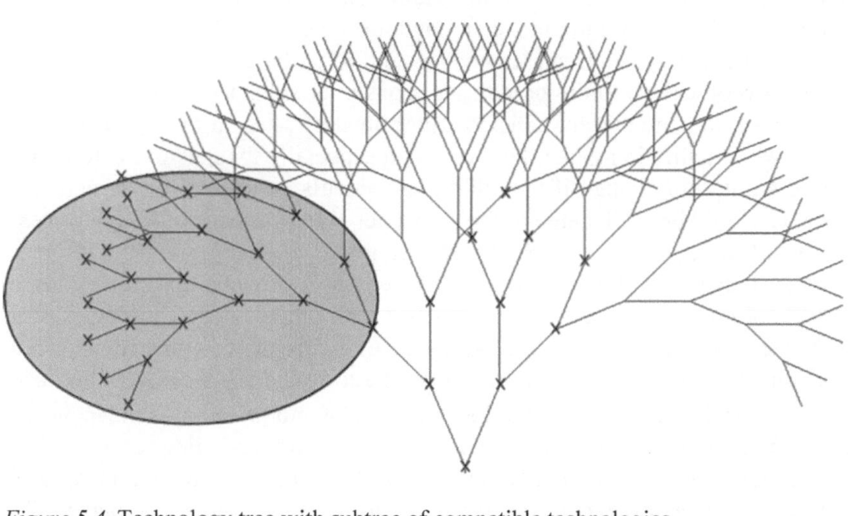

Figure 5.4 Technology tree with subtree of compatible technologies.

labour–output ratio $a_{j,TEC}$ but rather the actual labour–output ratio after the network effect is included. Of course, it is not realistic to assume that agents are capable of making this calculation for all possible technology types in real time and organize a collective coordinated transition to another technology type as soon as that one becomes slightly superior to the currently used one in terms of the standalone labour–output ratio. It would be heroic even to assume that agents are perfectly informed about newly developed technologies. They do, however, likely know the last period's usage shares $s_{TEC,t-1}$ and can use those to estimate the actual labour–output ratio as a basis for strategic decisions, hence

$$\frac{l_i}{y_i} = l_i = \frac{a_{j,TEC}}{s_{TEC,t-1}}. \tag{5.12}$$

Common examples of this type of production-side network effects include large-scale infrastructure like the technical specifications of water and energy supply and the track width for railroads, etc.,[8] but also methods of production requiring specific resources or services, supporting the development of certain skills among the employees, and similar patterns.

The specific form of the network effect as given by equation (5.10) is unrealistically strong; however, it is not the specific form that matters but that the impact of the network effect on labour demand fulfills property (5.11).[9] In turn, the increased efficiency of production with technologies with a large network increases the amount of labour available for less efficient plants, and consequently the number of plants that can be operated at a time, which, assuming that labour is scarce (see above) also leads to increasing output by the agent (and the economy as a whole) resulting from the network effect.

It is worth noting that apart from assumption (5.8) which is merely a side argument, the model does not make any assumptions about the allocation of the additional output (and consequently value may then be consumed or used for subsequent production) to the producer or consumer side. This allocation is not a central issue for the current consideration; however, assumption (5.8) will be retained, while underlining that the model and especially the network externality structure would not change if the additional benefits generated by the network externality would be reallocated to the consumer side by means of falling prices.

The resulting dynamic of growth and technological change is very different from what happens without the network effect. It is obvious that most widely used technology types become more efficient than others. All technologies of a (hypothetical) very common technology type *TEC** experience a greatly enhanced real labour–output ratio with the network effect applied $\frac{1}{s_{TEC^*}} a_{j,TEC^*}$ because s_{TEC^*} is larger than for other technology types. This modification is independent of the standalone labour–output ratio a_{j,TEC^*} of the corresponding technologies. As a consequence, the labour requirements of these technologies are lower and therefore more technologies of this type tend to be among the active part of the economy, even if their standalone labour–output ratio a_{j,TEC^*} is not as good as that of other technologies of more rarely used technology types. This in turn increases the share of the technology type *TEC** in total actively used technologies, s_{TEC^*}, again. The positive feedback loop is thus complete, which implies the presence of a positive network externality resulting from this network effect. From the point of view of the vendors, any entity that has plants of the common technology type available will likely succeed in putting these plants into active production; further, any entity that has a technology of the favourable type available will, if possible, construct plants using this technology. The industry will face a technological lock-in, with this type of technology being the successful industry standard. Consequently, the development of both the industry structure and the technology of production will be path dependent and non-ergodic.

However, as long as the lock-in is not total[10] the situation may be resolved – if only to arrive at another lock-in with another technology type. As soon as a technology is researched the standalone labour–output ratio of which is still better than the actual labour–output ratio of most technologies of the widely used type TEC^*, entities will start to switch to this technology. With network externalities that are weaker than the one defined in equation (5.10) as discussed above, this is not unlikely. The transition occurs very similarly to the transitions predicted by evolutionary models of technological change (e.g. Nelson and Winter (Nelson, 1968; Nelson and Winter, 1974, 1982a)). While the transition occurs, the scope of the network effect for any one of the technologies reduces, thus lowering total production. This effect may account very well for growth cycles; the idea follows those discussed by Perez and Freeman (Perez, 1983; Freeman and Perez, 1988) as well as Silverberg, Lehnert and Verspagen (Silverberg and Lehnert, 1993, 1994; Silverberg and Verspagen, 1996). For details on the empirical scope and exact form of this phenomenon, see the simulation model in section 5.3.4.

5.1.3 Network externalities of consumption

Another kind of network effects are those resulting from more efficient or more beneficial consumption of the goods in question. This is probably the most common example for network effects; it includes communication networks, all kinds of standardized infrastructure, more technical standards like the common size of vehicles and the width of streets, but also the more social-behavioral traffic laws. All those have in common that the users experience a greatly increased utility from the consumption of a good employing a more common type of technology than one using a rare type of technology. For instance, a communication network tends to be more useful for the individual customer, the more people are connected to that network. As detailed in the previous chapter, this type of network effect may also be limited to a more or less direct neighbourhood, in turn changing the utility patterns and adoption dynamics. It will, however, be shown that this type of network effect is formally equivalent to production-side network effects as discussed in the previous section. Therefore, the current study does not treat network effects rising from consumption as a separate type of network effects, but rather as a special case of network effects from production. In a way, the consumption network effect simply causes an increased utility to arise or be produced for the consumers as a result of a larger network of users of the same technology. Given sufficient market power, this utility is also appropriable by the production side.

For simplicity, consider a global network effect only. To model this network effect, the normalization of output as given by equations (5.5) and (5.6) has to be dropped in favour of a production function that takes the network effect into account. This time, the direct impact of the network effect is not to decrease the costs or factor requirements but to increase the output value (higher utility, higher reservation price). For instance, using a linear network externality which again follows the shares of the type of technology (as used in production) changes equations (5.5) and (5.6) to

$$s_{TEC} \times y_i = s_{TEC} \times k_i = \sum_{j=1}^{k_i} \frac{1}{a_{j,TEC}} \times l_{i,j} \tag{5.13}$$

and

$$s_{TEC} \times y_{i,j} = s_{TEC} \times 1 = \frac{1}{a_{j,TEC}} \times l_{i,j}, \tag{5.14}$$

Applying an allocation of labour that takes the network effect into account, the labour demand function changes to

$$\frac{l_i}{y_i} = l_i = \frac{a_{j,TEC}}{s_{TEC}}. \tag{5.15}$$

This is the same equation as for the production-side network effect as shown above. It was also discussed above that this type of network effect formally leads to a positive network externality and lock-in effects as well as all their consequences, including a path dependent development of the industry and technology. As the two network effects are formally equivalent, the same holds for the consumption-side network effects currently under consideration. The only important difference is that the benefit resulting from consumption-side network externalities originates at the consumer side. It may or may not be reallocated to the producer side via prices and profits or via private investments. This, however, is not the primary concern of the current analysis.

Therefore, production-side and consumption-side network effects shall be treated as subtypes of one form of network effects that exercise a profound influence on growth and technological change. Both lead to the production of greater utility or greater value as a result of the wider spread of the employed type of technology. The effects they induce into industry dynamics as well as paths of technological change and economic growth are expected to be very similar. It is appropriate to term this class of network effects as *network effects of production.*

5.1.4 Network externalities of research and innovation

A radically different type of network effects are the network effects resulting directly from research and innovation. This kind of network effect influences patterns of technological change directly and as a consequence also the paths of technological change in the economy, in the sector or in other entities of the socio-economic systems.

In this case, following the analysis in Kauffman *et al.* (2000) and Silverberg and Verspagen (2005, 2007) the probability of successful research increases as a result of wide use of the currently best available technology $a_{t,i,TEC}$, where the network effect again depends on the share of the respective technology type *TEC* in the technologies of active plants s_{TEC}. As an example consider the network externality modifies the success probability in (5.9) such that equation (5.9) becomes

$$a_{t+1,i} = \begin{cases} \rho_{inno} \quad a_{t,i,TEC} & \text{with probability } 1-(1-p)\frac{1}{s_{TEC}} \\ a_{t,i,TEC} & \text{otherwise} \end{cases}. \tag{5.16}$$

Note that the modified probability is still a valid probability value (between 0 and 1)

$$0 < 1-(1-p)\frac{1}{s_{TEC}} < 1$$

and increases in s_{TEC}

$$\frac{\partial 1-(1-p)\frac{1}{s_{TEC}}}{\partial s_{TEC}} > 0. \tag{5.17}$$

This is again just an example, the relevant property of which is given by condition (5.17) and should hold for all network effects of this type. Also note that the example does not restrict the type of the technology resulting from the research that is now more likely to be successful; a feature that acknowledges that innovative research may also lead to developments that are incompatible with the current state of the art. Therefore, there will not be a reduction in variety of technology types,[11] the network effect just influences the rate of innovation. Coordination of a significant part of the population of economic entities would always spark research followed eventually by the development of incompatible technologies, then the slow switching to these technologies, and the reduction of the network effect.

It is evident that the network effect causes an increase in output in this case as well – though this time indirectly by sparking the research into better technologies that in turn have lower labour requirements and (analogous to production-side network effects) result in the available labour being able to operate more plants and thus increase output.

A model of *network effects of research* could also be designed by defining a system on a Kauffman-type technology landscape (see section 3.4.4 for details of this concept) with mostly open information. Research is conducted in the vicinity of the currently used technology. As soon as a much better technology is discovered and becomes public knowledge, all agents (firms for instance) switch to this technology. As a consequence, the vicinity of this technology is explored by a number of agents simultaneously while all findings become open information (thus, the entities cooperate implicitly) and further possible improvements are revealed at a much higher rate. This is exactly the phenomenon discussed by Kauffman *et al.* (2000) as a possible cause of growth cycles – however, without any reference to network externalities. Indeed, in this case it might also be a matter of interpretation whether the network externality causes the boom in innovations or if the initial innovation simply causes both an increase in follow-up innovations and a lock-in-like coordination of technologies.

The example also raises the question of openness of information and its impact on network effects, network externalities, technological change and economic growth in general; a question, that will be addressed below.

5.1.5 Conclusion on stylized models of network externalities

In this section, the basic models of two fundamental types of network effects, resulting network externalities and their effects on growth and technological change have been discussed. Network effects have been found to lead to increased production of value either directly (in the case of network effects of production) or indirectly (in the case of network effects of research) with increased size of a network using the same type of technology. Thus, they induce positive network externalities by reducing the labour requirements of production, which allows the production of greater quantities (if labour, or, more broadly speaking, also human capital, is scarce) and possibly by generating greater revenues that may be reinvested. These network externalities lead to temporary lock-ins and a non-ergodic, path dependent development of technology and industry structure (this will be discussed in more detail in section 5.2.2). The eventual break-out of the lock-in is possible because of the technological development. That is, another technology will most likely eventually become superior, at which point a change in the technological paradigm occurs. This mechanism is relatively simple but justified by examples from real-world economic systems[12] and powerful in its effect on the model where it results in breaking up of the lock-in – this will be discussed in more detail in the simulation section, section 5.3.4. Some extent of industry monopolization may be expected. As pointed out in evolutionary economic models, e.g. Nelson and Winter (1982a), entrepreneurship is motivated by profit opportunities resulting from temporary comparative advantages. This, in turn, leads to larger resources for reinvestment and thus to a slight shift in industry shares to the first comers in using the technology on which the lock-in is focused. Assuming that other agents follow (which depends on the openness of information on technology), this effect will, however, not be large and cease as soon as the lock-in is broken up.[13]

Though the examples are very stylized, they allow assessment of the kind of influence network externalities exercise on the technological and socio-economic development. This brings the current argument one step closer to investigating the phenomenon with a more realistic model, which will be done in section 5.3.4 using computer simulation. This simulation model will rely heavily on the model discussed in this section. Some issues that were neglected for the sake of simplicity in this section, first and foremost the method of information transmission, research by imitation, or, more illustratively open knowledge and secret and tacit knowledge, will be modelled much more explicitly in the simulation model. Also, the simulation will provide data from simulation runs with rather realistic settings with a large number of agents (firms) both with and without network effects, and with both shared and non-shared technological knowledge.

Before going into such detail, however, some further theoretical considerations will be reviewed, now focusing on the macro effect of network externalities (sections 5.2.1 and 5.2.3) and the properties of the micro-to-macro transition (sections 5.2.2 and 5.2.4). In addition, existing efforts to model technological change as discussed in Chapter 3 and their possible applicability to the current context shall be reviewed again (section 5.3.2).

5.2 Microfoundations of growth and the emergence of macrostructure in systems of technological innovation: aspects of the micro–macro transition

5.2.1 Macro-level phase transitions as a result of path dependence on the micro-level

A model of the macro-perspective

Network effects resulting from research may especially lead to a more global form of positive feedback mechanisms. When seen in a broader sense than the one considered in the model presented in the previous section, increased research as a result of network effects may give rise to the development of derivative industries – industries secondary to the original one that caused the network effect. As an example, the development of personal computers eventually facilitated not only an industry sector of personal computer operating systems but also very dynamic sectors developing office software, computer games, network applications, and furthermore, online platforms, online search engines, etc.

Consider such a development seen not from the micro but from the global perspective, the development of several connected industries as a macroeconomic effect. A large enough network of a specific technology type that causes sufficient network externalities to develop derivative industries results in rapid economic growth of the sector. Returning to the example, this may be exemplified by the boom of the 'new digital economy' during the 1990s. If the network effects are not large enough or it abates at some point in the development of the structure of derivative industries, the macro-level growth process stagnates. What remains is the usual development of slow growth caused by the expansion of capacity, which may be the result of effects from the environment of this core industry such as growing demand, increasing labour force, growing investments, efficiency gains in other sectors, etc.

This structure is resembled by the model as discussed in the last section of the previous chapter, section 4.7. The model described growth as a result of a network effect depending on output y_t at time t produced using a specific technology and simply indicated by the size of this production y_t. The model is given by the recurrence equation (4.67)

$$y_{t+1} = \frac{y_t^{\gamma}}{\gamma}.$$

(where γ is a parameter) with the solutions (4.68)

$$y_1^* = 0$$
$$y_2^* = \sqrt[\gamma-1]{\gamma}$$

of which the first is stable for $\gamma > 1$ while the second is stable for $\gamma < 1$. The second solution stands for the decline of the sector until it ceases to exist; the first solution is a stable equilibrium with positive production. For $\gamma > 1$ the sector either breaks down towards the trivial first equilibrium (sector size 0) or expands infinitely with increasing velocity, giving rise to typical path dependent processes in this system which is appropriately characterized as non-ergodic. It was noted above (section 4.7) that infinite growth is not exactly a realistic scenario. However (as also noted above), this may easily be changed by internalizing the parameter γ. The complete situation is illustrated in Figure 4.17.

Consider γ to be a simplified one-dimensional but inverse term for the technology level – that is, the technology level is higher, the lower the value of γ and vice versa. In a way, γ corresponds to the labour–output ratio a in the model in the previous section just with the difference that a was for technical reasons limited to values between 0 and 1 while γ is not. Assume further that the technology level grows slowly as long as the system is in the area above the non-trivial equilibrium ($y > y_2^* = \sqrt[\gamma-1]{\gamma}$) and falls slowly in the area below ($y < y_2^* = \sqrt[\gamma-1]{\gamma}$). To give a more specific example,[14]

$$\gamma_{t+1} = \begin{cases} \frac{\gamma_t}{2} + \frac{y_t - \sqrt[\gamma_t - 1]{\gamma_t}}{y_t} \frac{\gamma_t}{2} & \text{for } y_t > \sqrt[\gamma_t-1]{\gamma_t} \\ \gamma_t & \text{for } y_t = \sqrt[\gamma_t-1]{\gamma_t} \\ \gamma_t + 0.01 & \text{for } y_t < \sqrt[\gamma_t-1]{\gamma_t} \end{cases} . \qquad (5.18)$$

which forms a system of recurrence equations together with the development equation of output, equation (4.67) (as already given above).

$$y_{t+1} = \frac{y_t^{\gamma}}{\gamma}.$$

To analyze the dynamics of this system of recurrence equations analytically, the eigenvalues would have to be computed using the Jacobian matrix of the system. This is not possible since equation (5.18) is not continuous. Even if a continuous equation could be found to approximate the system with sufficient accuracy, the equation would be a highly complicated one making it doubtful whether the eigenvalues can be computed analytically. Since the dynamics resulting from this system in the relevant area of the phase space (positive y_t and γ_t, especially low positive values of y_t and γ_t) are relatively plain and steady, the system shall simply be investigated by simulating its development for a number of initial values (y_0 and γ_0) evenly distributed in a grid-pattern over the area $0 < y_0 \leq 5$, $0 < \gamma_t \leq 4$. The simulation was conducted using a Python program; the source code is available from the author on request.

As had been expected, it can be seen that the trivial equilibrium (of the underlying simple system without the technology level internalized) $y_t = 0$ is locally stable for $\gamma > 1$, attracting the trajectories from the area below the other equilibrium (which is unstable in this area of phase space), that is $y_t < \sqrt[\gamma_t - 1]{\gamma_t}$, while the trajectories above this latter equilibrium $y_t > \sqrt[\gamma_t - 1]{\gamma_t}$ move to the upper left (y_t increasing, γ_t shrinking[15]) until they transgress the $\gamma_t = 1$ line, where the simple system (without the technology level internalized) experiences a phase transition and the non-trivial equilibrium $y_t = \sqrt[\gamma_t - 1]{\gamma_t}$ becomes stable. Consequently, the trajectories approach this equilibrium from the right or from above (as do of course those trajectories with lower values of y_t in this area from below). The discreteness of the system (as a system of recurrence equations) leads to intersections of trajectories in some areas of fast movement as well as the trajectories approaching the non-trivial (upper) equilibrium with movements oscillating around it – this would not happen for the continuous system[16] (real-world systems are continuous) which is to be approximated here. At which point a trajectory approaches the upper equilibrium depends on the initial conditions, i.e. where the trajectory started. Thus, the system is indeed path dependent, both the technology level and the production tending to be higher (y_t high, γ_t low) the higher the system had been above the upper equilibrium. The final approach of the trajectories toward the upper equilibrium when the trajectory passed the phase transition at $\gamma_t = 1$ may be interpreted as the final breakdown of the expanding regime of the industry sector, i.e. as soon as the – in reality certainly much more complex – phase transition had occurred, the system ceases to expand into further derivative industries, as no further value-adding or efficiency-increasing derivative industries can be established. The growth dynamics of the now stagnating industry will subsequently change to match the new nature of the industry in production, development and investment, which may well lead to a certain restructuring and decline in output as seen in the model system in Figure 5.5 for the trajectories in the upper left after having passing the $\gamma_t = 1$ phase transition event.

Macro-level phase transitions in economics

The scenario as discussed in this section may a little more illustratively be imagined as a process of rapid growth in a complex economic system, which includes stochastic elements that cause the system to shift between the trajectories of the corresponding deterministic system. It is certain that the rapid growth will not last forever as it relies on the sufficiently fast and functional adaptation of organization and infrastructure to fit the needs of the incorporation of yet another derivative

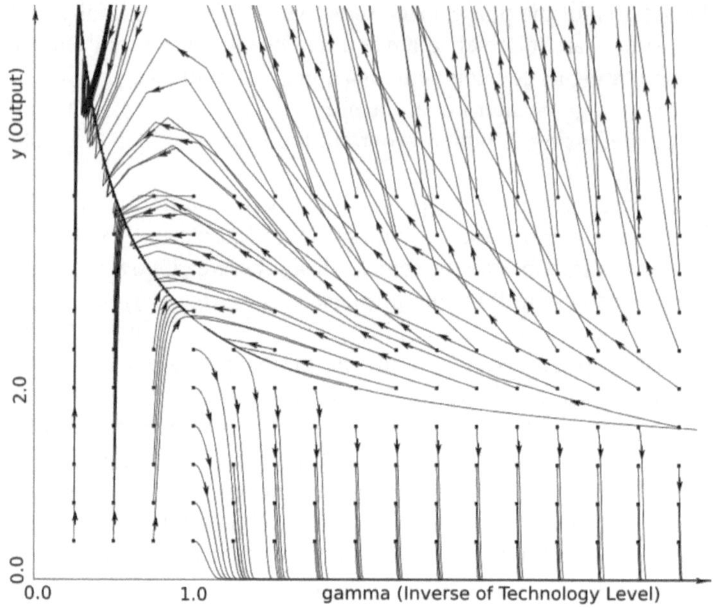

Figure 5.5 Phase portrait of the model system (trajectories of the model for a number of different initial conditions). The image was created using a Python program (source code available from the author on request), the squares indicate the starting points for the different initial conditions, the lines and arrows denote the trajectories.

industry into the system. It is, however, ex ante unclear what happens when the regime of fast growth comes to an end. Either, the economy can return to rapid growth after a period of reorganization of the underlying infrastructure with the next major scientific advances. Or it may also stagnate at its current size or collapse entirely.

The stagnation of the rapid growth regime does not mean that there is no growth at all. A variety of effects in macro-variables may still lead to either real or nominal growth, either in per capita terms or just for total macro-variables. For example, assuming with Schmookler's (1966) hypothesis that the demand side has at least a long-term effect on innovation and thus also growth, changes in the population size as well as the society's preferences may lead to a simple expansion (or certainly also contraction) of production, or (closer to Schmookler's argument) to an adaptation of the technology in use which in turn effects the efficiency and the scope of production.

There are a number of real-world examples. As mentioned above, the so-called 'new digital economy' is the most recent and the best documented one. Though the interpretation of the data is debated (see e.g. Liagouras (2005), Carlsson (2004) and Egyedi and Heijnen (2005)), it is clear that first communications and computer technology, and subsequently the software sector and internet related sectors have emerged rapidly over the past decades, generating economic sectors of a huge scale that have simply not existed before, and have revolutionized the mode of operation of most if not all other industries by applying computer technology and internet related communication. For a more detailed analysis of the 'new digital economy' under the aspect of network externalities and their effect on growth and technological change, see Chapter 6. The internet economy is not the only example; the emergence of the electricity supply net as a simple and widely used means of power supply may have had a similar effect as did the steam engine, the railway, the industrial exploitation of electromagnetic waves (radio and other communications media, radiotelegraphy, etc.), and others have probably had similar effects. Those core innovations, sometimes called *general purpose technologies* (Carlsson, 2004), led to the emergence of new industries and periods of growth clearly at least within the industry sector or the respective region, but arguably also on a national or global level. This is what the current model suggests. It is, however, difficult to show empirically, since the phenomenon, including secondary effects arising from the emerging technology, is supposed to be a global one. Therefore, it is not clearly separable from other general growth effects while, of course, the growth of the respective sector that happens partly as net economic growth, partly at the expense of superseded technologies can be shown. Possible evidence of long waves has been discussed (see section 3.5) and so have possible explanations focusing on technology paradigm changes (Perez, 1983; Freeman and Perez, 1988; de Liso and Metcalfe, 1996) and even evidence that suggests alternating periods of slow technological change and periods with a concentration of innovations (de Groot and Franses, 2009). The possible relation to network effects that is under consideration here has, however, to the knowledge of the current author not yet been studied in detail.

5.2.2 Network externalities and the emergence of order

Stable patterns resulting from network externalities

The two previous sections have detailed two different stylized models for network externality induced effects on economic growth and technological change. While one model includes the micro-level to a certain degree including stylized agents owning a number of production plants and using specific technologies, the other one focused on the feedback dynamics on the macro-level. The current section, focusing on the first model but not ignoring the dynamics of the macro-level model, shall analyze the theoretical properties of this approach in general and its results in particular.

The model results in the coordination of the whole or much of the sector on one dominant type of technology. The lock-in may or may not be broken at a later time. This follows from the processes of research into better technologies of the currently dominating technology type on the one hand and all competing types on the other hand. With the latter process being suficiently ahead of the former, the lock-in would be broken. The expected time to this point may, as will be explained in the following, be finite or infinite. Technological development is cumulative (see equation (5.9)). Neglecting the network effect for a moment, the innovation process is formally simply a group of modified random walk processes: for each new technology, either a technology of the dominant technology type is added to the system, hence increasing the potential efficiency of the dominant technology (the modified random walk process moves into one direction), or another technology is invented (literally the process moves in the other direction). The trigger for break-out from the lock-in is when one of the technology types becomes more efficient than the currently dominating one (in spite of the network externality) – formally, when one of the modified random walk processes transgresses a certain value. In turn, this property can be used to estimate the time from lock-in to the subsequent break-out (or rather the statistical distribution of these time spans). The problem is the same as the time it takes a random walk process starting at zero to collapse to zero again, thus a random walk with *gamblers, ruin*. These time spans are power law distributed[17] – therefore it may be expected to follow a similar distribution for the current problem.

Let there be a continuum of possible period lengths from one technological paradigm change (the next starting with 1 (the unit is not relevant) and running through ∞, to be split into equal sections. Let the sections be characterized by their average value (of period length) ξ and let the empirical relative frequency of technological regimes falling into category ξ (i.e. having on average a dominance time of ξ) be $p(\xi)$ (frequency of ξ). A pure power law distribution would suggest a relation of the period length and the relative frequency of

$$p(\xi) = \Upsilon \times \xi^{-\alpha} \tag{5.19}$$

where Υ and α are positive parameters. This would mean that a period length increased by factor h (e.g. twice as long) always results in a frequency reduced by the same factor h (e.g. half as often). However, the lock-in implies a certain deviation of the technological efficiency and thus a distance from the point at which the modified random walk processes meet and the lock-in is broken. Therefore, as soon as the lock-in is in place, there is a minimum time length before this point can be reached, for time lengths greater than this minimum time length, the empirical frequency will at first increase until a maximum is reached. This part of the distribution does of course not follow the power law; it is, however, common for empirical power law distributions to hold only for values greater than some minimum.[18] The process without the network externality or with a very weak externality would be very close to the actual power law distribution. For stronger network externalities, the power law relation continues to hold for the 'tail' of the distribution, the part for time values greater than the maximum. Especially, the scale free reduction of probability for an increased time length is preserved.

Whether the distribution has a finite or infinite mean (the exponent of the power law α being smaller or larger than 2) – that is, whether it is empirically likely that the lock-in is broken within finite time – depends on the strength of the network effect.

The pattern as such – the emergence of a comparatively superior technology type, the subsequent lock-in, the possible break-up and the immediate reestablishment of the lock-in structure now with another dominant technology type – is stable. It happens (as will also be seen in the simulations in section 5.3.4) independently of the simulation run, time, and most importantly independently of the agents involved, independently of the identity[19] of the superior technology and the identity of its technology type. In a way, the elements of the system 'flow' through a stable structure, which exists only on the macro-level of the model but which nevertheless has an effect on the micro-level entities via the network externality matching the definition of a dissipative structure.

Dissipative structures and emergent properties

A *dissipative system* in physical terms is a system with contractive dynamics, which means dissipative systems generally converge towards fixed points or other attractor structures. More broadly, and more importantly to the current context, the trajectories converge, reducing initial differences with time. This also implies that the effect of perturbations, whether unique shocks or stochastic influences, *dissipates* in time. In information theoretic terms, such perturbations are information with respect to the system's history and the dissipative system's loose information with time, while by contrast in conservative systems information is conserved.[20] *Entropy* evolves correspondingly to information; it dwindles away from dissipative systems while it is preserved in conservative ones.

Prigogine in his seminal paper (Prigogine, 1975) and later in an influential book co-authored with Stengers (Prigogine and Stengers, 1984) has introduced the concept of *dissipative structure* in physical, chemical and by implication also more complex social systems. Dissipative structure does not constitute an entire system but rather a structure in a system that as such is stable. The term refers especially to structures that are stable while the surrounding system is not, in a way they are dissipative subsystems in conservative or even chaotic systems.

This is precisely what is the case for the model of production-side network effects above. The system as such is doubtlessly conservative: as stochastic influences enter the system by means of the stochastic success rate of efficiency increasing research, they are preserved. However, the technology type standardization resulting from the network externality is stable. All agents having a technology of the dominant type available will employ this technology. Moreover, when this lock-in is broken, the system will tend to standardize rather rapidly to a new dominant technology type. In effect, the standard dominance follows the properties of a dissipative structure. Dynamically, for weaker forms of the network effect – such that the break-out from the lock-in is sufficiently likely – the pattern of repeated standardization, lock-in and break-out is itself a stable pattern, though potentially with varying time intervals between the technology-type changes. In real-world systems this corresponds to technology paradigm changes. It is also a dissipative structure. Furthermore, dissipative structures co-evolving to this one may exist, e.g. in the development of the technological progress, the form of expansion of production capacity, industry structure, etc. These effects are not immediately obvious from the theoretical model; however, the simulation in section 5.3.4 provides data from simulation runs suggesting that the said effects do exist.

Such dissipative patterns tend to exist on a meta-level of complex systems – in this case an agent-based model. They are generally not directly linked to the dynamics of single elements of the system but rather arise from the interaction of those elements. For instance, an individual agent in the model above, say a firm, has no direct influence on the lock-in at the system-level, though it likely follows the lock-in process as the network effect of the dominant technology gets stronger as its usage share grows. The same is true for the dynamic path-breaking, break-out, and re-lock-in, thus technology-type change processes: the development path of the individual firm is not identical to the system-level trajectory, though the system-level network effect 'pulls' all the individual level trajectories into a common direction in turn forming a system-level order. The system-level order is said to *emerge* dynamically in the process of the systems development. The effect of redirection of individual level dynamics by system level structures, which in turn have emerged as a consequence of the interaction of individual level mechanisms, is a classical example of *downward causation*. The form, size, and type of the resulting emergent pattern and even the incidence of the emergence itself is (or may be) subject to path dependence. It occupies a set of unconnected parts of the system's phase space – which of these attractors is approached, i.e. which technology type is successful, is generally not ex ante predictable. In this case, the attractor as a stable fixed point is subject to changes in time and may likely dissolve at some time – when the break-out from the lock-in occurs. The trajectories move in turn to another part of the system. However, the stability at a meta-level persists in that both the properties of the coming new standard and the form and speed of the transition are predictable by then.

In short, the model considered in this chapter contains a typical dissipative structure and leads to persistent patterns, emergence and emergent structural properties on the system level. Returning to real-world examples, the emergent properties of growth and technological change processes with network externalities include repeated standardization (lock-in) and intermediate periods of growing variety; it alternates between periods of rapid technological change (when the standardization begins), rapid growth (an effect of the standardization) and subsequent stagnation as well as potentially even periods of decline (a possible result of the decline of network externalities and the corresponding benefits). As the system is, however, too complex for exhaustive analytical investigation, details will be given in the simulation in section 5.3.4.

5.2.3 Resonant oscillations

In the previous sections it was established that specific network effects cause network externalities with a strong effect on economic growth as a whole, and the technological development of the sector or region or even the entire economy in particular. The growth effect in one sector takes (as discussed in sections 5.1

and 5.2.2) the form of alternating phases of rapid and slow or stagnating growth in economic development. The corresponding technological system co-evolves with periods of increased innovation alternating with phases of stagnation. The length of these periods is not particularly regular; rather it is to be expected to follow a power law distribution beyond a certain minimum, the relative frequency of technological lock-ins lasting that long decreasing along the 'fat tail' of the distribution (see previous section). For the micro-to-macro transition, the link between the arguments of section 5.1 and 5.2.1, one question is of particular interest. How does the system behave on an aggregated level, including several economic sectors all following the same kind of process as specified in section 5.1? The answer does of course depend on whether the processes between the sectors are linked and the standards tied or whether the sectors are completely unrelated.

Without the tying effect, the aggregated level comprised of several or many industry sectors is simply the sum of a number of time series that oscillate between slower and more rapid growth. Not considering possible demand-supply feedbacks or reinvestment of revenues that may also cause feedback mechanisms, the oscillations are independent from one another. Though some turbulences remain, any perturbations including cycles and oscillations should dissipate at an aggregated level. Some sectors are always in the rapid growth regime, others in a phase of stagnation, and since these states are not correlated, the relative number of slowly and rapidly growing industries should approximately remain constant according to the law of large numbers, at least if the number of industries considered is sufficiently large.

This does not hold for tied industries where network effects transfer boom effects from one sector to others. In that case, the oscillations may start to 'resonate'. This resonance between two individual sectors is of course not lasting but limited to one regime change once triggering the same oscillation in the other industry. However, this occurs between potentially all involved industries. It may trigger cascades and may even amplify with resonance. Theoretically, the effect is similar to one-time boosts in the aggregated data that may trigger further boosts. The effect persists for some time and dissipates thereafter. This in turn, may be seen as a naturally occurring moving average effect: there is a stream of random perturbations in growth, ε_t, the current growth is, however, not only a function of these perturbations but also of the perturbations over a number (LAG) of past periods resulting in a time series of growth rates

$$Growth_t = \frac{\sum_{i=t-LAG}^{t} \varepsilon_i}{LAG} \tag{5.20}$$

which is exactly the moving average mechanism. As discussed in another context above in section 3.5.1, this leads to short-term perturbations leveling out while longer cycles are amplified (Slutzky–Yule effect). Notwithstanding, emergent properties, including downward causation as described in the previous section, may still hold, in turn forcing the single industries into the oscillation pattern of the aggregated system.

This theoretical idea is, of course, rather abstract while the system under consideration itself is very complex. The development of a sector according to the pattern of transitions between slow and rapid growth is not trivial even if the system is reduced to an output and a technological efficiency term as well as mechanisms for research and network externalities. The corresponding path of output development – the variable currently under investigation – with occasional transitions between rapid and slow growth is not easily reproducible as a simple function. An agent-based approach is arguably still one of the easiest ways to produce such a time series. Agent-based models in turn are by magnitudes more complicated than simple functions. To describe such a system analytically, one or several development equations would have to be introduced for each agent, yielding a system of very high degree that is most likely not analytically solvable. It is therefore justified to use simulation as a means of analyzing the dynamics and the resulting patterns.

Consider the following setting for the simulation model. As the emphasis is currently on the interaction of several sectors and not on the workings of a sector itself, the sectors are designed as simply as possible using a standard mechanism of evolutionary growth theory as employed in Nelson and Winter's (1974; 1982a) models in a similar way. Agents produce according to their technology level a which is again the labour–output ratio. The lower this variable is, the better the available technology and the higher the output generated by the agent. a increases with research similar to equation (5.9) above

$$a_{t+1} = \begin{cases} (0.8+\varepsilon_t) & a_t \text{ with probability } 0.0005 \\ a_t & \text{otherwise} \end{cases} \tag{5.21}$$

where ε is a random disturbance with mean 0. In turn, successful research grants large efficiency gains (by $\frac{1}{5}$) but is highly unlikely to succeed and thus happens rarely. Once invented, technologies are semi-open: other agents have a probability of 0.005 (much higher than the probability of successful research) to learn about them. This might be interpreted as open knowledge or even open source policies, as successful imitation (as Nelson and Winter do (Nelson and Winter, 1974, 1982a)), or also as a research-side network externality (as discussed in section 5.1). Once an innovation is made, successive innovations follow that meet the needs of other vendors to integrate the radically new findings of the original innovation into their mode of production. One way or another, the structure of rare true innovations occurring only from time to time by only a few agents and the subsequent slow adaptation of the remaining ones is a powerful device to generate alternations of periods of slow and rapid growth and circular patterns in the growth rates.[21] (See also section 5.3.2.) Note that up until now the industry sectors are not tied; the aggregated level should not show any cycles or oscillations. An example simulation run for this model with 10 sectors is shown in Figure 5.6; the source code both for this setting and the one below is available from the author on request.

Now, the sectors may be tied by introducing a research-side network externality into the model. This is also done in the simplest form possible by changing the

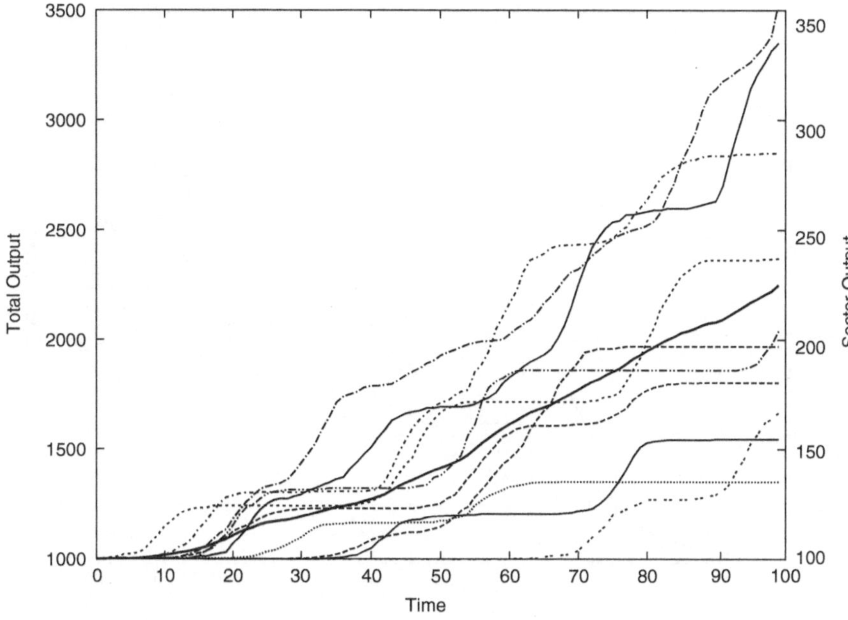

Figure 5.6 Output development without tying network externalities for 10 sectors separately (right scale) and at an aggregated level (bold line, left scale); simulation.

research success probability to depend positively on the aggregated level's output growth. If the economy grows rapidly, research success is more likely to occur, thus transferring rapid growth from one sector via the augmented growth of output at the aggregated level to other industries. Let y_t denote output at the aggregated level. The output growth rate is

$$\dot{y_t} = \frac{y_t - y_{t-1}}{y_{t-1}} \tag{5.22}$$

Research is now modelled as

$$a_{t+1} = \begin{cases} (0.8 + \varepsilon_t) \quad a_t & \text{with probability } 0.0004 + 0.02 \times \dot{y_t} \\ \qquad a_t & \text{otherwise} \end{cases} . \tag{5.23}$$

The parameter values in equations (5.23) and (5.23) (0.0005, 0.0004 and 0.02) are chosen to yield an approximately equal average research success and, as a consequence, an approximately equal average growth at the aggregated level for both settings in order to make them comparable. An example simulation run with 10 sectors for this setting is shown in Figure 5.7.

It is immediately obvious from Figures 5.6 and 5.7 that for the setting with tying the oscillations of the aggregated growth rate are larger and more visible,

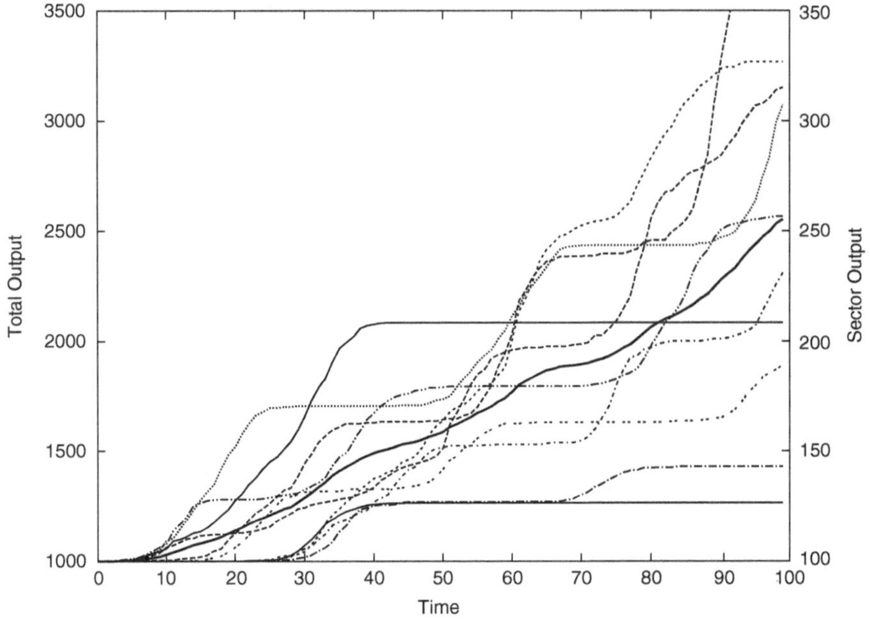

Figure 5.7 Output development for 10 tied sectors seperately (right scale) and at an aggregated level (bold line, left scale); simulation.

and that regime changes in the individual sectors (thin lines) tend to coincide in time in the setting with tying while they do not without tying of industries. Figure 5.8 shows the aggregated level growth rates for both settings. The regime without tying also shows some kind of apparently periodic disturbance; however, the periodic nature is much clearer and stronger in the tied regime. This becomes even more obvious when considering a larger setting with 100 industry sectors, the growth rate development for which are shown in Figure 5.9. To investigate the cycles further, consider the autocorrelation spectrum for the two mentioned example runs as depicted in Figure 5.10. It shows that there is some kind of circular pattern with the autocorrelation coefficients switching between strongly negative and strongly positive values for growing lags.[22] However, it is also apparent that the pattern is not a simple wave form – it appears that for lags between 10 and 30 periods there may be both positive and negative autocorrelations while the pattern settles to a more clear form for larger values.

Theoretical considerations at the beginning of this section suggested that the tying of standards of different industries, and hence the tying of these industries, may lead to significant phenomena at a macro-level. Specifically, tying of standards caused by network effects may yield sector-level waves and disturbances to resonate with other sectors and result in a mechanism similar to the Slutzky–Yule effect in simple moving averages that leads to cycles at

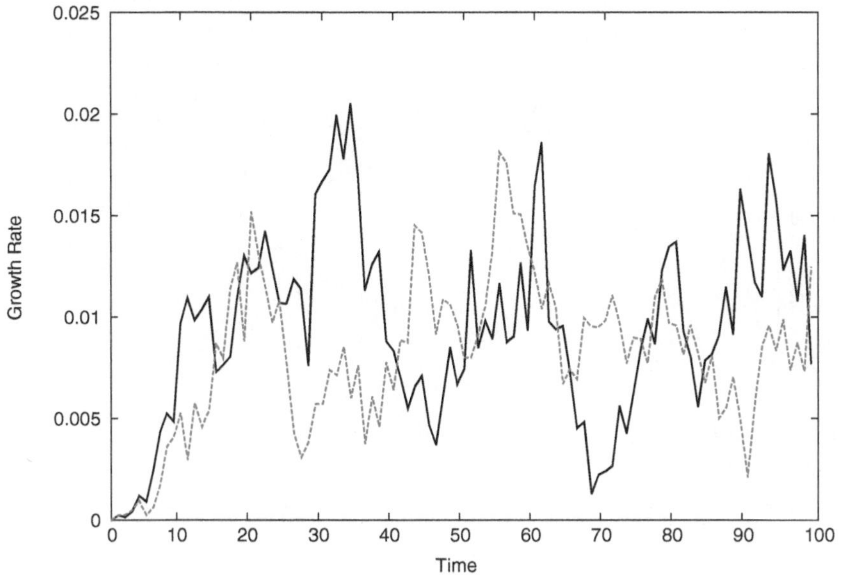

Figure 5.8 Growth rates of the aggregated level with (black line) and without (grey line) tying network externalities (10 sectors).

an aggregated level. In effect, network externalities may at least in part be responsible for macroeconomic cycles, a phenomenon that despite many efforts is not sufficiently explained (see section 3.5 in the chapter on growth theories). This hypothesis was confirmed by the results of the computer simulation as discussed in this section.

It should be noted that the simulation in this section merely modelled research-side network effects for simplicity, and it did so in a very simple way to avoid interference by other unrelated effects.[23] Many other effects may produce positive feedback loops that may cause or amplify cycle patterns as well as other network externalities. Resonances may in reality be much stronger than in this carefully balanced model. For instance, the dynamics of newly emerging sectors while other sectors shrink and loose their central position for the economy have not been included. The same is true for dynamics of production, revenue and investment, and dynamics of correlated investment decisions (Keynes' 'animal spirits' (Keynes, 1936)) that may also result in catastrophic crashes, but certainly have an impact on growth and technological development even if they do not lead to financial crashes. Further phenomena that may have such an impact on macro-level growth and development and that have not been modelled in this section include the dynamics of labour availability, supply and demand relations between sectors, Forrester's 'bootstrap effects' (Forrester, 1977), as well as political influences such as economic policy and education policy.

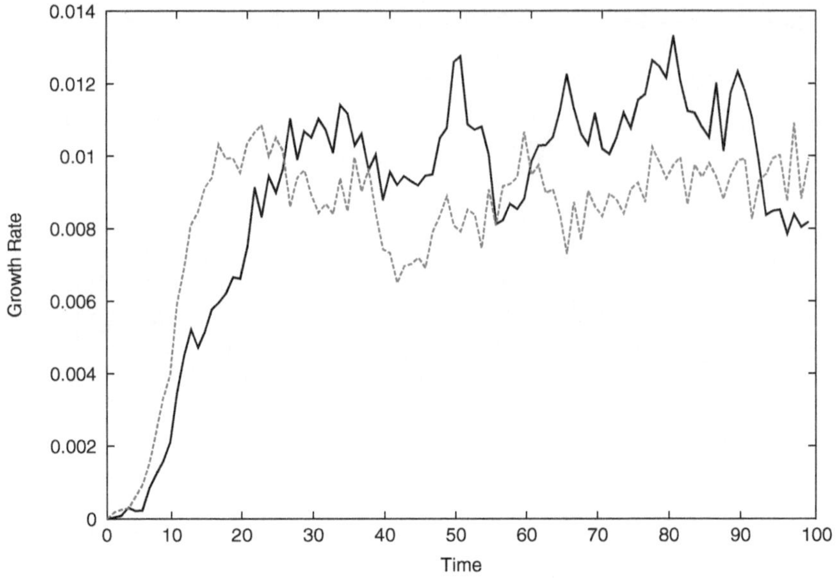

Figure 5.9 Growth rates of the aggregated level with (black line) and without (grey line) tying network externalities (100 sectors).

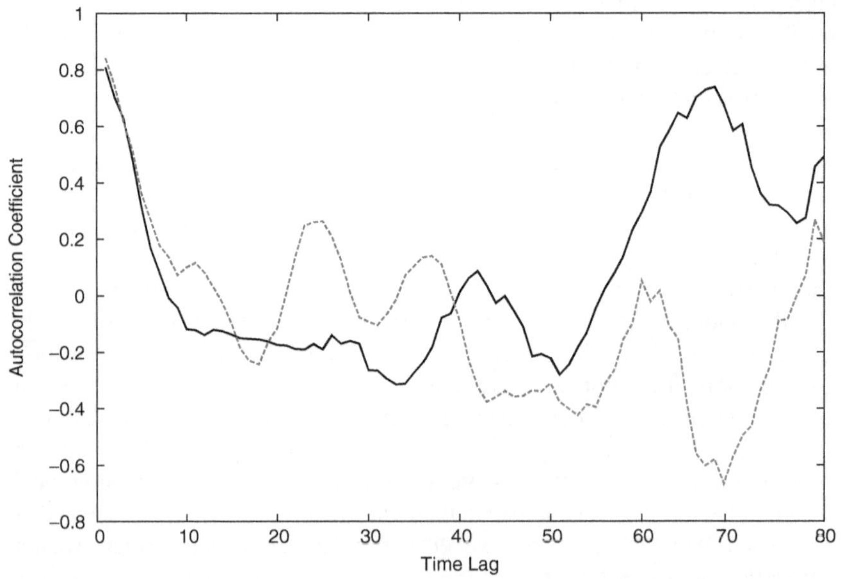

Figure 5.10 Autocorrelation spectrum of the aggregated level growth rates with (black line) and without (grey line) tying network externalities (10 sectors).

5.2.4 Research-side network externalities and power law distributions

So far, technological progress has been treated in a rather abstract and oversimplified way just using an efficiency parameter and a technology type assignment. The previous sections discussed the micro-to-macro transition, the emergent and path dependent nature, the stochastic distribution of fundamental changes in the technological system as well as the possible cross-industry correlation of rapid growth periods and the macro-effect of those. The nature of the underlying technology landscape has so far been neglected.

This aspect may, however, be an important one for various reasons. It may have an impact on the speed and stochastic distribution of the technologies. It may add additional aspects of technological path dependence. However, the most important issue is probably the effect it has on the industry structure. Starting with the theoretical analysis from section 5.2.1 that a rapidly growing technology may give rise to secondary sectors, tertiary sectors and all kinds of derivative industries[24] the most likely picture is one of a 'technology tree' as depicted in Figure 5.4. Several derivative technologies are developed from one successful technology – all of them are subject to standard setting processes and hence to network externalities. The structure does not strictly follow a tree pattern since several technologies (including the respective industry standards) may be combined to form a new one; the overall structure, however, follows the pattern that more specialized technologies rely on more basic ones and accordingly on more basic standards. The network size of successful standards as well as their effect on and importance for the socio-economic system most likely vary. Standards may also be defined for rarely used derivative technologies, but the successful technologies and therefore standards and lock-ins that matter on an industry level are a mere fraction of those. The standards may simultaneously or independently[25] be changed but it does not happen too often since the current lock-in (the exclusive usage of the current industry standard) has to be overcome.

Let b be the number of feasible and successful technologies derivative to some basic technology $\overline{b}$ including $\overline{b}$ itself.[26] Consider the technology tree for simplicity to be strictly hierarchic such that any technology is directly derivative of only one other technology. However, any technology may have any number of derivative feasible and successful technologies. Thus, any of the derivative technologies counted in b for $\overline{b}$, have their own (smaller) values b that cover only the respective 'subtree' of the technology tree. Consider the number of feasible and successful derivative technologies b of a technology $\overline{b}$ as a measure for its size, its success, and possibly also the revenue a commercial vendor would be able to extract from a commonly accepted standard of this technology. The distribution of this variable $p(b)$ is thus of special significance.

Assume, contrary to other approaches (like Kaufman *et al.*'s 'technology landscapes' (Kauffman *et al.*, 2000), or Silverberg and Verspagen's 'perlocation model' (Silverberg and Verspagen, 2005, 2007)), an infinite space of theoretically possible derivative technologies to any technology. However, let the technology tree unfold iteratively and according to a very simple process. Every technology has the same probability of gaining a new directly derivative technology, i.e. not a derivative of a derivative technology but a direct offshoot. Hence, as the size of derivative industries, b, of a technology $\overline{b}$ increases (compared to the remainder of the system), so does its likelihood of gaining a new derivative technology (now including derivatives of derivative technologies, etc.). This may be seen as the representation of research-side network externalities. The more successful research has been conducted in some specific part of the technological system (e.g. a subgraph of the original technology tree that only encompasses a specific technology and its derivatives), the more likely is further innovation in this part.

Suppose that the system is initialized with just one technology at time $t = 1$ and that only one technology is added at a timestep (which means that the total size of the system at time t is equal to t). Note that this process is similar to the Yule process, which describes the increase of genetic diversity and is known to produce beta distributions which in turn behave like power law distributions for large values (Newman, 2005). The following mathematical analysis will follow the algorithm Newman used to discuss the Yule process (Newman, 2005: p. 340).

Define $m_{b,t}$ as the number of technologies with exactly b derivative technologies (including itself as defined above) at time t. For any particular technology $\overline{b}$ belonging to this group, the probability of gaining another derivative technology at time $t+1$ is $\frac{b}{t}$ since there are t technologies currently in the system, b of which are derivative to $\overline{b}$ and therefore they and their possible future derivative technologies contribute to the size of the network of directly and indirectly derivative technologies of $\overline{b}$. Hence

$$b_{t+1}(\overline{b}) = \begin{cases} b_t(\overline{b}) + 1 & \text{with probability } \frac{b}{t} \\ b_t(\overline{b}) & \text{with probability } \frac{t-b}{t} \end{cases}. \tag{5.24}$$

Furthermore, the probability that one of the technologies with exactly b derivative technologies (to which $\overline{b}$ belongs) will gain one further derivative at time t is $\frac{b}{t} m_{b,t}$ (since there are $m_{b,t}$ such technologies). This is also the expected value of how much the variable $m_{b,t}$ decreases because technologies with *subtree size* (of the network of derivative technologies) $m_{b,t}$ gain another derivative and increase that number to $b+1$. At the same time, however, the technologies with a smaller value $b-1$ face the probability $\frac{b-1}{t} m_{b-1,t}$ to extend their network of derivative technologies and thus increase to $m_{b,t+1}$. Hence, the development equation for $m_{b,t}$ (for $b > 1$) may be written

$$m_{b,t+1} = m_{b,t} - \frac{b}{t} m_{b,t} + \frac{b-1}{t} m_{b,t} \tag{5.25}$$

So far, the development equation is written as an equation of the absolute frequency of technologies with subtree size b, $m_{b,t}$. However, for convenience it shall be rewritten with the relative frequency

$$p_{b,t} = \frac{m_{b,t}}{t}, \tag{5.26}$$

hence,

$$(t+1)p_{b,t+1} = tp_{b,t} - bp_{b,t} + (b-1)p_{b-1,t}. \tag{5.27}$$

The case $b = 1$ is an exception since, with every period, one technology is added with certainty, hence equations (5.25) and (5.27) become

$$m_{1,t+1} = m_{1,t} - \frac{1}{t}m_{1,t} + 1 \tag{5.28}$$

and

$$(t+1)p_{1,t+1} = tp_{1,t} - p_1 + 1. \tag{5.29}$$

Thus, the system is completely defined by equations (5.27) and (5.29). Now, let $t \to \infty$ and consider the steady state of the system in relative frequencies

$$p_{b,t} = p_{b,t+1} \tag{5.30}$$

This steady state, if existing, denotes the time-independent distribution $p(b)$ in this model that is approached when t is sufficiently large and, in turn, the distribution of the impact of technologies on the socio-economic system this model predicts. Equation (5.29) becomes (index t eliminated since it is meaningless for the steady state)

$$(t+1)p_1 = tp_1 - p_1 + 1 \tag{5.31}$$

and resolved for p_1

$$p_1 = \frac{1}{2}. \tag{5.32}$$

Equation (5.27) becomes (again, time indexes eliminated)

$$(t+1)p_b = tp_b - bp_b + (b-1)p_{b-1} \tag{5.33}$$

which, resoved for p_b yields

$$p_b = \frac{b-1}{b+1}p_{b-1}. \tag{5.34}$$

Substituting equation (5.34) for $b-1$ into itself gives an equation of p_b as a function of p_{b-2}; after repeating the substitution until p_1 is reached, the equation becomes

$$p_b = \frac{2(b-1)!}{(b+1)!}p_1 = \frac{2(b-1)!}{(b-1)!b(b+1)}p_1 = \frac{2}{b(b+1)}p_1. \tag{5.35}$$

Substituting the solution (5.32) for p_1 in equation (5.35) resolves the system as

$$p_b = \frac{1}{b^2 + b}. \tag{5.36}$$

The distribution of the technology subtree sizes follows accordingly

$$p(b) \sim \frac{1}{b^2 + b} \tag{5.37}$$

which in the tail of the distribution (that is, for sufficiently large values of b) approaches a power law distribution with exponent $\alpha = 2$

$$\widetilde{p(b)} \sim b^{-2}. \tag{5.38}$$

The described model is simulated using a Python simulation program (source code available from the author on request) for 60,000 technologies. The results resemble the analytically derived distribution functions (5.38) and (for values of about $b > 5$) also (5.37) closely. (See Figure 5.11.)

It is not the intention of the current section to claim that real-world technology's influences follow this model exactly. Rather, the model should be seen as a very simplified representation of technological development, especially with respect to the interrelations of technologies and research-side network effects (as explained above) in this process. It predicts a highly skewed distribution which in its tail is most likely also scale free[27] (i.e. a power law distribution). Still, it is highly doubtful whether the impact of a technology on the socio-economic system or the technology subtree size can be measured that cleanly. Real-world technologies do not show such a clear hierarchy with technologies being derivative to exactly one other technology, but are part of a complex and changing system. Empirical studies suggest, however, that variables related to such a value as the technology subtree size are indeed highly asymmetric and often do follow power laws. First and foremost, firm size, which may partly follow the success the enterprise had with a technology developed there, is highly skewed (Axtell, 2001; Newman, 2005; Andriani and McKelvey, 2007; Cordes *et al.*, 2011) and is said to follow a power law (Axtell, 2001). Different explanations have been suggested for this phenomenon (among others, the greater flexibility of small firms and the costs of opportunistic behaviour in small or large groups and firms (Cordes *et al.*, 2011)). The current section argues that research-side network effects in connection with the development of technology trees of derivative technologies in times of rapid growth may be another approach to this phenomenon. Other related variables that have been analyzed to follow power laws are the size of industrial agglomerations (Andriani and McKelvey, 2007) as well as social agglomerations (Newman, 2005) and social networks and organizations (Andriani and McKelvey, 2007).

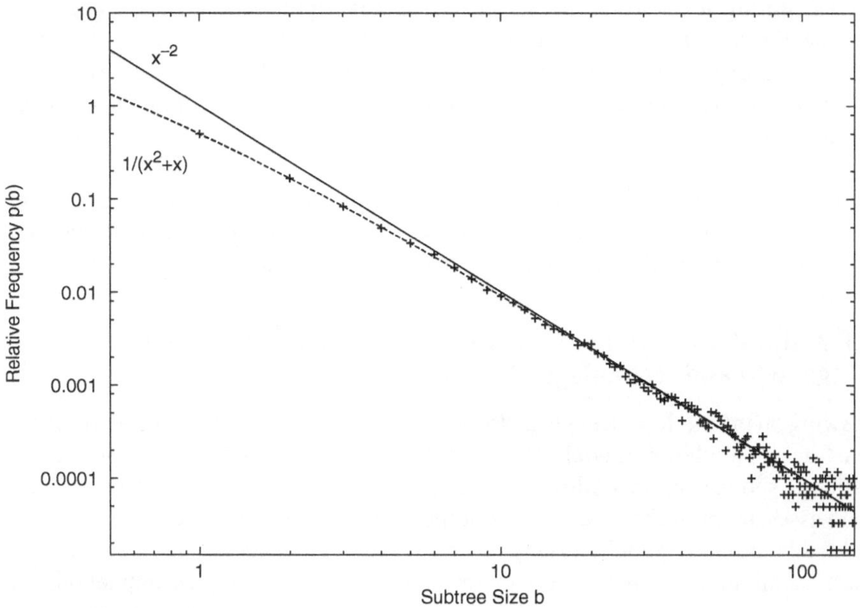

Figure 5.11 Distribution of subtree sizes in a technology tree with 60,000 technologies (scatter plot). Additionally, the distribution functions $p(b) \sim \frac{1}{b^2+b}$ and $\widetilde{p(b)} \sim b^{-2}$ are shown.

5.2.5 Towards a theory of microfoundations in models of technological change

Several aspects of the micro–macro relation in economic systems with network externalities and especially with respect to growth processes and technological development have been discussed in this section. Four particularly interesting aspects were studied in detail:

1. Sectors with network externalities that alter between periods of slow and rapid growth lead to path dependence at an aggregated level.
2. The macro-level of such a model may sustain persistent stable patterns, also called dissipative structures that emerge from the complex micro-level.
3. The expected period lengths between transitions to slower or more rapid regimes of technological change are highly skewed and possibly power law distributed. They may further be coordinated between sectors (especially interdependent sectors), which at an aggregated level leads to a structure similar to a naturally occurring moving average smoothening. With the Slutzky–Yule effect, this leads to cycles at the macro-level that may at least in part explain cycles in macroeconomic data (i.e. business cycles or also long waves if they do indeed exist).

4. The 'size' or impact of successful technologies on the socio-economic system (assuming it could be measured reliably) is highly skewed and the tail of the distribution may be scale free (approaches a power law). This may help to explain structural asymmetries in data like firm size, sizes of social and industrial agglomerations as well as sizes of organizations and social networks.

With these results established for the analysis of the micro-to-macro transition in growth and technological change in systems with network externalities, the following section will focus on the sector-level dynamics.

5.3 A simulation study of the microfoundations of economic growth and technological change

Drawing on the model discussed in section 5.1 and its implications, this section will construct a simulation model that internalizes effects of direct interdependence. As discussed earlier, this allows for an integration of micro-level effects as well as an assessment of the exact results of the involved network externalities.

At this point it should briefly be asked again (since it was discussed in more detail in the introduction to this chapter) which empirical phenomena are to be modelled. Real-world growth data show superlinear growth with persistent fluctuations, apparently alterations between slower and faster growth (that have been suggested to be in fact changes of the technology regime). The technological base of this economic system co-evolves, also undergoing transitions between phases of slow and rapid innovation, and the usual industry structure is oligopolistic.

The section will start by assessing the mechanisms suggested in the literature as possible explanations for this picture and continue by proposing a different mechanism based on network externalities, specifically the model considered in section 5.1. Accordingly, such an agent-based simulation model will be constructed and analyzed.

5.3.1 The explanatory range of pure macro-level models

The pure macro-level models of the Solow–Swan type, AK, EGT, or other systems similar to Harrod–Domar and related model classes are generally incapable of generating non-monotonic dynamics. The main variations are either the convergence towards a stable steady-state as in Solow's (1956; 1957) and Swan's (2002 [1956]) models, or monotonic steady growth with constant or converging growth rate as in the models of endogenous growth.[28] (Uzawa, 1965; Cass, 1965, 1966; Romer, 1986; Lucas, 1988) Stochastic perturbations may lead to temporal disturbances of the dominant growth regime as suggested by Lucas (1972, 1975, 1981) but do not produce lasting effects and produce much less persistent circular or otherwise disequilibrium patterns since the dominant growth regime is stable and the dynamics will simply fall back to the stable path[29] (Chen, 2002).

A deterministic system with inherent instability to generate cycles or oscillations requires at least one complex eigenvalue.[30] Such a differential or difference equation system must be at least a second-order system[31] – or equivalently a system with two equations – but is not difficult to construct. A simple linear system with the product of the elements of the main diagonal being negative

$$\prod_i j_{i,i} < 0$$

($j_{i,i}$ denoting the ith main diagonal element of the Jacobian matrix or transformation matrix $\mathcal{J}$) is sufficient to ensure circular dynamics. If, furthermore, the dynamics are not contracting,[32] the cycles do not collapse either and are sustained indefinitely. This is, in essence, what Samuelson contributed to growth theory with his 1939 model (Samuelson, 1939a,b). His model gives a two-dimensional recurrence equation system using a Keynesian macroeconomic interpretation. Depending on the parameters, the model is either expanding or contracting, and either with or without cycles. If there are cycles, however, they are perfectly regular and stable. Though the model is elegant and the approach is insightful in its interpretation, it seems oddly stiff from today's perspective with its broader understanding of more elaborate models with an explicit inclusion of the micro-level.[33]

The progress of dynamic theory sparked further contributions relying on increasingly complex models. The start had been made with the *Samuelson oscillator* (Samuelson, 1939a,b) and some vaguely dynamic approaches to growth theory (Harrod, 1939; Kaldor, 1940) in the early era of Keynesian thought. While the mainstream of growth theory subsequently set out to prove that even growth processes are an equilibrium situation (starting with Solow (1956, 1957)), Keynesian, Marxian, Post-Keynesian and Kaldorian economists added an economic *Lotka–Volterra predator–prey system* (Goodwin, 1967), chaotic systems with strange attractors (Lorenz, 1987) and complex models with periods of stability and catastrophic collapses (Keen, 1995). However, all of these mechanisms are limited to continuous flow along relatively simple trajectories of relatively simple systems defined by relatively few macro-level equations. Though perturbations may add further aspects, they are limited to at most shifting the system between different attractors.

Agent-based modelling on the other hand, adds even for rather simple settings (consider the Nelson–Winter system (Nelson and Winter, 1974; Nelson *et al.*, 1976; Nelson and Winter, 1982a)) dissipative structure and emergent stability that may last for a while and become unstable again. These dynamics rely on system of much higher dimensionality than pure macro-level settings and are capable of generating very different dynamics. Furthermore, they are arguably much closer to economic reality by not just considering some abstract or even arbitrary macro-level but starting from the micro-level and investigating whether or not aggregated-level structures emerge.

5.3.2 Simulation models and evolutionary growth theory

The single most central mechanism used in evolutionary growth theory is Nelson and Winter's (1974; 1976; 1982a) formalization of the Schumpeterian entrepreneurship motivated by temporal efficiency advantages if not monopoly rents that result from successful innovation. (See section 3.4.1.) A setting with several firms is considered, with each of them characterized by their technological efficiency (or simply their technology level). The firms are allowed to engage in innovative research (resulting in true innovation, a better technology, if successful) and imitative research (copying a competitor's better technology) according to their available capital. The mechanism works by setting the success-probability of innovative research relatively low compared to imitative research. Hence, truly new technologies are implemented rarely, giving the first-mover a considerable advantage; imitators follow, swiftly collapsing the initial advantage of the first-mover. The first-mover retains the revenue collected in the time between her innovation and the imitators eventual catch-up in the form of increased capital that may, of course, be used to increase the chances of her future research. The model may contain regular slow growth based on the expansion of available capital. Periods of fast growth correspond to the catch-up process of the imitators. A single first-mover does not increase the economy's performance very deeply; however, as the new technology diffuses, the aggregated effect is strong. As soon as most of the firms have completed their catch-up, the growth rates return to normal levels of slow growth until the next innovation is accomplished.

This mechanism was discussed briefly in section 3.4.1 in the context of Nelson and Winter's growth model and evolutionary growth theory. It is applied to a very basic model of production quite similar to the one considered in section 5.1 above, but without employing a vintage system of construction and depreciation of plants – instead, capital is another input of the production equation and the state of the art technology can be employed immediately as soon as it is invented for the full scope of the production. As the model is agent-based, the equations are defined for the micro-level. An agent (firm) i using technology level $(a_{L,i}, a_{K,i})$ with labour productivity term $a_{L,i}$ and capital productivity term $a_{K,i}$ as well as capital k_i produces according to

$$y_i = \frac{k_i}{a_{K,i}} \tag{5.39}$$

and requiring a labour force of

$$l_i = \frac{a_{L,i}}{a_{K,i}} k_i. \tag{5.40}$$

The idea of defining the technology level as a two-dimensional term may serve the purpose of making the model more realistic and more easy to give an economic interpretation. It fails, however, to add any new aspect to the model. Research works to decrease both terms to increase labour or capital productivity, respectively. In effect, the capital productivity term allows production of larger quantities while the labour productivity term allows production of the same quantity at lower cost. Both values are monotonic; if they were not, the two-dimensionality could introduce diversity of technologies in use with a now complex technology landscape. However, as they are, the most successful firms will just be those that are able to attain a balanced increase of both capital and labour productivity (a balanced decrease of $a_{K,i}$ and $a_{L,i}$). The model's dynamic properties remain intact if the technology level is reduced to a one-dimensional term, most conveniently by setting $a_{K,i} = 1 \quad \forall i$. This transformation will be used below to study the Nelson–Winter mechanism in a less complex form.

Nelson and Winter's model also has two different mechanisms to introduce scarcity.[34] On the one hand, labour is limited either absolutely (with an amount of work hours that are available to be employed in production or similar[35]) or relatively, with the wage level increasing with employment (a Phillips curve) and hence a declining revenue function. On the other hand, demand may be limited again either absolutely (a maximum level of consumption, thus maximum value of sales) or relatively, with the demand function decreasing in price. Again, one of these mechanisms suffices to make the mechanism work.

Nelson and Winter further define research as depending on available capital. This shifts the dynamics of the model just to tend to faster monopolization since with this assumption, capital-rich firms have a decisive advantage in inventing and operating newer and better technologies. The same is true for the capital efficiency parameter $a_{K,i}$ since successful firms would not just produce more cheaply but also more. It will be shown below why this monopolization effect may not be desirable and may even reduce a part of the potential of this very powerful and simple mechanism.

Reducing the required scarcity in the model to scarce labour (with a Phillips curve), and the technology level reduced to labour productivity, that is, in this case, the labour–output ratio $a_{L,i}$ the equations above are reduced to

$$y_i = k_i \tag{5.41}$$

and

$$l_i = a_{L,i} k_i. \tag{5.42}$$

The demand side is thus not part of this simplified version of the model. Furthermore, research may be reduced to a simple stochastic term for both innovative research and imitation (with attempts to imitate succeeding with a higher probability than innovative research).[36]

To investigate the behaviour of the model, it is simulated for 128 agents and for 2200 time periods (source code available from the author on request). Figures 5.12 to 5.17 show the dynamics produced by the Nelson–Winter mechanism both with respect to growth effects and the industry structure. In Figures 5.12 and 5.13 – that were already discussed above in section 3.4.1 – it can be seen that the model initially resembles the expected behaviour very closely: growth

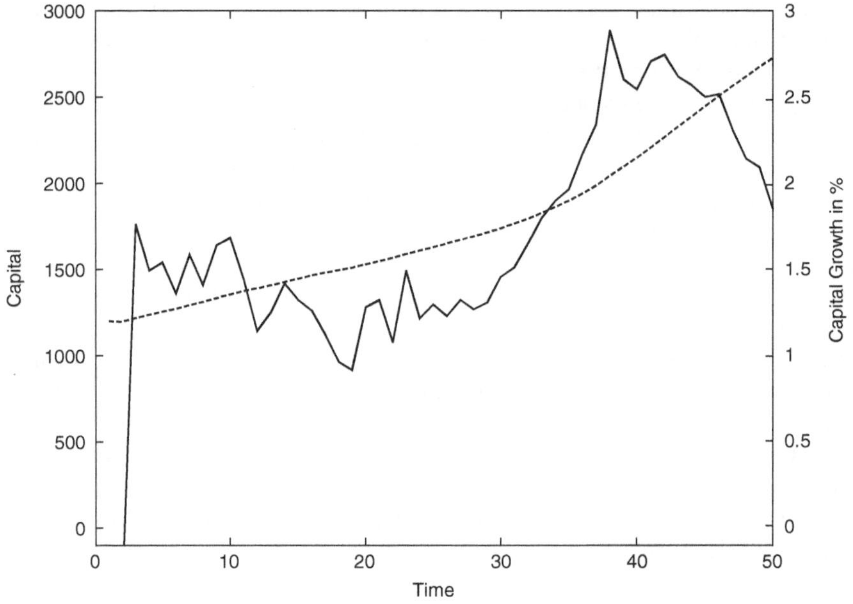

Figure 5.12 Capital growth (output grows proportionally) as generated by the Nelson–Winter mechanism using a Python simulation. The dashed line indicates absolute capital (left scale; unit values arbitrary, starting with 1200.0), the growth rate (right scale) is given by the solid line; first 50 iterations.

is present, it is superlinear, growth rates are oscillating smoothly and are not converging to zero as in linear growth. The oscillations lead to a circular behaviour between slow and fast growth. In fact, the alternation of the growth rate involves an obviously stochastic influence, since fast growth is only made possible by successful innovative research which is stochastically irregular and non-periodic. The transition between regimes of slow and fast growth is further gradual and without large sudden turbulence. The concurrent evolution of the industry structure shows the emergence of an asymmetric industrial organization as the increasing normalized Herfindahl index[37] (depicted in Figure 5.13) indicates. This might not be unrealistic; in the later development of the current model it does, however, prove to be a dubious blessing. Figures 5.14 to 5.17 show the development of the model for a longer time, 500 (Figures 5.14 and 5.15) and 2200 (Figures 5.16 and 5.17) iterations compared to just 50 iterations in Figures 5.13 and 5.13 discussed above. As the monopolization of the industry progresses rapidly, the pattern of growth and development changes. As may be seen in Figures 5.14 and 5.15, there is no smooth transition between higher and lower growth rates but rather occasional steep increases in productivity, and thus output, while normal growth rates are negligible. In fact, if there is just one large firm in the industry alongside with several very small vendors, growth occurs always and only in the period when a

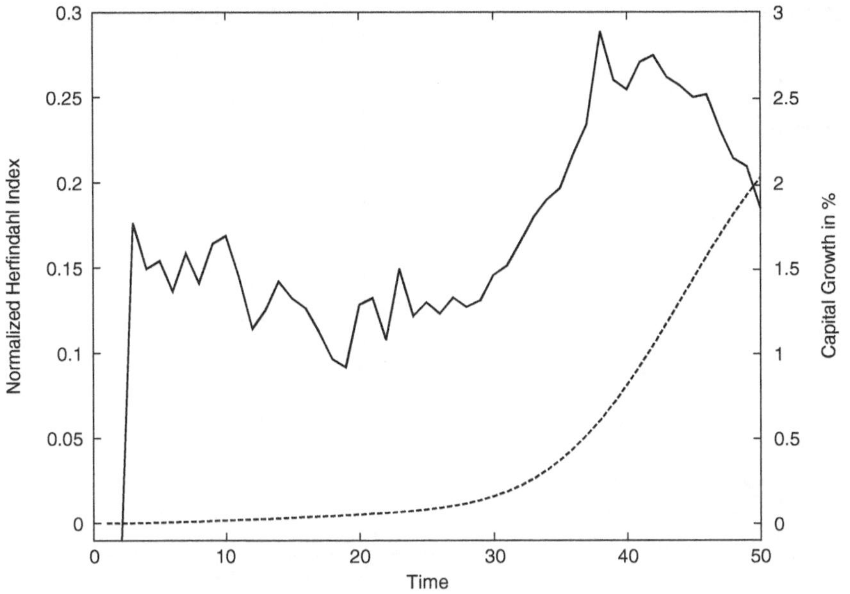

Figure 5.13 Growth rates (capital growth) as generated by the Nelson–Winter mechanism using a Python simulation. The solid line gives the growth rate (right scale) while the dashed line shows the normalized Herfindahl index of the industry (left scale); first 50 iterations.

better technology becomes available for the large firm whether by innovation or by imitation. This pattern continues and the monopolization is eventually completed, the normalized Herfindahl index approaching 1 as seen in Figures 5.16 and 5.17. The effects neglected for this simplified version of the model, specifically price competition at the output market and research depending on capital, would tend to increase this tendency toward monopolization.

It should, however, be noted that Nelson and Winter were not only interested in economic growth but also in a wide variety of other effects. They developed a powerful device that does produce evolutionary growth patterns for a number of turns and which was a large step towards proper microfoundations of growth theory. They then turned to study other special aspects of this model in order to establish a comprehensive Schumpeterian theory of economic growth and innovation that is very different from the neoclassical growth theory that was dominant at the time (and arguably still is today) (Nelson and Winter, 1982a).

The parameters of Nelson and Winter's mechanism may be adjusted to extend the period for which the model behaves smoothly and before complete monopolization occurs. However, it is common for not just Nelson and Winter's model but also for other evolutionary growth models that the industry structure

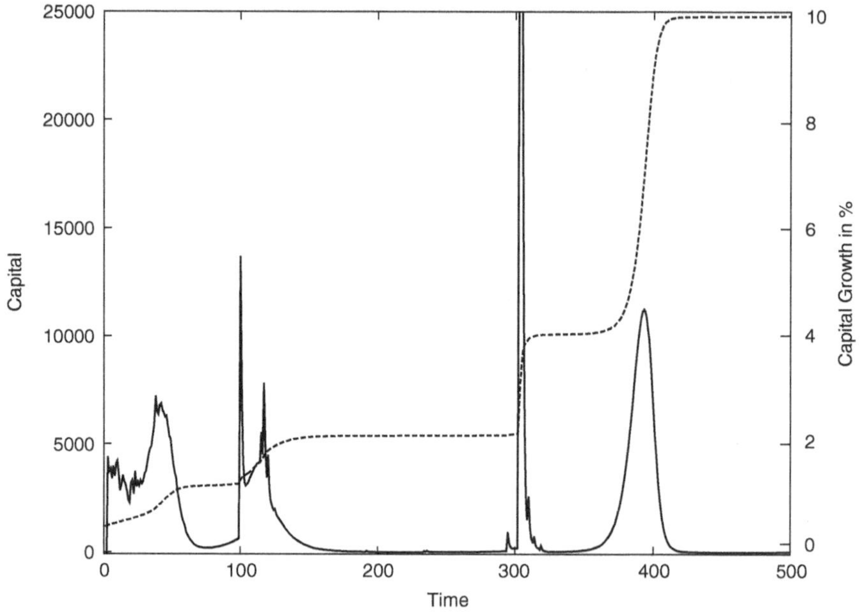

Figure 5.14 Capital growth (output grows proportionally) as generated by the Nelson–Winter mechanism using a Python simulation. The dashed line indicates absolute capital (left scale; unit values arbitrary, starting with 1200.0), the growth rate (right scale) is given by the solid line; first 500 iterations.

tends towards monopolization and the dynamics of the models change at some point. (See e.g. the models by Silverberg and Lehnert (1993, 1994).)

5.3.3 Central aspects of the Nelson–Winter mechanism: inertia and openness

The increased revenue caused by comparative advantages through better technology provides the core of Nelson and Winter's Schumpeterian theoretical foundation of the mechanism (Nelson and Winter, 1974, 1982a). However, Nelson and Winter's mechanism works also without including a capital-stock and a revenue and re-investment dynamic. The smooth transition between slow and fast growth regimes is generated merely by significant technological improvements made by one party and the subsequent diffusion of the new technology throughout the industry. Removing the revenue and re-investment part certainly leaves the model reduced to a mere shell, a mechanism that lacks its theoretical justification. However, it may also be argued that adding a – admittedly common and easily conceivable – revenue and re-investment loop provides just one possible context for the mechanism that is as good as any other. Instead of exploring different possible interpretations of the Nelson–Winter mechanism, it shall suffice to discuss

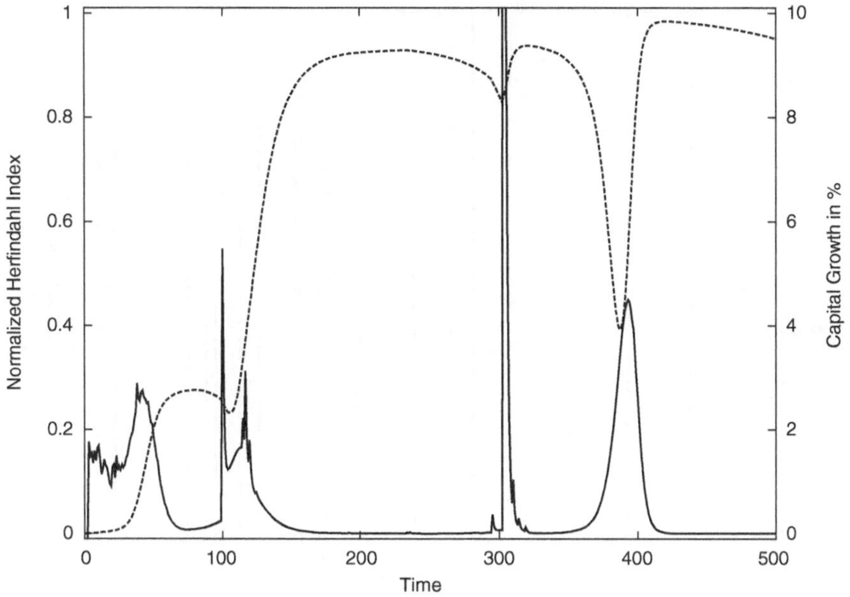

Figure 5.15 Growth rates (capital growth) as generated by the Nelson–Winter mechanism using a Python simulation. The solid line gives the growth rate (right scale) while the dashed line shows the normalized Herfindahl index of the industry (left scale); first 500 iterations.

the bare mechanism and its properties, and, with some instructive conclusions, return to the task of constructing another device based on network effects and the model introduced in section 5.1 above.

The breakdown of smooth transitions in Nelson and Winter's model after an initial period is caused by the quick and accelerating increase of the monopolization (measured using the normalized Herfindahl index) in the sector. This in turn is brought about by vast differences in revenue between successful and less successful firms and the effect of this inequality on research, etc. Not including this in the model leaves the mechanism of technological transition intact. What remains is a stochastic process for innovative research with rather low success rates and the swift though not immediate diffusion of the technology caused by imitation processes with much higher likelihood of success. In fact, such a device was used to generate the sector level time series in section 5.2.3.

That the diffusion of the new technology is caused by imitation is another illustrative interpretation of the mechanism. What is necessary to ensure a smooth transition is merely a certain persistence of the established technology, and that the new technology is readily available, though possibly after a time lag or just for some firms or with just with a certain probability. The Nelson–Winter mechanism guarantees the former requirement by limiting innovative

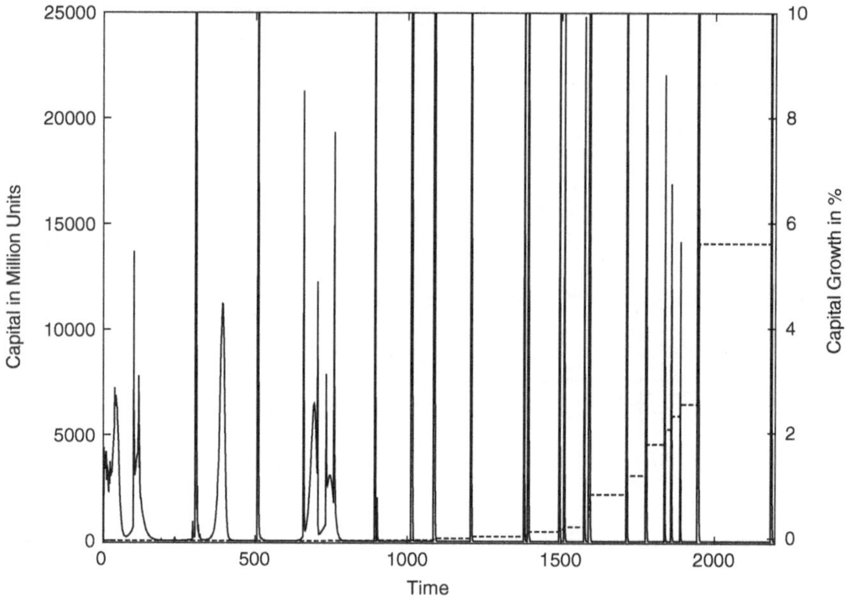

Figure 5.16 Capital growth (output grows proportionally) as generated by the Nelson–Winter mechanism using a Python simulation. The dashed line indicates absolute capital (left scale; unit values arbitrary, starting with 1200.0), the growth rate (right scale) is given by the solid line; entire simulation including all 2200 iterations.

technologies to exclusive use for the innovating firm at first. The latter is ascertained by allowing imitation (with a certain probability) by any other firm willing to engage in 'imitative research' without any means for the innovator to protect knowledge of the technology.

The first requirement – a certain inertia of the industry when adopting technological improvements – may also and quite illustratively be attained by using a vintage model. Production relies on physical facilities (in recent times maybe also virtual facilities in the form of digital social networking, etc.) that must be replaced or at least adapted to employ a new technology which in turn takes time. Conlisk (1989) used such a model; the model above in section 5.1 has been formulated in this way and the simulation model in the following section will also follow this structure.

The swift diffusion of a new technology may also be described as openness of such a technology. This requirement is by no means limited to the Nelson–Winter mechanism; in fact, research is done faster if results are shared and may be relied on by other entities for further improvement. In a way, this introduces an implicit research-side network externality: if findings are openly or more or less openly available for all agents, the expected rate of improvement increases with

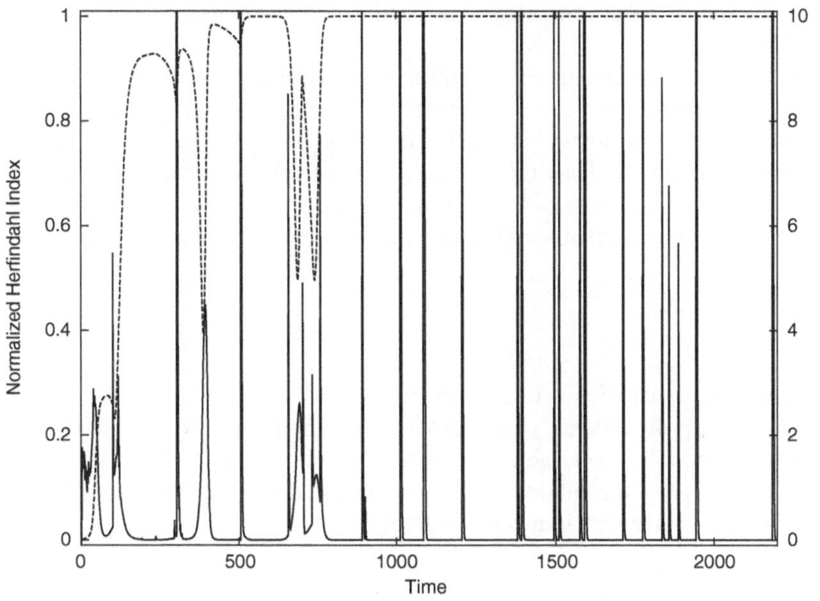

Figure 5.17 Growth rates (capital growth) as generated by the Nelson–Winter mechanism using a Python simulation. The solid line gives the growth rate (right scale) while the dashed line shows the normalized Herfindahl index of the industry (left scale); first 2200 iterations.

the number of agents. If research on the other hand is strictly excludable (firms can protect their technologies from imitation), the technological progress happens for each firm separately and generally slowly. The sufficient openness of new technologies does, however, not have to take the form of stochastic imitation. It may also be an explicit openness, as open standards or open source technologies[38] that are available to the entire industry. In reality, not all agents may be inclined to make their findings openly available (but some are, such as public institutions). A relatively small share of agents committed to openness will, however, be sufficient to keep the mechanism functioning, especially as they benefit from the above-mentioned research-side network externality among each other, and are thus able to conduct successful research much faster than the rest of the industry – though the rest of the industry will still benefit from their open research findings.[39]

5.3.4 Simulation of growth: the benchmark simulation model

This section discusses a simulation model constructed along the lines of the model of growth and technological change in section 5.1. The theoretical model was constructed specifically to study network externalities. However, this will not immediately be integrated into the simulation model. First, the general dynamics of the model shall be analyzed to provide a benchmark for the later assessment of network externalities; then, as a second stage, the model will also include network externalities.

In the following, the details of the simulation model are discussed. Note that this follows the model of section 5.1 closely. The simulation programs (written in Python; source code available from the author on request). The following paragraphs are merely the description of the simulation program; equations should be seen as the illustration of how the program works, not as a mathematical model (since the underlying mathematical model has been elaborated and discussed in section 5.1).

FIRMS

Consider an industry consisting of a number of firms,[40] the specific firm (or similar entities) identified with index i. This number is for simplicity kept constant in time t. Any firm commands a – variable – number of production plants $n_{i,t}$ denoted by index $j = 1, ..., n_{i,t}$, has a capital stock $k_{i,t}$ available at time t, and the technologies at its command at time t are indicated by a set $\Theta_{i,t}$.

TECHNOLOGIES

Θ_t is the complete set of operational technologies at time t; most likely none of the i firms has all these technologies Θ_t at its command; rather the technologies available for the firms are subsets $\Theta_{i,t} \subset \Theta_t$ (though theoretically, the subset may encompass the entire set Θ_t). Any technology in the set Θ_t, $\theta \in \Theta_t$ is defined by (1) its standalone labour–output ratio $a(\theta)$ and (2) its technology type $TEC(\theta)$. As long as network externalities are not included in the model, the technology type is meaningless. The labour–output ratio allows production of an amount of output $\overline{y}$ with labour input $l(\theta) = a_\theta \overline{y}$.

For the later setting below, which includes network effects, let $\widetilde{a}(a(\theta), TEC(\theta))$ denote the effective labour–output ratios including changes caused by network effects. (For the setting without network effects, let $\widetilde{a} = a$.)

PRODUCTION PLANTS

The firms start without any production plants at time $t = 0$, hence $n_{i,0} = 0 \quad \forall i$. However, production plants may be constructed at a – fixed – cost of c_{Plant}. Firms have sufficient funds to immediately construct one plant (such that the firm will subsequently be able to generate revenues), $k_{i,0} = c_{Plant} \quad \forall i$.[41] Furthermore, it is assumed that demand is unlimited, i.e. the industry under consideration is merely a part of the economic system and other parts of the system generate sufficient wealth to pay for and consume any possible output of the industry in question. Hence, firms will always try to expand their production capacity by generating as many new plants as they can afford. The plants are built using the owner firm's best currently available technology (i.e. that with the best standalone labour–output ratio a in the case without network effects, and otherwise the technology with the best total labour–output ratio after including changes caused by network effects $\widetilde{a}$). It is assumed that they do not save for building better plants at a later time. Plants decay over a time of $t_{Plant\ \ TTL}$[42] after which the plant is rendered unusable; for the time before this, the plant is operational and may be used in production if there is sufficient labour to operate it. The output is assumed to be proportional to the revenue; it consists of a revenue part and a wage part, with the ratio of the two (revenue to wage) remaining constant at all times for the sake of simplicity. The output of the plants is normalized to produce a revenue of 0.25 at any time the plant is active. That is, suppose all $n_{i,t}$ plants of a firm i are operating at time t, firm i earns a revenue (equal to the increase in the stock of available capital of that firm) of $k_{i,t} = \frac{1}{4}n_{i,t}$.

LABOUR

Labour is limited at level L_0, which for simplicity is assumed to remain constant.[43] At any point in time, labour is allocated efficiently to operate the most efficient plants of the industry. Consequently, a – possibly empty – subset of the plants of every firm i, i.e. $n_{i,t,active} \leq n_{i,t}$ is active at any time t. This subset tends to be (absolutely, not necessarily relatively) larger, the smaller the labour–output ratios of the available technologies are, and, in other words, the better the technologies in use are.

PRODUCTION, OUTPUT AND CAPITAL

Consequently, the industry produces at time t according to $\sum_i y_{i,t} \propto \sum_i k_{i,t} = \frac{1}{4}\sum_i n_{i,t,active}$.[44] Knowing that $\sum_i y_{i,t}$ is proportional to $\sum_i k_{i,t}$, it will be sufficient to keep track of $k_{i,t}$ in the simulations below. Note that if some firm gains a comparative advantage by having better technologies available than the competitors, thus generating greater revenue, the firm will be able to make larger investments and further increase its revenue. Analogous to Nelson and Winter's (1982a) (and also Silverberg and Lehnert's (1993; 1994)) models, monopolization is possible and not unlikely; therefore, the normalized Herfindahl index of the current industry structure will also be measured in the simulations below.

RESEARCH

All firms i engage in innovative research (and innovative research only) which succeeds with a probability $p = 0.1$. Further, innovative research is assumed to be cumulative; any technology is first assigned a random improvement value $\rho \sim \sqrt[32]{Uniform(0, 1)}$, which is most likely close to one. (See Figure 5.18.) Then, for any other technology known to the innovator firm $\theta \in \Theta_{i,t}$, there is a probability (0.5 in the simulations) that the new technology is compatible with this technology, which would generate a combined standalone labour–output ratio (i.e. technology

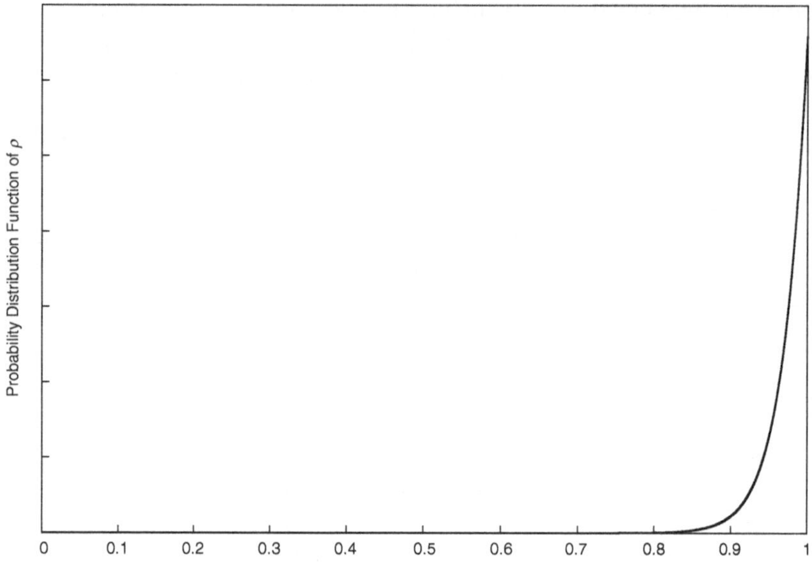

Figure 5.18 Probability distribution of ρ.

level) of $\overline{a} = \rho \times a(\theta)$. Since both ρ and $a(\theta)$ are smaller than 1, the result $\overline{a}$ is even smaller, though not by much, than $a(\theta)$ since ρ is by definition close to 1. In theory, there could be as many possible $\overline{a}$ values as there are technologies known to the firm; the firm chooses the smallest among them $min(\overline{a})$ and integrates the resulting technology into the set of the known technologies, its standalone labour–output ratio now being $a(\theta_{new}) = min(\overline{a})$. The success probability is considerably higher than in both Nelson and Winter's models and the simulations in the previous sections of the current chapter. Note, however, that the innovation is combined with at most one previously known technology;[45] hence progress is still very slow but less unsteady than in Nelson and Winter's mechanism.

In the setting with network externalities to be considered later in section 5.3.6, the technology type of the new technology is randomly drawn and does not depend on which technology the innovation may have been combined with (since backward compatibility is not necessarily given in reality either).

OPEN TECHNOLOGIES

Furthermore, it is assumed that some of the actors (or firms) i make their technologies always openly available. In that case, any technology developed by such an actor would immediately also become part of the set of available technologies of all the other firms. The share of firms inclined to develop 'open standards' will be varied in the simulation below to study the effect of openness, as well as the edge cases with no open technologies and with 100% open technologies.

Performance, output variables, and expected behaviour of the model

The current simulation model differs from those discussed in the previous sections in that it is much more extensive, complex and costly in terms of both computation time, processor speed, and memory storage requirements as well as programming and program structure. Generally, the parameters for the simulation runs have been chosen to both be economic in performance and yield robust results. The latter is achieved by:

- Running the simulation for a sufficiently large number of firms, though the 128 used in the simulations are not exactly large by standards that would apply for professional equipment. Considerably larger population sizes would have either increased the computation time or limited the flexibility to set other parameters to settings that cause larger system requirements.
- Running the simulation for a sufficiently large number of iterations. This is, however, limited by other parameter settings. For example, for runs with 100% open technologies (see below), the quick technological progress causes both the number of technologies and the output and subsequently the number of production facilities (plants) to rise to large values.[46] Accordingly, the simulation has been pursued for a different total number of iterations for different settings of the remaining parameters.
- Conducting the same simulation run several times with the same parameter settings to average out stochastic influences. This is particularly useful if path dependence is involved: the simulation might behave very differently for different runs though the initial settings had been identical. In this case, with a sufficient number of simulation runs, the distribution of the outcomes, the interdependence of variables during the progress of the simulation, and their effect on the (path dependent) outcome may be identified.

In the course of the simulation, the program keeps track of and records the development of a number of variables, categorized into variables concerning (1) output and growth, (2) technological progress and (3) industry structure: capital (output and growth variable), labour demand and (as labour supply is fixed) the employment rate (output and growth variable), best, worst and average *employed* technology; i.e. best, worst and average labour–output rates of the currently active production facilities (technological progress variable). For the setting with network effects, these variables include the network effect. Furthermore, it records the best available open technology (a technological progress variable), the normalized Herfindahl index of active firms.[47] Finally, the industry share of the largest competitor is measured; again this variable develops almost parallel to the normalized Herfindahl index (industry structure variable).

As is not difficult to see, the technology level rises (the labour–output ratio decreases) over the course of a simulation since innovations occur with positive probability. Furthermore, as open technologies are available for every firm, including those not contributing to those open technologies, it is reasonable to expect more rapid progress and faster output growth the more open technologies are being developed. The simulation runs have yet to answer the question whether the contributing firms will simply be driven out of the industry by their competitors or not (see below).

While the theoretical number of firms remains constant over all iterations,[48] the number of production facilities (plants) rises at first and than takes a course that is determined by the size of the firm's revenues, and the number of available technologies increases steadily. The revenues and consequently also the capital and the output follow the development of the number of plants by definition.

Finally, the industry structure measures may be expected to show an initial development towards a slightly asymmetric industry structure (since the system starts with perfectly equal industry shares). The further development is not clear and could depend on initial values (the share of open technology producing firms, the presence or absence of network externalities) as well as stochastic influences if the system is path dependent.

The questions to be answered by the simulation runs conducted below, are therefore:

1. How does the presence or absence of open technologies change the development of the simulation – in terms of output data, technology data and industry structure data?
2. How fast does the system grow in terms of output variables? Are any random perturbations, circular patterns, or otherwise noticable structures identifiable? Are patterns of economic growth as observed in the real-world matched by the simulation?
3. Under which conditions does a monopolization of the industry occur?

The current simulation will be followed by a very similar simulation study discussed at the end of this chapter which introduces technology types and labour-saving network effects as theoretically analyzed in section 5.2. Besides the obvious increase in efficiency, this modification will (as discussed in section 5.3) result in different structural properties, growth patterns and macro-behaviour. The current benchmark case will serve as a point of comparison for those findings.

5.3.5 A benchmark for the simulation: the behaviour of the model without network externalities

Figures 5.19 to 5.21 depict the development of simulation runs of the specified simulation model for different shares of firms committed to the development of open technologies. Capital revenue (and proportional to that the output) is indicated by a bold black curve (which is not true to scale). Lowest, average and highest labour–output ratio in active production are represented by three thin dashed black lines; in runs with open technologies, the best-practice open technology (that with the lowest labour–output ratio) is given by a thin solid black line. Further, two grey curves represent the development of the industry structure; the lower one is the normalized Herfindahl index, the upper one is the industry share of the largest firm.

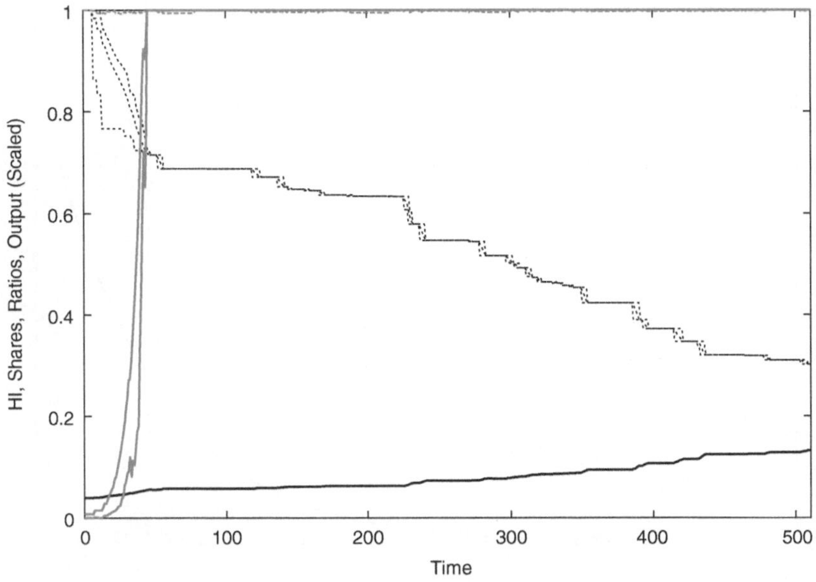

Figure 5.19 Simulation run of the model without open technologies (and without network externalities); 512 iterations. The grey curves measure the development of the industry structure (the upper one the industry share of the largest competitor, the lower one the normalized Herfindahl index of the industry only including active firms). The falling thin lines denote the decreasing labour–output ratios; the dashed lines stand for best, worst and average, the solid one is the best open technology. Finally, the steadily rising bold black line indicates the output level (not true to scale; scaled to $\frac{1}{800}$) and the thin dashed grey line shows the employment share (here always close to 1.0).

Unsurprisingly, in a setting without open technology (depicted in Figure 5.19), technological progress is modest and as seen from the course of the average labour–output ratio (the middle thin dashed black curve) almost linear in the initial stage. At some point after 40–50 iterations, this regime of slow but steady progress breaks down and reverts to a horizontal line with only occasional improvements. This coincides with the complete monopolization of the industry. In fact, there is only a single competitor left after 50 iterations and actual technological improvement is limited to the infrequent research successes of this firm. Labour remains scarce, employment at 100% and capital and output grows very slowly. Figure 5.19 indicates that this dynamic apparently prevails for the entire simulation no matter how many periods the simulation lasts.

A very different case is that with 100% open technologies – i.e. the entire technological knowledge is available to any firm (Figure 5.20). Research progresses much faster than without open technologies; in this case the theoretical speed of research may be seen from the grey curve (the best-practice open technology and,

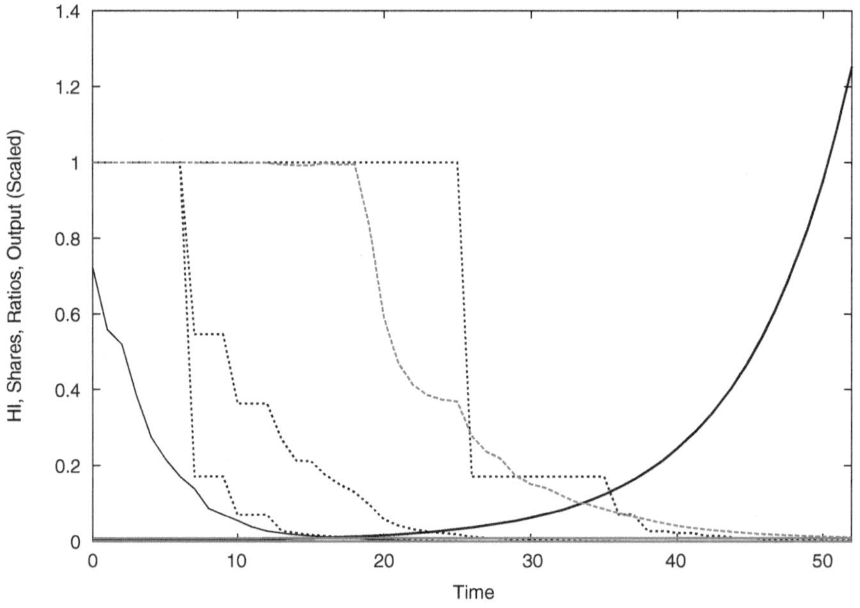

Figure 5.20 Simulation run of the model with 100% open technologies (but without network externalities); 52 iterations. Line types as explained in Figure 5.19; output not true to scale, scaled to $\frac{1}{20,000}$.

since all technologies are open, the overall best-practice technology). Innovations that have been put to practical use lag behind theoretical research since new plants are first employed in the period after construction, though by then, an even better technology has most likely been developed; furthermore, the firms do not always command sufficient capital to construct new plants. Indeed, technological progress occurs fast enough for the labour–output ratio to decrease more rapidly than the production capacity grows – hence after about 20 periods, the employment rate (in this figure indicated by a dotted line) begins to fall. Output and capital grow exponentially, and no noticeable monopolization occurs in the industry in the course of the first 52 time periods.[49]

Figure 5.21 shows an intermediate setting with just a small part of the industry making their technologies openly available. Rapid innovation prevails and though non-open standards may now be temporarily technologically better than open technologies, the technological progress is obviously driven by the best-practice open technology (the thin solid black line). As temporary competitive advantages through better technology are now possible, it is not surprising that the level of the industry share of the largest firm rises notably to a level of between 5 and 10%; the normalized Herfindahl index also increases; however, this is far from monopolization as observed in the setting without open technologies; growth of

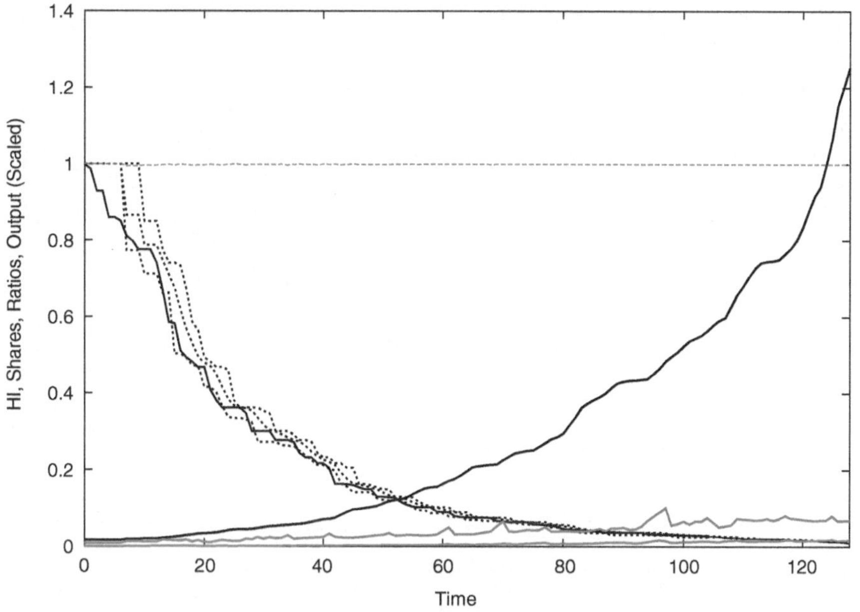

Figure 5.21 Simulation run of the model with 10% open technologies (but without network externalities); 128 iterations. Line types as explained in Figure 5.19; output not true to scale, scaled to $\frac{1}{2000}$)

capital and output remains exponential but labour also remains the scarce factor and employment is consequently at 100%. As depicted in the logarithmic plot in Figure 5.22, the capital growth (and consequently output growth) for the different levels of shares of open technologies are linear in a logarithmic plot (hence exponential in a simple plot) with different slopes. It follows that for output terms $y_{OT=1.0}$ (100% open technologies) and $y_{OT=0.1}$ (10% open technologies) there exists a factor h such that

$$\log y_{OT=1.0} = h \times \log y_{OT=0.1} \tag{5.43}$$

$$y_{OT=1.0} = y_{OT=0.1}^{h} \tag{5.44}$$

Hence, the impact of the (partial) openness of technology on growth is of exponential magnitude in this model.

Note that the simulation runs conducted so far are merely example runs; the statistical robustness of the displayed behaviour has to be established by repeating the simulation sufficiently often. For more trivial settings (no open technologies and all technologies open), this confirms the same patterns of development, as also indicated by the very fast lock-in for the setting without open technologies and the very clear readings of the industry structure variables. The mixed setting

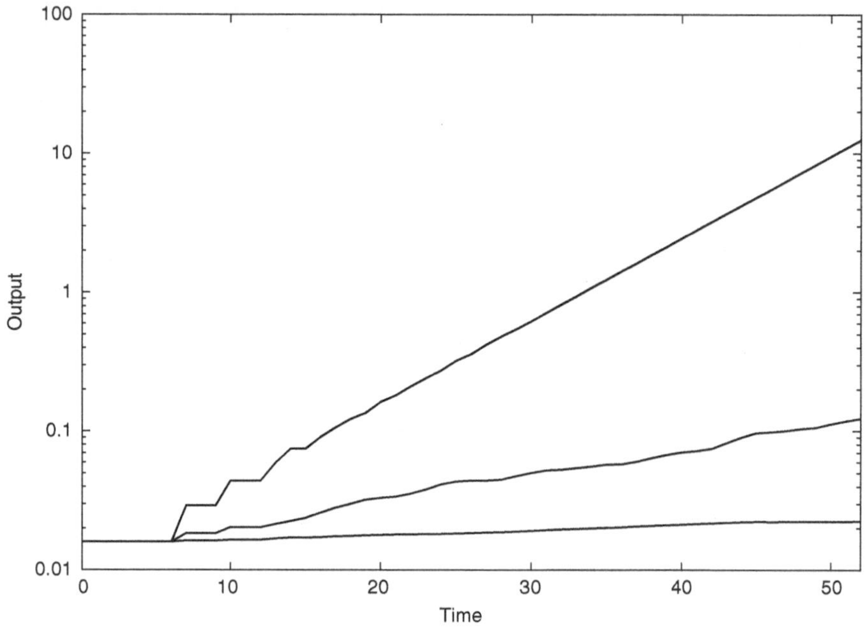

Figure 5.22 Development of output in example simulation runs of the model with 100%, 10% and without open technologies (without network externalities); output not true to scale, scaled to $\frac{1}{2000}$.

(10% open technologies), however, shows some divergence between different runs. Figures 5.23 and 5.24 show the development of output data and normalized Herfindahl indices in 20 independent runs of the simulation model with this setting; each for 96 iterations. There is clearly a cluster encompassing two thirds to three quarters of the simulation runs that develops comparable to the example simulation run, the normalized Herfindahl index rising to about 0.1 and the output data growing at a modest rate. For all 20 of the runs including the outliers, growth is well below the growth rates of the setting with 100% open technologies and well above the setting without open technologies. However, there are some outliers with high normalized Herfindahl indices (dotted in Figures 5.23 and 5.24), in fact, for two runs (but only two out of 20), the industry is monopolized completely. For these, the growth rates fall after monopolization (as expected[50]) and the output after the 96 iterations remains comparably small. There are also some outliers with very low normalized Herfindahl indices (dashed in Figures 5.23 and 5.24); these prove to grow faster than the average run. It is obvious that the development of the system is path dependent.

Returning to the questions posed for this investigation by computer simulation at the end of the previous section, it can be concluded that (1) complete or partial openness of technologies is crucial for inducing superlinear growth. This is

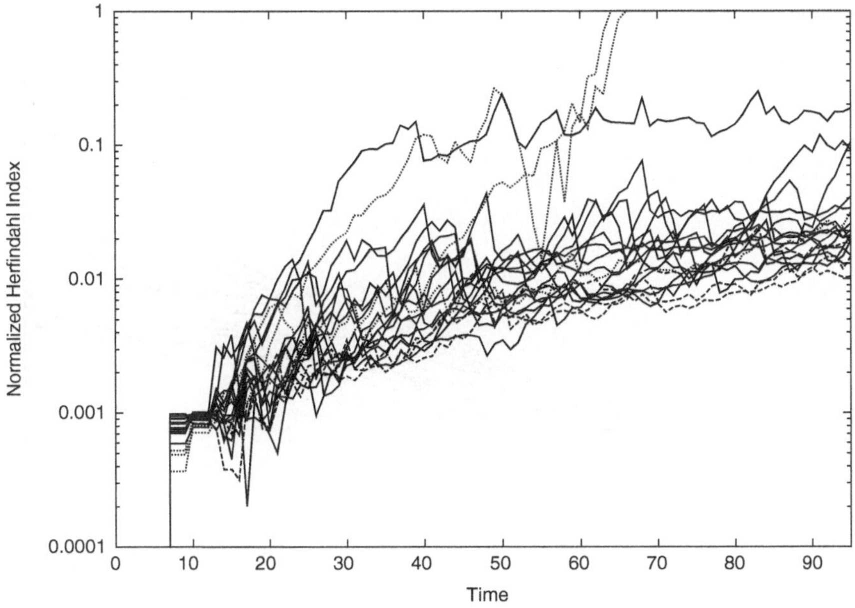

Figure 5.23 Industry structure in the model without network externalities: normalized Herfindahl index in 20 simulation runs (the same as in Figure 5.24) of the model with 10% open technologies (without network externalities). The simulation runs are depicted in the same line types (solid black, dashed black, dashed grey) as in Figure 5.24.

consistent with the results of the investigation of the Nelson–Winter mechanism above. It underlines the importance of network effects for economic growth as openness of technology is clearly a mechanism that speeds up technological progress with the size of the network; superlinear growth is caused by open technologies in the current model and may be caused by open knowledge and normal diffusion of technological knowledge in real-world systems. (2), growth is generally exponential though growth rates vary slowly; some data series may be interpreted as cycles between slow and rapid growth. This is probably caused by the stochastic occurrence of more significant technological improvements and their possible subsequent diffusion just as in the Nelson–Winter mechanism. However, the resemblance of real-world data is limited. Finally (3), the industry monopolizes rapidly if technological research is exclusive as this may grant lasting competitive advantages; with open technologies, the open community with their combined and cooperative research efforts is faster and more successful than other firms. Though other firms do also benefit from open technologies, it is unlikely for a non-open best-practice technology to maintain the technological advantage for a longer period of time. Also note that this setting may be seen and investigated as a public good problem, an *n*-person prisoners' dilemma supergame in that industry, though this is not the focus of the current investigation.[51]

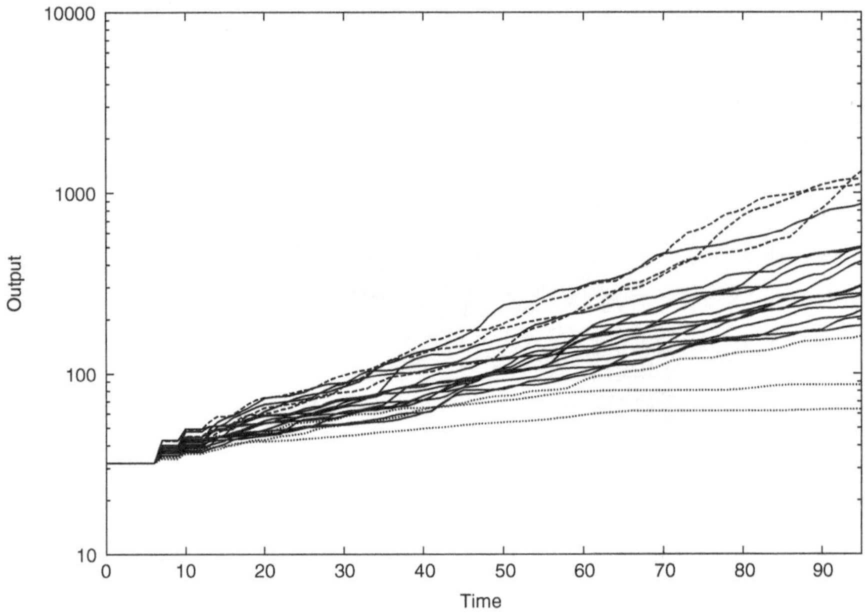

Figure 5.24 Development of the output in 20 simulation runs (the same as in Figure 5.23) of the model with 10% open technologies (without network externalities). The simulation runs are depicted in the same line types (solid black, dashed black, dashed grey) as in Figure 5.23.

5.3.6 A simulation of growth and network effects: the simulation model

In addition to the simulations conducted above, the simulation model will now be extended to include network effects. As already explained above, the network effects are assumed to be production-side network effects and to work by reducing the labour output ratio of commonly used technologies relative to that of rarely used technologies. Suppose the share of a technology type *TEC* in the technologies used in *currently active plants* in the industry is s_{TEC}. The labour–output ratio $\widetilde{a}(\theta)$ of a technology belonging to this type ($TEC(\theta)$ with industry share $s_{TEC(\theta)}$) is now computed from the standalone labour–output ratio $a(\theta)$ according to $\widetilde{a}(\theta) = \frac{(5-s_{TEC(\theta)})a(\theta)}{4}$. Note that in this case, in order not to change the dynamics of the model fundamentally by vast increases in efficiency, the network effect is not included by enhancing the technology level of commonly used technologies but rather by reducing it for rarely used technologies.[52]

In order to be able to simulate a few more periods for this more important setting with the available computation power, the speed of technological progress is reduced by increasing the order of the root in the function for the technological improvement per innovation from $\rho \sim \sqrt[32]{Uniform(0,1)}$ (see above) to $\rho \sim \sqrt[64]{Uniform(0,1)}$. This does not change the general shape of the function (as depicted in Figure 5.18) but merely shifts the steep increase further to the right and thus closer to 1. With this, the expected value of ρ grows also closer to 1 which means that the expected extent of any successful technological improvement is reduced.

In the simulation runs below, the number of technology types is set to 21. The low technology level the agents start with (labour–output ratio 1.0) is a technology type of its own that is, however, not researchable – so to speak a primitive technology. The technologies developed by the firms are randomly assigned to one of the remaining 20 types. As all agents start with the mentioned primitive technology, the system starts with a lock-in, which will, however, eventually become unfavourable since new technologies can by definition not be integrated into this network (may not belong to this technology type). To keep track of the lock-ins and break-outs in the course of these simulations, the program additionally records the normalized Herfindahl index of the distribution of the technology types (additionally to the firm's industry share Herfindahl index as detailed above).

This structure leads to the lock-in of the industry, i.e. to one of the technology types being the exclusively used standard. However, as research is not affected, and especially since the lock-in has no effects on the technology types of newly researched innovative technologies, the eventual break-out is also possible (as discussed above in sections 5.1 and 5.2.2).

In the following, a number of simulation runs will be conducted with the modified simulation model again with a share of 10% of firms that make their research openly available (in turn, 10% open technologies). The data collected in these runs will be used for theoretical and statistical analysis to investigate the impact of the network effect on the development of the industry structure, growth and technological progress. Note that as the research function has been changed (the speed of research slowed down) the absolute values of capital and output growth are not comparable. However, the shapes of the development function, the presence or absence of regular patterns, etc., are. Further, several example runs are conducted with more and less than 10% open technologies to establish how the effect of open technologies changes with network externalities.

Specifically, the questions to be addressed are

1. Is the network externality measurable, are there lock-ins, and (as the answers to these question are presumably yes) how does they effect the technologies in use and the patterns of technological change?
2. Does the network effect change the pattern of the development of output (and, proportionally to that, capital) growth?
3. Is the development of the industry structure affected (monopolization), and, if so, in which way?
4. Does the network externality, as theoretically predicted, increase the effects of path dependence that were already present without network externality?
5. Is there any interplay between the network effect and the effect of the openness of technologies?

5.3.7 A simulation of growth and network effects: the simulation results

General behaviour of the simulation model with network effects

Figure 5.25 shows a largely representative example run of the simulation model with network effects. The form, line style, and notation is equivalent to the figures discussed in the previous section. Furthermore, the dashed bold line indicates the normalized Herfindahl index of technology type shares in use in the industry. In turn, the industry starts locked-in to the trivial technology as explained above; subsequently the technology type Herfindahl index drops occasionally (six times in the 256 iterations of the example simulation run) to lower levels, only to subsequently rise again to 1.0. This indicates when break-outs from the current lock-in occurs. As suggested in the theoretical discussion above, the industry locks-in to another (or more unlikely, the same) technology type; usually that occurs almost immediately. Technological changes between different technology types

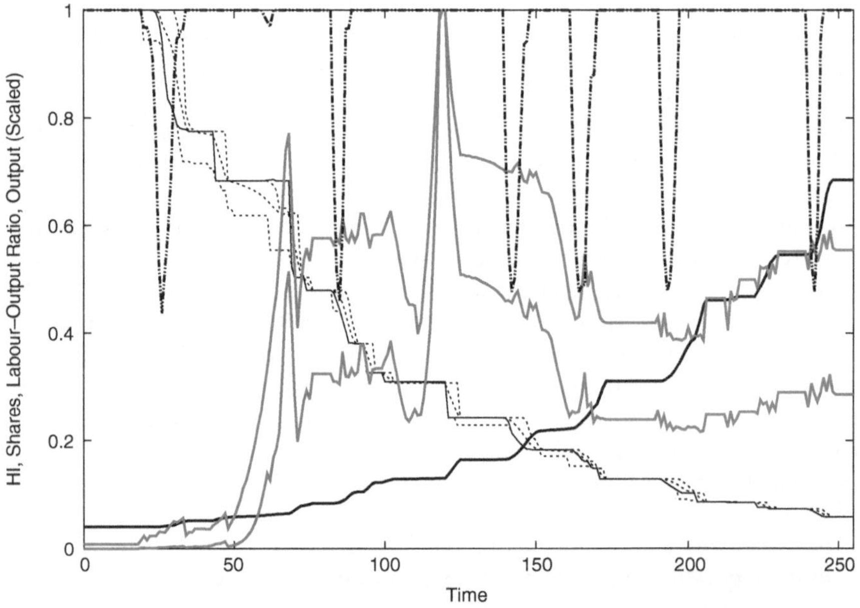

Figure 5.25 Simulation run of the model with network effects and 10% open technologies. The bold grey lines measure the development of the industry structure (the upper one the industry share of the largest competitor, the lower one the normalized Herfindahl index of the industry including only active firms). The dashed bold black line tracks the lock-ins using the normalized Herfindahl index for the technology types. (1.0 means locked-in, deviations below 1.0 indicate temporal break-outs.) The falling thin lines denote the decreasing labour–output ratios; the dashed lines stand for best, worst and average, the solid line is the best open technology. Finally, the steadily rising bold black line indicates the output level; output not true to scale, scaled to $\frac{1}{800}$.

(in turn, the mentioned break-outs and re-lock-ins) obviously do not occur on a regular basis but rather are randomly distributed over time. They are accompanied by productivity increases – presumably since the new technology type is of greater productivity at the time.

The labour–output ratio drops with occasional sharp decreases, almost step-wise (which is especially striking when compared to the same setting without network effects as shown in Figure 5.21). That is due to the lock-in patterns: While the lock-in is in place, only improvements of the same technology type[53] can be implemented.

As mentioned, technological improvements (reductions in the average labour–output ratio) coincide with output (and capital) growth. Consequently, the output/capital function is also much less continuous, now featuring clear (though not sudden as in a monopolized industry) transitions between regimes of slow and rapid growth. This pattern must be expected to produce oscillating growth rates and therefore possibly growth cycles; this will be investigated in more detail below.

The most interesting difference to the setting without network effects, however, is the development of the industry structure – indicated in both the share of the largest firm (the upper bold grey curve in Figure 5.25) and the normalized Herfindahl index of the firm's industry shares (the lower bold grey curve in Figure 5.25). Other than without network effects (see Figure 5.21), the industry structure does not converge to 1.0 (monopolization), nor does it remain close to 0.0 (equal sharing). In the example run, the normalized Herfindahl index increases to 5%, then jumps to about 35%, where it remains steady for a number of iterations. The level is changed for at least two more times. Within these periods, the level is relatively stable. Briefly, the normalized Herfindahl index even jumps to 100% but the competitive industry structure recovers; this coincides with the best-practice open technology (thin solid black curve) being equal to the lowest labour–output ratio in use. It seems that the technological openness is able to prevent complete lock-in, at least for some cases[54] with network effects. Furthermore, the network externality seems to drive the industry structure towards oligopoly even though monopolization seems to become less likely than in the above case without network effects. These oligopoly structures also seem to be pretty stable.

The example run is naturally not sufficient to assess the robustness of these results. In order also to get an idea of how likely such an outcome is and which other outcomes are possible or likely, this simulation with these parameters has been repeated 20 times, the results of this will be investigated in detail in the following sections. However, the example run already allowed a glimpse of the differences to the setting without network effects and provides some hints about which phenomena should be analyzed with a larger sample (i.e. repeated simulation runs).

Robustness of the systems behaviour and path dependence

From comparisons of the different simulation runs (Figures 5.26 and 5.27) it becomes immediately clear that the development of this system differs widely from simulation run to simulation run. For some, the attained level of output is

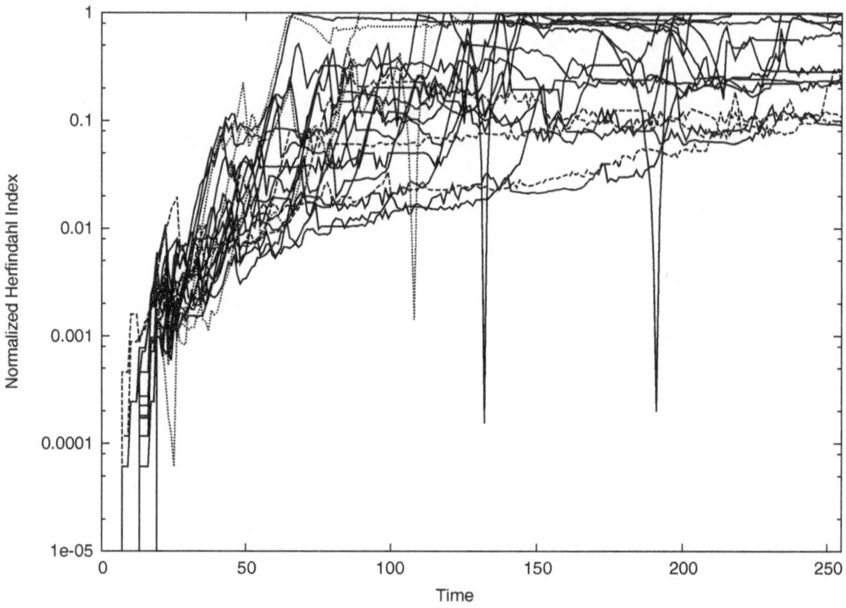

Figure 5.26 Industry structure in the model with network effects: normalized Herfindahl index in 21 simulation runs (the same as in Figure 5.27) of the model with network effects (10% open technologies). The simulation runs are depicted in the same line types (solid black, dashed black, dashed grey) as in Figure 5.27.

by magnitudes larger than for others. And again, as analyzed above for the setting without network effects, this seems to be related to the industry structure. The nature of this relation, however, will be discussed in the section on the industry structure below. Almost half of the simulation runs result eventually in complete and final monopolization of the industry – though this occurs at different times and the trajectories taken before differ as well. Another quarter of the sample shows an oligopoly with a very asymmetric industry structure, as seen from a very high (though below 1.0) normalized Herfindahl index. In four runs the industry stays rather competitive though the normalized Herfindahl index is at about 10% even there, and only stabilizes after reaching this level. And finally, in two runs the system was almost monopolized at a time but restored to a more or less competitive state, an effect that has already been discussed with respect to the example run above though the phenomenon is much less prevalent there. The number of break-out and re-lock-in events differs as well, between 4 and 11 such events within 256 iterations.

This indicates that the system is heavily path dependent, and a number of different results are possible; averaging of the sample is probably of no help in analyzing the model's general behaviour. However, some structures and patterns seem to be stable over the sample:

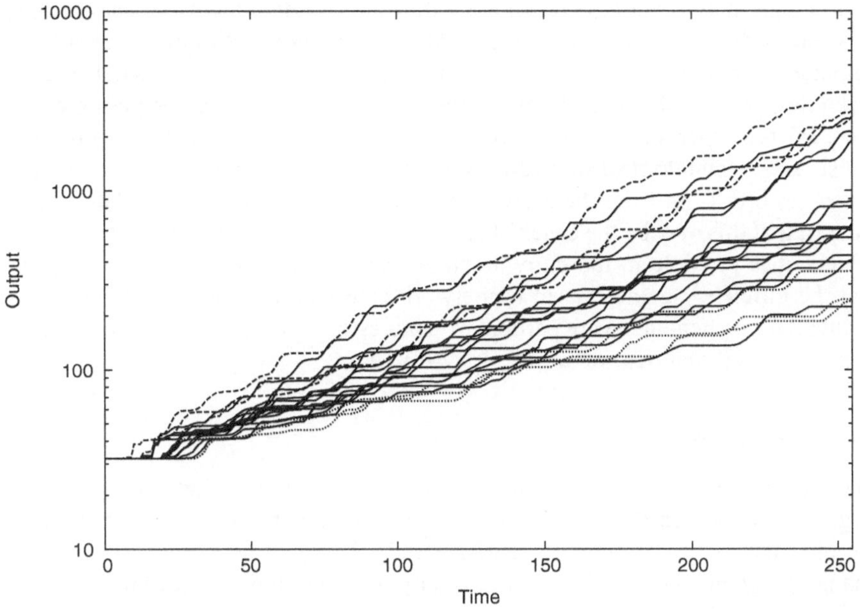

Figure 5.27 Development of the output in 21 simulation runs (the same as in Figure 5.26) of the model with network effects (10% open technologies). The simulation runs are depicted in the same line types (solid black, dashed black, dashed grey) as in Figure 5.26.

- Technological improvements and growth occur step-wise, with a slow growth regime at a constant technology level and a regime of rapid growth accompanied by technological change. This is consistent with what had been predicted in section 5.1.
- Technological improvements are sometimes but not always accompanied by break-out and re-lock-in events.
- The Herfindahl index shifts between levels at which it is relatively stable. This indicates that the industry structure is undergoing periods of rapid change and periods of greater stability.

Other than in the setting without network effects, however, there is no clear evidence here that the development of output (Figure 5.27) and the industry structure (Figure 5.26) are linked. It seems that a larger degree of monopolization leads to lower output and vice versa (as above) but further analysis is needed to establish that (see below).

Technological change

As mentioned before, technology tends to improve step-wise in this setting, being prevented from implementing the great majority of innovations by lock-ins.

However, on average (e.g. smoothed over sufficiently many periods for the steps to no longer play a role) the technology trajectories retain the exponential shape they had without network externalities. Major technological improvements often coincide with break-out and re-lock-in events, i.e. the system switches to a technology type with currently much better technologies. It may be possible that there exists a supercycle of technology type switches covering longer periods than the steps of technological innovations. Only very large improvements allow the system to break-out from the current lock-in. Data from several simulation runs suggest that this may be the case. Other than in the above setting without network effects, it is possible in the current setting that technologies lose efficiency, in that the efficiency gains caused by network externalities reduce with break-out events. These phenomena are particularly interesting, as they show the interaction of network externalities and the industry structure.

Industry structure

The industry structure tends to show some degree of oligopolization – but with large variations according to the respective development trajectory. Furthermore, the industry structure seems to remain steady for some periods of time before changing again. The changes are usually but not always accompanied by technological developments. If there is no obvious connection to successful innovation or break-out from the current lock-in, changes in the industry structure may still be (and probably are) late effects of earlier innovations, etc.

For all runs, the system swiftly approaches a level of about 0.1 on the normalized Herfindahl index, which indicates a fast and considerable oligopolization. The setting seems to work strongly towards an asymmetric industry structure. After this level is reached, however, the simulations vary between different values between 0.1 and 1.0 (i.e. also different degrees of oligopolization) without necessarily converging towards monopoly.

Output and growth patterns

The level of output and capital production follows the (average) labour–output ratio of the currently employed technologies in the current setting, i.e. with the current network effects. Hence, output grows at different speeds, alternating between a slow regime and occasional occurrences of rapid growth, the latter always accompanied by technological progress. Sometimes, these fast growth phases also coincide with a break-out from the current technological lock-in. More accurately, the break-out does not cause the rapid growth, but the following re-lock-in into a technology type with vastly better labour–output ratios does. Figure 5.29 compares the growth patterns in the current setting with and, in the previous setting, without network externalities. The presence of different regimes in the current pattern is most obvious while without network effects and consequently without technological lock-ins, growth seems more random and more evenly distributed over time.

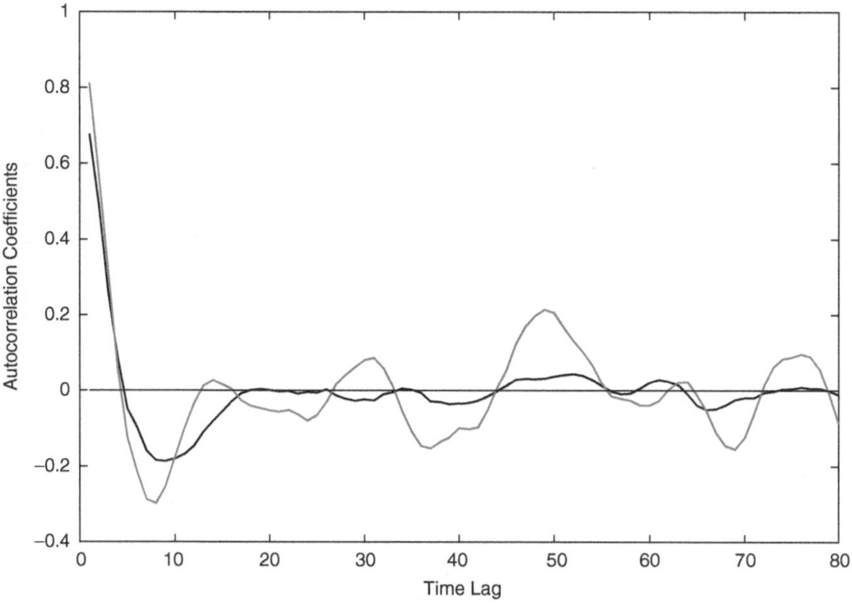

Figure 5.28 Autocorrelation spectrum of the growth rates of output in the 21 simulation runs (bold curve) and in for comparison to autocorrelations in non-aggregated data also the autocorrelation spectrum of a single example run as grey line.

The continuing changes between slow and fast growth indicate that there may be growth cycles or even different cycles of several periods. Figure 5.28 depicts the autocorrelation spectrum of both the growth rates of an individual simulation run (thin line) and the average of the autocorrelation of the growth rates of the whole sample (bold line). It may be expected that strong circular patterns would still be visible on average over many simulations while path dependent developments differ from run to run. In fact, one short cycle is noticeable. The bold line (averages) shows strong autocorrelations for very short time lags indicating that the growth rates change slowly; it also shows a strong negative correlation with a time lag of about 10 iterations. This might indicate a cycle with an average period length of 10. The thin line (single run) shows further waves indicating the presence of still strong correlations over larger periods of time. These, however, seem to average out to 0 when considering the entire sample as the average function indicates. It is therefore quite likely that the short-term autocorrelations really stand for rather stochastic perturbations (i.e. naturally stochastically distributed technological improvements) that induce a characteristic pattern of growth and technological change. The distribution of these perturbations (significant innovations) over time, however, does not follow a regular pattern (as expected from section 5.1). Note that the current simulation discusses the industry level dynamics; irregular industry

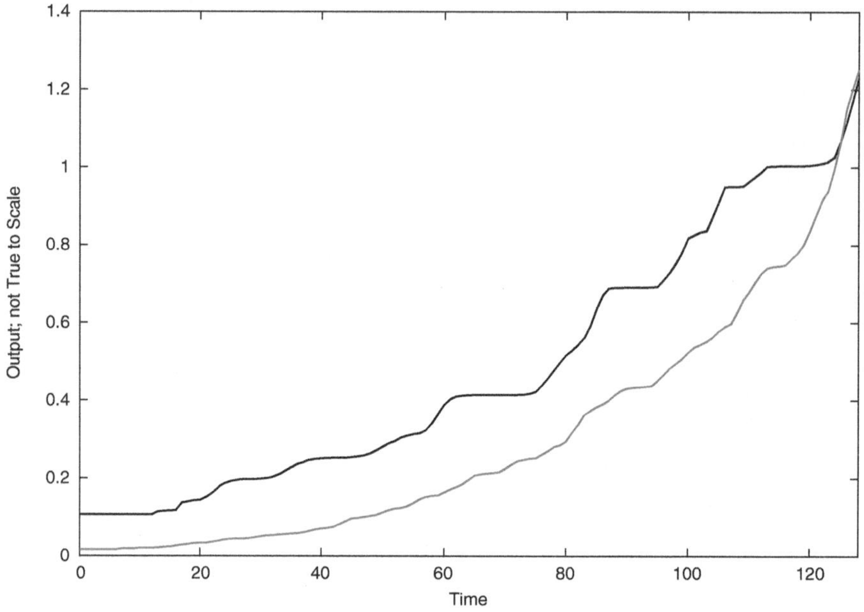

Figure 5.29 Shape of the output development in the model without (black line) and with network effects; both runs with 10% open technologies. Output levels are not true scale, the graphs are scaled to be of approximately the same scope (without network effects scaled to $\frac{1}{2000}$, with network effects scaled to $\frac{1}{180}$). Note that the research functions differ; direct numerical comparison is therefore meaningless.

level growth patterns may well introduce more regular aggregated level patterns as discussed in section 5.2.3.

In some cases, one firm achieves a monopoly over the industry at some point in time. In these cases the usual smooth transition between slower and more rapid growth of output and capital production reverts to a pattern of sharp sudden increases in otherwise static levels.

The dream of full employment and the monopolist's despair

There are some odd spikes in the industry structure data just before some change in the level of monopolization (indicated by the normalized Herfindahl index) occurs (Figure 5.26). What is odd about this is that these spikes point in the wrong direction; if the industry structure becomes more asymmetric, more monopolized, this is sometimes preceded by a brief moment of increased competition and vice versa. Additionally, there are some rare occasions where a monopolist loses her monopoly again. This turns out to result from the same effect. It results from the scarcity device in the model; this was discussed in some detail in section 5.3.2 with respect to the Nelson–Winter model, which has two different ways to introduce scarcity, either by limiting labour and thereby production capacity or by limiting demand and thereby revenue. The current model chose an absolute limit to labour L_0. Growth occurs mainly by increasing labour efficiency such that L_0 can be used to produce larger quantities and in fact by operating more plants. That means, a rising technology level enables technologically backward plants to re-enter an operational state and possibly revives firms that were driven out of the industry but that still own some technologically backward production plants. However, if the technological knowhow of these firms does not improve, while they can produce with these facilities, they will be driven out of the industry again as soon as the 'market-leader' has constructed sufficiently many new plants to employ all available labour. If they are successful, however, the former monopolist may end up as one ordinary small competitor among many.

Note that this mechanism works for all vintage models regardless where or how scarcity is designed in the model. For instance, if caused by a drop in prices (that in turn was the consequence of recent innovations) the demand for a specific type of consumption goods rises steeply and the innovator having insufficient production capacity is unable to match the additional demand, some other (maybe technologically backward) competitors may step in. As a result, the sector becomes less monopolized and the innovator loses some of her industry share.

It is very likely that this mechanism would work for the real-world as well. One might expect that in real-world systems, such a development would also result in quick diversification of the product range of this industry, possibly giving rise to secondary, tertiary and further derivative industries. This is the very mechanism discussed in section 5.2.1 and may give rise to macro-level developments as discussed there.

Open technologies with network effects

Figures 5.30 and 5.31 contrast four runs of the current model with network externalities for different shares of open technologies (or, more accurately, of firms contributing open technologies), i.e. 0%, 10%, 20% and 30%. The findings of the above simulation without network effects are largely confirmed. Output (and proportionally capital) generally rises exponentially, just with different exponents for different shares of open technologies. Technological progress largely follows an exponential function as well (though with negative exponent when studied in the form of labour–output ratios) and happens faster the larger the share of open technologies is.

For technical reasons (limited computation time), just the first 180 (for 20% open technologies) or 160 (for 30% open technologies) turns are simulated to be compared here. Up to this stage, however, it can be seen that the industry structure approaches the 0.1 level apparently slower for more than 10% open technologies, but otherwise behaves similarly in these cases; without open technology, as already seen in the case without network effects, the industry

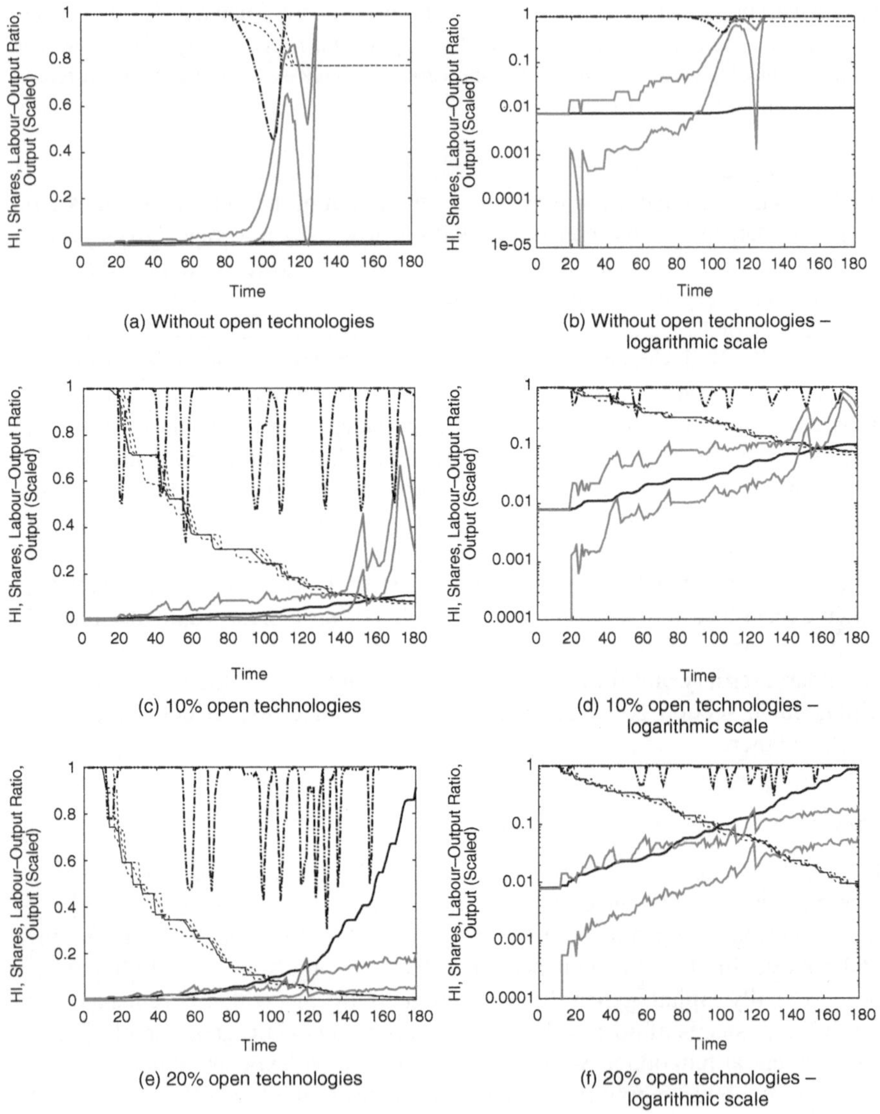

Figure 5.30 Simulation runs of the model with network effects and 0%, 10% and 20% open technologies. Line types as explained in Figure 5.25; output not true to scale, scaled to $\frac{1}{4000}$.

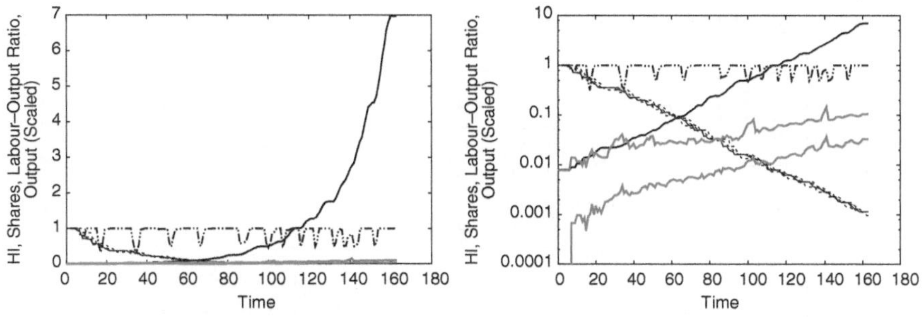

(a) 30% open technologies

(b) 30% open technologies – logarithmic scale

Figure 5.31 Simulation runs of the model with network effects and 30% open technologies. Line types as explained in Figure 5.25; output not true to scale, scaled to $\frac{1}{4000}$.

is quickly and irrevocably monopolized. This again confirms the above result that open technologies decrease the risk of monopolization and tend to increase competitiveness.

Additionally, break-out events seem to happen more often the higher the share of open technologies, specifically only once within 180 iterations without open technologies, and 9 times and 11 times for the case with 20% and 30% open technologies respectively. Most likely, this is just a result of the accelerated technological progress.

Rapid and slow growth: the roots of the realized development trajectory

Table 5.1 is the matrix of Pearson correlation coefficients between the variables output (Y), average technology level, i.e. average labour–output ratio (a) and the normalized Herfindahl index (HI), each at three different times: once in the beginning, after 32 iterations; once after 96 iterations; and finally at the end of the simulation runs, after 256 iterations. Furthermore, two more variables are included, the number of break-out events within the 256 iterations and the multiplied normalized Herfindahl indices after 96 iterations and at the end. The latter variable is used as a proxy variable for the development of the industry structure in time (as opposed to its state at merely one point in time, as the single normalized Herfindahl indices after both 96 and 256 iterations are). The sample is composed of the 21 simulation runs with network effects and 10% open technologies as discussed above; hence the correlation coefficients have statistically 19 degrees of freedom (two less than the sample size) and the significance of these correlation coefficients is computed using the one-sided Student's t distribution with the t-value[55]

$$\bar{t} = \nu \frac{d.f.}{1 - \nu^2}$$

Table 5.1 Matrix of correlation coefficients between normalized Herfindahl indices, output, the labour–output ratio at different times (after 32, 96 and 256 iterations) as well as the number of lock-in/break-out events within 256 iterations for 21 simulation runs.

		$a_{t=32}$	$HI_{t=32}$	$Y_{t=96}$	$a_{t=96}$	$HI_{t=96}$	$Y_{t=256}$	$a_{t=256}$	$HI_{t=256}$	*Number of break-outs*	$HI_{t=96} \times HI_{t=256}$
Output iteration 32	$Y_{t=32}$	−0.974***	0.165	0.674***	−0.718***	−0.491*	0.758***	−0.651***	−0.282	0.509**	−0.518**
Labour–output ratio iteration 32	$a_{t=32}$	1	−0.152	−0.635***	0.715***	0.520**	−0.683***	0.653***	0.231	−0.543**	0.577**
Normalized Herfindahl index iteration 32	$HI_{t=32}$		1	0.115	−0.215	−0.083	0.118	−0.337	−0.486*	−0.105	−0.297
Output iteration 96	$Y_{t=96}$			1	−0.933***	−0.437*	0.819***	−0.687***	−0.154	0.490*	−0.434*
Labour–output ratio iteration 96	$a_{t=96}$				1	0.505**	−0.819***	0.813***	0.221	−0.580**	0.577**
Normalized Herfindahl index iteration 96	$HI_{t=96}$					1	−0.432*	0.533**	−0.018	−0.452*	0.841***
Output iteration 256	$Y_{t=256}$						1	−0.786***	−0.386*	0.721***	−0.452*
Labour–output ratio iteration 256	$a_{t=256}$							1	0.475*	−0.739***	0.719***
Normalized Herfindahl index iteration 256	$HI_{t=256}$								1	−0.182	0.330
Number of lock-in/break-out periods	♯ Break-Outs									1	−0.440*
Herfindahl indices iteration 32 and 256 combined	$HI_{t=96} \times HI_{t=256}$										1

***Significant at 99.9% confidence level (t-value> 3.58, one sided t-test, 19 degrees of freedom)
**Significant at 99% confidence level (t-value> 2.54, one sided t-test, 19 degrees of freedom)
*Significant at 95% confidence level (t-value> 1.73, one sided t-test, 19 degrees of freedom)

The matrix shows, unsurprisingly, that output and technology are autocorrelated. Further, output and technology levels are highly correlated both in the same period and between periods. This merely underlines that the processes of growth and technological change that are currently under consideration are indeed path dependent: early growth and technology in the system (if only the first 32 periods) shapes the long-term development. The Herfindahl indices seem to fall out of this pattern; there is not even a positive autocorrelation (indeed with a low confidence level, the Herfindahl indices would be significantly negatively autocorrelated). However, as seen in the analysis of details above, the industry structure plays a major role in its development. In fact, however, as the Herfindahl index may jump between very different levels, it is not sufficient to consider the index at one point in time. A measure that combines the normalized Herfindahl indices after 96 iterations and at the end of the simulation (after 256 iterations) performs better. This value will still be between 0 and 1; it is low, if the normalized Herfindahl stays low over the entire course of the simulation; if the system is close to a monopoly after either 96 or 256 iterations the value will be much higher. The result of the multiplication of the two normalized Herfindahl indices is strongly correlated to the technology level at each of the three time stages; it correlates also reliably to the output – though this correlation seems to fall with time. Furthermore, the relation between this measure for the long-term development of the industry structure and output is quadratic (since the two Herfindahl indices were multiplied) as shown in Figure 5.32.

The number of break-outs is also strongly correlated to both the combined Herfindahl indices (the long-term industry structure measure) and the output and technology levels at almost all times. However, as Figures 5.33 and 5.34 show, the linear regression with the long-term industry measure is barely different from that with the normalized Herfindahl index after 96 iterations. It seems that the earlier Herfindahl index has some (negative) effect on the subsequent number of break-out events. Finally, Figure 5.35 shows the positive relation of the number of lock-ins and the output level attained at the end of the simulation.

Conclusion

Returning to the initially posed questions, it can be concluded that

1. The network externality is present in that usually only one technology type is used (basically a lock-in as predicted) and technological change takes a more periodic shape as many innovations belonging to the wrong technology type cannot immediately be implemented.
2. Output and capital growth follows accordingly, equally switching between periods of slow and rapid growth.
3. Significant innovations induce recurring wave patterns; this does, however, occur on a stochastic, not a strictly regular basis.

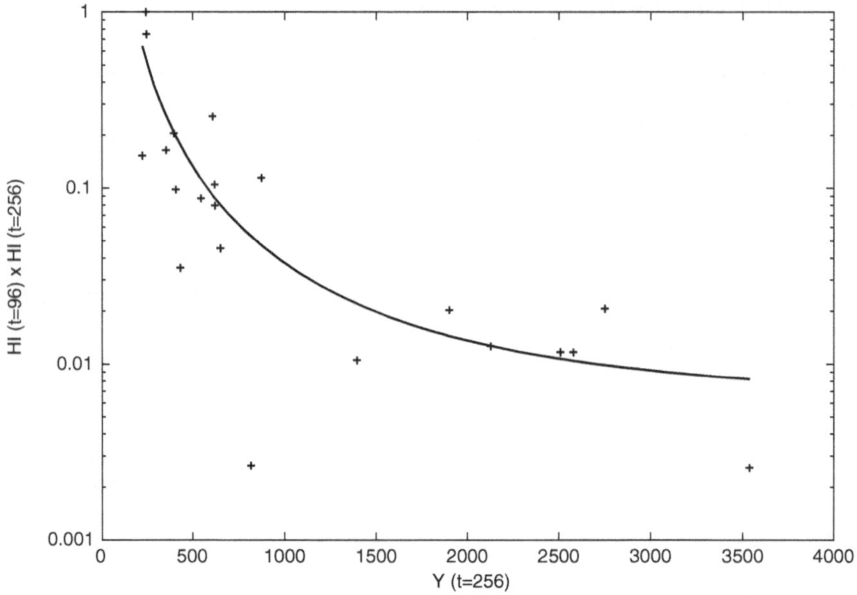

Figure 5.32 Scatter plot of output after 256 iterations against the multiplicatively combined normalized Herfindahl indices after 96 and 256 iterations; data from 21 simulation runs, with network externality and 10% open technologies. The solid line indicates the exponential fit of the data; function $HI_{t=96} \times HI_{t=256} = 31764.8Y_{t=256}^{-2} + 0.00567362$.

4. The network externality seems to work towards an asymmetric (oligopolistic) industry structure, but also has a stabilizing effect. The industry structure remains at a level of monopolization for considerable time.
5. There is evidence that path dependence is stronger than in settings without network effects, the trajectories may also be much more complex.
6. The network externality leaves unchanged the effects found above to be caused by open technologies. However, there are more phenomena that are related to the presence of open technologies (or, more widely, technological diffusion); they may play a part in preventing monopolization and may lead to more frequent break-out events from the respective current lock-in.

5.3.8 Implications of the simulation model

To summarize, this section has developed and analyzed a simulation model following the theoretical model laid out in section 5.1. The simulation technique was loosely based on the conclusions drawn from the Nelson–Winter mechanism of technological change and economic growth. The simulation was successful in reproducing systems of technological change that produce (1) superlinear

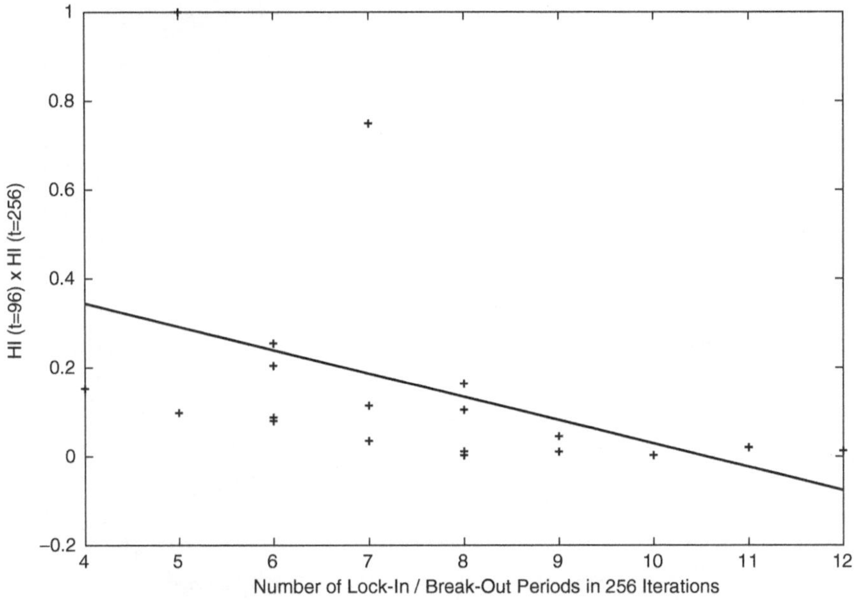

Figure 5.33 Scatter plot of the number of lock-in/break-out periods within 256 iterations against the multiplicatively combined normalized Herfindahl indices after 96 and 256 iterations; data from 21 simulation runs, with network externality and 10% open technologies. The solid line indicates the linear fit of the data with function $HI_{t=96} \times HI_{t=256} = -0.0525085$ (*# Break-outs*) $+ 0.55417$.

growth, (2) heavy path dependence, (3) an oligopolistic industry structure with (4) continuing smooth changes between slow and rapid growth, and (5) a series of technological lock-ins caused by network effects. The dynamics of the model are driven by network effects that lead to the said lock-in patterns (though break-out from the lock-ins is usually possible and does occur) and a certain degree of openness of technological knowledge. While technological diffusion or openness of technologies plays a key role in Nelson and Winter's model as well, the current model follows an alternative approach by including network effects and building the microfoundations of the growth model on the properties of technology, and accordingly the presence of network externalities and technological lock-in.

Compared to the most influential models of evolutionary growth theory, i.e. the models by Nelson and Winter (Nelson and Winter, 1974; Nelson *et al.*, 1976; Nelson and Winter, 1982a,b); Conlisk (1989) and Silverberg and Lehnert (1993, 1994), the current model is a rather simple one. It does not try to provide a general model of all branches of the economic system or to include possible other feedback mechanisms such as via the financial sector, wages, price level, etc. Rather, it is reduced to the minimal structure that allows the device to work and the necessary functions to track and record its development.

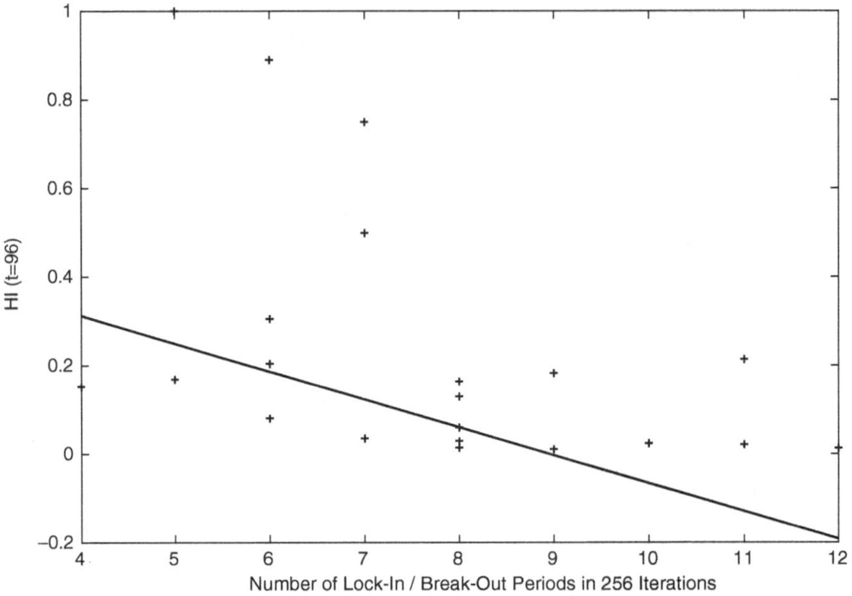

Figure 5.34 Scatter plot of the number of lock-in/break-out events within 256 iterations against the normalized Herfindahl indices after 96 iterations; data from 21 simulation runs, with network externality and 10% open technologies. The solid line indicates the linear fit of the data with function $HI_{t=96} = -0.0630576\,(\#\,\mathit{Break\text{-}outs}) + 0.719061$.

5.4 Technological change with network externalities: a conclusion

While a number of fruitful approaches to economic growth – some of them with interesting and promising microfoundations – are to be found in the literature (see Chapter 3), we are far from fully understanding the workings of growth and technological change. Different traditions of growth theory provide us with a number of hints, modelling tools, probable interrelations and interesting mechanisms. It is very clear that economic growth depends strongly – among other things – on technological progress. The economic analysis of network effects suggests that especially in relation to technological development, standardization plays a major part. Standardization in turn, is subject to strong network effects, leading to technological and organizational lock-ins; a consequence that changes the dynamics of the system to a state with multiple equilibria, possibly without finite equilibrium, and thus persistent growth patterns, and continuing development in far-from-equilibrium states.

There are two different types of network externalities that may influence economic growth. First, regular economic processes may be conducted more efficiently as a result of wide usage. Second, network effects may lead to more

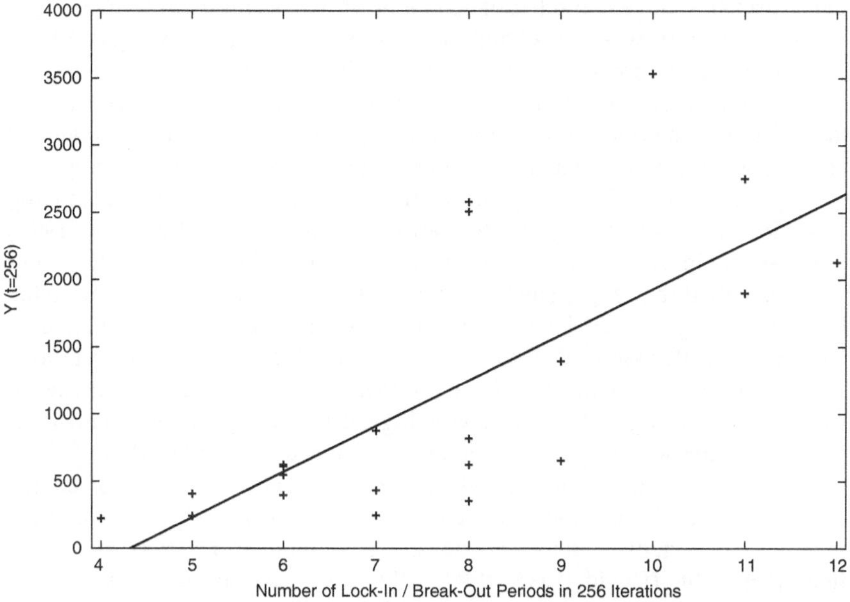

Figure 5.35 Scatter plot of the number of lock-in/break-out periods within 256 iterations against output after 256 iterations; data from 21 simulation runs, with network externality and 10% open technologies. The solid line indicates the linear fit of the data with function $Y_{t=256} = 339.523\ (\#\ \textit{Break-outs}) - 1466.34$.

rapid technological progress. Both types may coexist. The analyses conducted in this chapter suggest that a system of economic growth with network-externality-caused lock-ins is most likely to be subject to structural emergence. This in turn creates stable patterns at an aggregated level notwithstanding possible turbulence and random influence on the micro-level. Openness of technological knowledge – or equivalently rather rapid diffusion of innovations – is an important element in modelling technological change and economic growth with growth cycles. Lock-in effects based on network externalities of technology may explain persistent growth with – irregular – switching between regimes of slow and rapid growth without leading to certain monopolization (though the mechanism suggests a development towards an oligopolistic industry structure). This alternating pattern may be interpreted as growth cycles, but would more conveniently be described as irregular waves.

So far, these results apply mostly to the sector level. However, aggregation over a number of sectors each of which is subject to these changes between a rapid and a slow regime of technological change and growth (termed resonant oscillations above) results in a 'natural' Slutzky–Yule effect that may be responsible for circular patterns in economic growth. Research-side network effects may result in a power law distribution of sector size and use of technologies and standards and – considering that technological standards are often distributed by only one vendor – firm size. Several models of economic growth suggest positive feedback loops at the industry sector level and their consequences at an aggregated level as a cause for growth cycles and far-from-equilibrium dynamics in economic growth in general (Forrester, 1977; Lorenz, 1987). Structural emergence may further play an important part in giving the aggregated level of the economic system a certain stability that in turn may stabilize micro-level dynamics (an effect termed *downward causation* in the theory of emergence). This reasoning is very close to theories of group-formation (Axelrod, 1984; Lindgren, 1997) and meso-economics (Elsner and Heinrich, 2009, 2011). Indeed, the proper term for what is called the 'aggregated level' in this chapter is a meso-level. Stable patterns at this aggregated level are what allows us to define an economic or social (or political) macro-level in the first place. Even the modern state as a certainly cohesive economic, social and political entity only came into existence (historically) recently and driven by technological and organizational micro-level developments during the early modern period (Anderson, 1991 [1983]).

Much of the debate about network externalities and lock-ins as discussed in section 4.5 has been undertaken under the implicit assumption that these phenomena are strictly bad for growth and technological progress and to be avoided if possible (David, 1985, 1992, 2005; David and Bunn, 1988; Arthur, 1989; David and Steinmueller, 1994; Liebowitz and Margolis, 1990, 1994, 1995; Stango, 2004; Stango and Greenstein, 2008; Hall and Günther, 2011). This may not be the case. Technological standardization is certainly a process with heavy network externalities and lock-in effects. It may be and is frequently the case that further progress is hindered by the current lock-in (and this has also been the case frequently in the simulation above); however, only the lock-in grants access to the benefit of the network effects. Acknowledging that the whole of technological development is path dependent, only lock-ins allow the industry sector to expand to make further progress possible, though, as a side-effect, it also prevents immediate employment of the results of this further progress. This does, of course, not contradict David's famous argument that the successful technology is not necessarily (and usually not) the technically most desirable standard (David, 1985). In fact, this is much more complicated in real-world systems. For instance, for technologies that require an extensive infrastructure in order to be used properly, the successful technology may not only be a random one among the competitors or the one that was introduced fastest, but may also be one that relies on pre-existing infrastructure more than others, thus being easier to implement. This in turn may lead to this infrastructure being used for purposes it was never intended for. (Detailed examples will be given in the case study in Chapter 6.)

6 Case study

The information and communication technology sector and economic growth

After discussing theoretical and mathematical aspects of economic growth, technology and network externalities largely without practical illustrative examples, this chapter will offer a case study by analyzing one of the fastest growing industries of the recent decades, the *information and communication technology* (*ICT*) sector.[1] Proponents of the theory of technological paradigms have identified this industry as the carrier of the current technological paradigm starting in the 1980s (Perez, 1983; Freeman and Perez, 1988), a view that has, as shall be shown over the course of the next sections, so far lived up to these expectations. Both with heavy network effects (since ICT is mostly about networks of computers and computer technology) and rapid technological change and growth, the ICT industry is also a natural first choice for studying real-world evidence of network externalities, lock-ins and the resulting growth effects.

Unfortunately, despite the rather comprehensive documentation of the history of ICT, empirical data is not available to the extent and with the ease that would be desirable and might be expected. There are three main reasons for this. First, ICT networks are of fundamentally open and decentralized design, which complicates central data collection, but is also one of the strengths of this industry and arguably – as shall be seen in the following sections – one of the crucial factors for its rapid growth. Second, ICT involves a substantial amount of purely digital goods (what Quah proclaimed as the 'weightless economy' (Quah, 1999b)), which leaves both vendors and users rather ambiguous about the exact number of copies of a certain piece of software or information in circulation. Finally, older industrial systems of accounting had – naturally – no common definition of what constitutes and what belongs to software, digital information or information and communication technology and what does not; this leads to different modes of data collection on the ICT, the international comparability of which, especially for the early years, is questionable.

The data discussed in this section includes growth accounting data, particularly on ICT assets for 14 countries provided by the EU KLEMS Project (2011), data on software usage in the personal computer sector (from the Wikimedia Foundation (2011)) as well as the server sector (from the Netcraft SSL survey 2009 (Netcraft Ltd., 2009)), data on the development of the number of hosts on the internet from the Internet Systems Consortium (ISC, 2011), as well as data from the US Computer Emergency Response Team (US CERT 2009) and the US National Vulnerabilities Database (US NIST, 2011) on computer security related incidents and vulnerabilities in ICT products. All of these data sources either only provide indirect information on the primary issues in the current work (i.e. network externalities and their effects on economic growth and technological change), or are probably strongly biased,[2] or both. They will, however, suffice to give some basic idea of how the issues discussed in this study so far behave in the real-world. Furthermore, the current section uses qualitative accounts on the development of ICT provided by a number of economic scholarly papers, as well as first-hand accounts by information scientists[3] and some more technical papers and documents.[4]

6.1 A brief history of information and communication technology

The history of the modern information technology[5] goes back to the mid twentieth century. While non-computerized communication technology is older and was well developed and common by the 1950s, it worked independently from new computer technology and later emerging computer networks. The integration of computer and communication technology on a larger scale did not take place before the advent of personal computers in the 1980s. Early computers and systems did not have much in common with what is known as computer technology today. The machines were huge, slow, and their capabilities were very limited. Expensiveness of the technology, lack of knowledge about it, and poor design of user interfaces due to the early development stage of the technology limited its use to large government institutions, large companies and major universities.

People were well aware of the new kind of the technology as well as of their capability to solve until then practically unsolvable (computation intensive) problems. Computer science pioneer Alan Turing dreamt of the development of artificial intelligence in the near future. The rise of the new technology must have been expected to at least some degree. On the other hand, programmers and engineers, as well as the developing institutions and potential users, reportedly and repeatedly failed to anticipate the imminent technologic revolutions and the intensity of the following boom. For example Ken Olsen, founder of the Digital Equipment Corporation stated in 1977 that 'there is no reason for any individual to have a computer in his home' (Cowan *et al.*, 2002). Although Olsen was probably not referring to PC-style computers but to mainframes or industrial computers, the statement reflects that he was certainly not aware of the large potential of the PC that was realized only a few years later. Cowan *et al.* (2002) have compiled a list of similar misjudgements by IT specialists.

And yet the computer technology of the 1960s and 1970s shaped the future development of the branch deeply. With a growing variety of vendors and machine types as well as more and more complex architecture, the need for widely recognized standards became apparent. Although at the time this was not so much for reasons of connectivity, but for reasons of compatibility, to make the replacement of current equipment (for modernization or expansion of processing capacity) possible at a minimum of costs, and to avoid depreciating existing knowledge and expertise in handling equipment of a particular design (Tanenbaum, 2001). Naturally, it is a major effort to make the same programs (or at least similar looking programs) run with technologically different chipsets. In turn, it would eventually become common practice to distinguish between *drivers* – and the hardware – and the *operating system* itself, which the drivers are designed to allow to run on different computers, and finally the user programs that merely have to be designed to work with a particular operating system without bothering with the implementation of the intended computations at the hardware level. Furthermore, the advantages of connected computer networks had been recognized for a long time and slowly became reality, starting in the late 1960s (Ziewitz and Brown, 2011). It was a major effort to establish a mechanism that would work with devices of different design, different operating systems and programs and would allow different computers to use the same physical communication channel at the same time for different connections. The technology chosen to solve this latter problem, *packet switching*, is to compile data to be sent from one device to another into a packet and add a header that describes – among other things – the destination device. For this, it was also necessary to design a universally accepted address system.

Although the first attempts often failed disatrously, by the end of the 1970s common standards for a universally compatible computer operating system (*Unix*), network protocols (*NCP – Network Control Program in 1970*, *TCP/IP – Transmission Control Protocol/Internet Protocol* in the late 1970s, formally adopted as a standard by US governmental institutions in 1980, and replaced NCP in the ARPANET in 1983), information exchange protocols (*e-mail – electronic mail* in 1972, *FTP – file transfer protocol* in 1969) and an operating network structure (the *ARPANET*[6]) were in place. Widely used programming languages had already emerged with *Fortran* and later *C* (later to be followed by *C++*) in the 1950s and 1960s.

In the 1980s some further standards were added, including in the network branch the *DNS* (*Domain Name System*), *UDP* (*User Datagram Protocol*), *EGP (Exterior Gateway Protocol)*[7] and *http* (*Hypertext Transfer Protocol*). In the operation system branch the *POSIX* (portable operating system interface for Unix) standard for Unix systems was specified in the 1980s, while the emerging sector of operating systems for personal computers was captured by the *Apple Mac* and the *Microsoft DOS/Windows* product line.

With the 1980s, IT-products became widely available. Hardware was more and more affordable, knowledge and ability to use them became more common. Connectivity to IT networks, thus what became known as the internet, was readily provided in universities and enterprises and eventually to the general public. This was due to the fact, that the technology was able to operate using standard telephone lines, thus taking advantage of a large well-functioning earlier network. This in turn enabled people with greater technical understanding to connect to the internet via telephone, and likewise set the stage for the commercial distribution

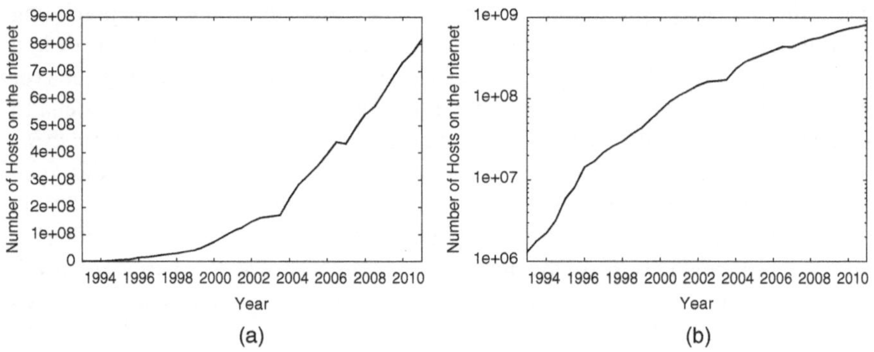

Figure 6.1 Growth of the internet in terms of connected hosts; right graph with logarithmic scale. Data from the Internet Systems Consortium (ISC, 2011).

of internet access. Immediately, the first online communities came into being, including mailing lists (in the late 1970s), online chats, online gaming (1980s) and finally the http-based world wide web (1990). The contemporary rise of offline applications like word processors, games, databases and graphic programming also contributed to the spread of of information technology in business and private life. The use of the internet rose exponentially, though it seems to have been slowed down to lower than exponential growth since the crisis of the 'new digital economy' in 2001 (see Figure 6.1).

These processes were accompanied by the gradual formation of particular standard setting organizations, as well as the growing awareness of the specific importance of standards and the possibility of controlling these standards.[8] These organizations, had very different backgrounds, from government institutions (e.g. *CERT*) to commercial coordination bodies (e.g. *RIPE*) and largely non-commercial organizations explicitly working on standard setting (e.g. *IEEE*), and finally to institutions that emerged from merely technical infrastructure entities (e.g. *IANA*). These organizations involve notably

- *IANA* (*Internet Assigned Numbers Authority*) established in 1972 by the primary network developer of the time, the DARPA (Defense Advanced Research Projects Agency), that belonged to the US Department of Defense. Its primary task is to maintain the unique naming of devices connected to the network in IPv4 namespace (see below). After the introduction of the DNS in the 1980s, the IANA started maintaining the DNS root as well (see below). Since 1988, IANA is a body sponsored directly by the US government.
- *IAB* (*Internet Advisory Board/Internet Architecture Board*) established in 1979 as a board formed by the chairs of several structures, that in turn were previously created by several organizations, including DARPA and other network developers.

- *IETF* (*Internet Engineering Task Force*) established in 1985–6 by the IAB to specify internet-related standards.
- *NSFNET* (*National Science Foundation Network*) was the network of the NSF that was to provide internet backbone services between 1981 and 1995, when it was decided to transform the hardware layer of the internet into a more distributed structure.
- *CERT* (*Computer Emergency Response Team*) established in 1988 in the USA to address the serious shortcomings of the security of the internet network design with respect to attacks from within (hacking). Since then, national CERTs have been established in most countries.
- *RIPE* (*Réseaux IP Européens*) established in 1989 as an independent body of several organizations involved in maintaining internet structures in Europe.

Qualitative data suggest a very rapid growth of the ICT sectors compared to other industries. It is, however, difficult to quantify this development reliably. First, products in the ICT involve a substantial part of non-material goods (e.g. software), i.e. what Quah calls the 'weightless economy' (Quah, 1999b). As a consequence, production costs for these are virtually limited to fixed costs, and since variable costs are negligible, vendors will not care how often the good was sold. What matters is just the gross revenue. If further network externalities are involved, there may be a strong incentive to give the first copies of the good away for free to create an initial user base. (For details on this strategy, see section 4.2.3.) There will also be no reason to keep track of the exact number of copies given away for free,[9] nor is there any way to tell if and how many of the copies given away for free are actually used. The same is even more true for software piracy including piracy that is directly or indirectly[10] tolerated by the vendors (see section 4.2.3). Second, common growth accounting systems were not prepared for the emergence of this sector, which would not just be a new and rapidly growing industry sector but would also revolutionize the way other sectors were organized.

Figures 6.2 to 6.5 show the development of the shares of the capital stock in ICT assets as well as (in Figures 6.8 and 6.5) the gross capital formation in ICT assets in all industry sectors in a set of developed countries. The data stems from the EU KLEMS project (EU KLEMS Project, 2011); the type of accounting, however, varies from country to country.[11] Consequently, different national growth accounting systems used different definitions of, for instance, what belongs to ICT investments and what does not. Especially for earlier times, this is obvious from the country level data as shown in Figures 6.2 to 6.5.[12] Hence, international comparisons of absolute values should be approached very carefully; however, it may be assumed that country level data is consistent in time, hence the development of the shares of capital invested in ICT is probably reliable.

The data (Figures 6.6 and 6.7) suggests that the growth of ICT (as indicated by the capital stock invested in ICT assets) is indeed exponential.[13] Furthermore, for many countries, there are periodic transitions from slower to faster growth and back. Roughly, three periods of rapid growth occur, in the mid 1980s, in the late 1990s, and after 2005. Country level data as shown in Figures 6.2, 6.3

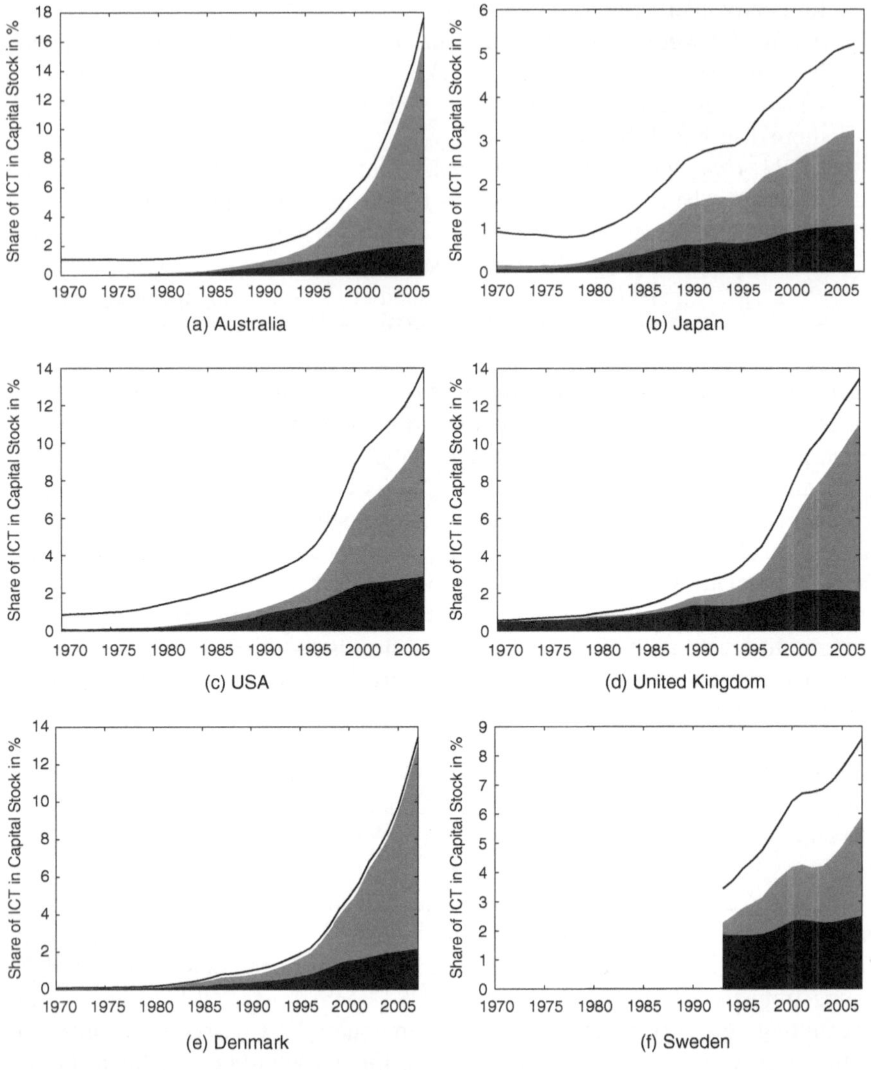

Figure 6.2 Share of software in capital stock (solid black), share of ICT without communication technology in capital stock (shaded), total share of ICT in capital stock (bold line); various countries. Real 1995 prices, data from EU KLEMS Project (2011).

and 6.4 confirms this pattern and shows that it does not originate in the (older and already more established) communication technologies: the graphs for ICT assets including and excluding communication technology develop proportionally in all countries except Slovenia and Finland.[14] This pattern is mirrored by the share of ICT assets in gross capital formation as shown in Figures 6.8 and 6.5.

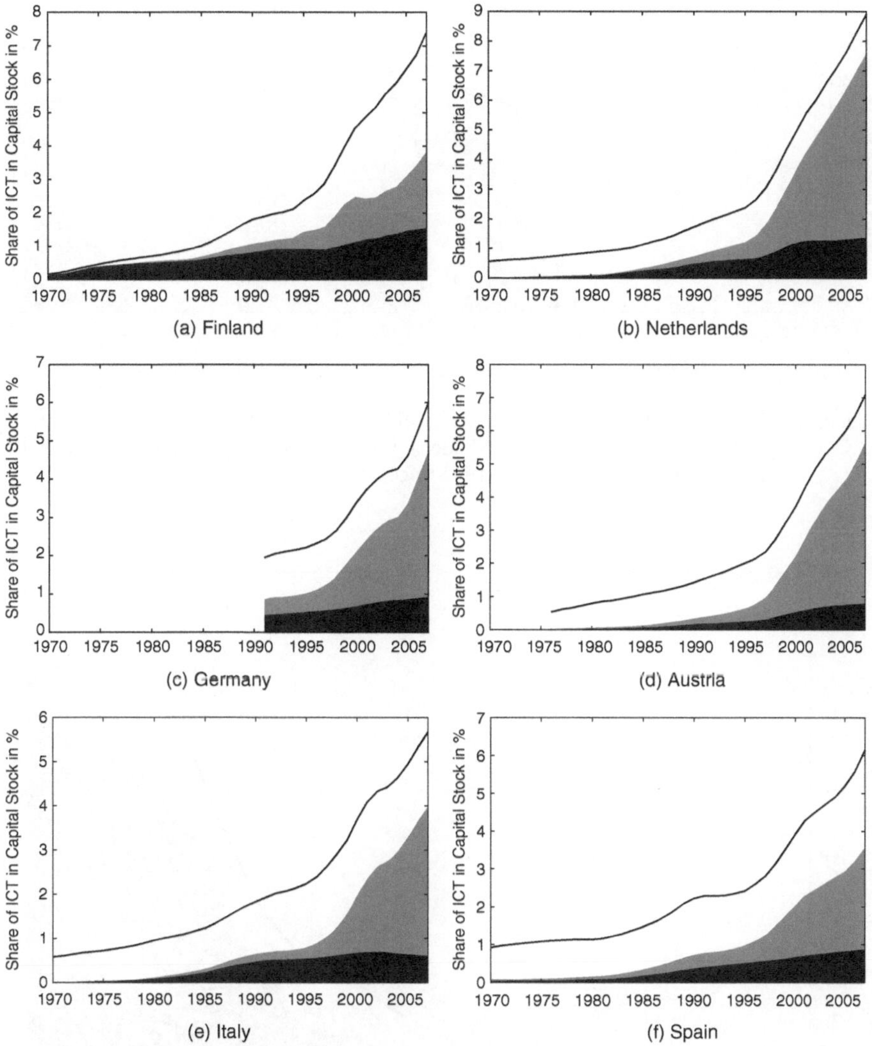

Figure 6.3 Share of software in capital stock (solid black), share of ICT without communication technology in capital stock (shaded), total share of ICT in capital stock (bold line); various countries. Real 1995 prices, data from EU KLEMS Project (2011).

Other indirectly related data confirm the rapid rise of ICT sectors. This includes the size of digital networks (see the growth of the number of hosts on the internet in Figure 6.1). While in the 1980s, hosts were mostly servers in universities, large firms, etc., which were maintained mainly for research and data exchange purposes, since the 1990s, most hosts have a more or less commercial nature

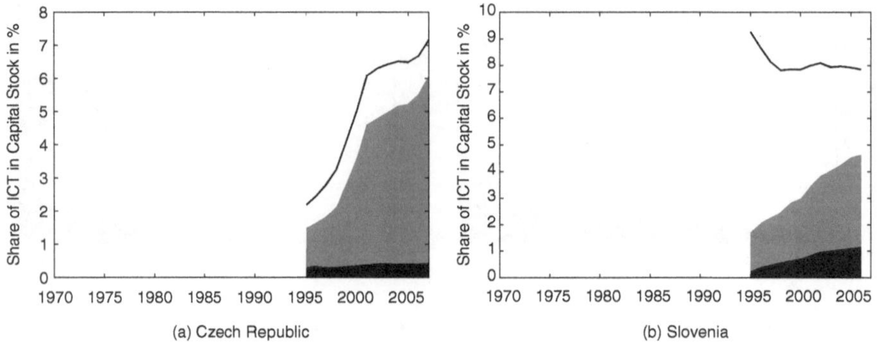

Figure 6.4 Share of software in capital stock (solid), share of ICT without communication technology in capital stock (shaded), total share of ICT in capital stock (bold line); various countries. Real 1995 prices, data from EU KLEMS Project (2011).

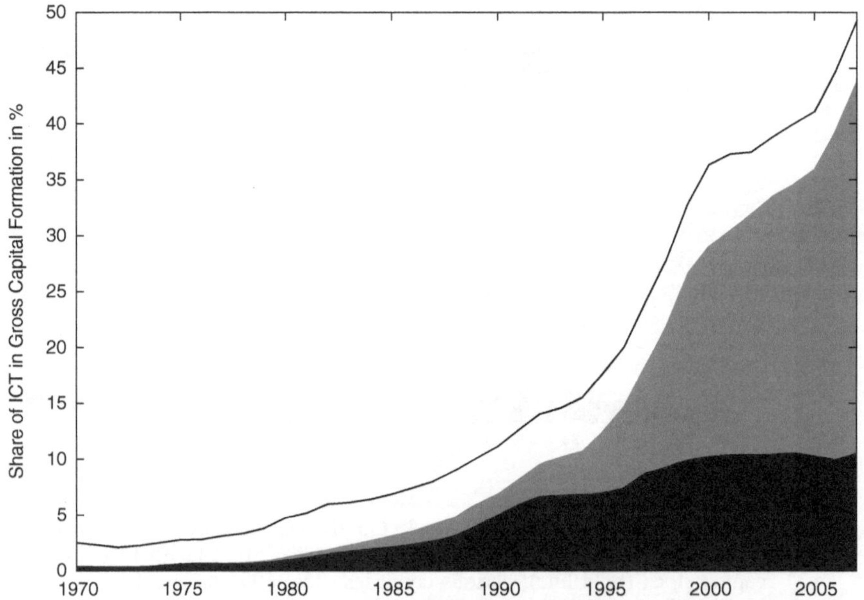

Figure 6.5 Share of ICT in gross capital formation in the USA. share of software in gross capital formation (solid black), share of ICT without communication technology in gross capital formation (shaded), total share of ICT in gross capital formation (bold line). Real 1995 prices, data from EU KLEMS Project (2011).

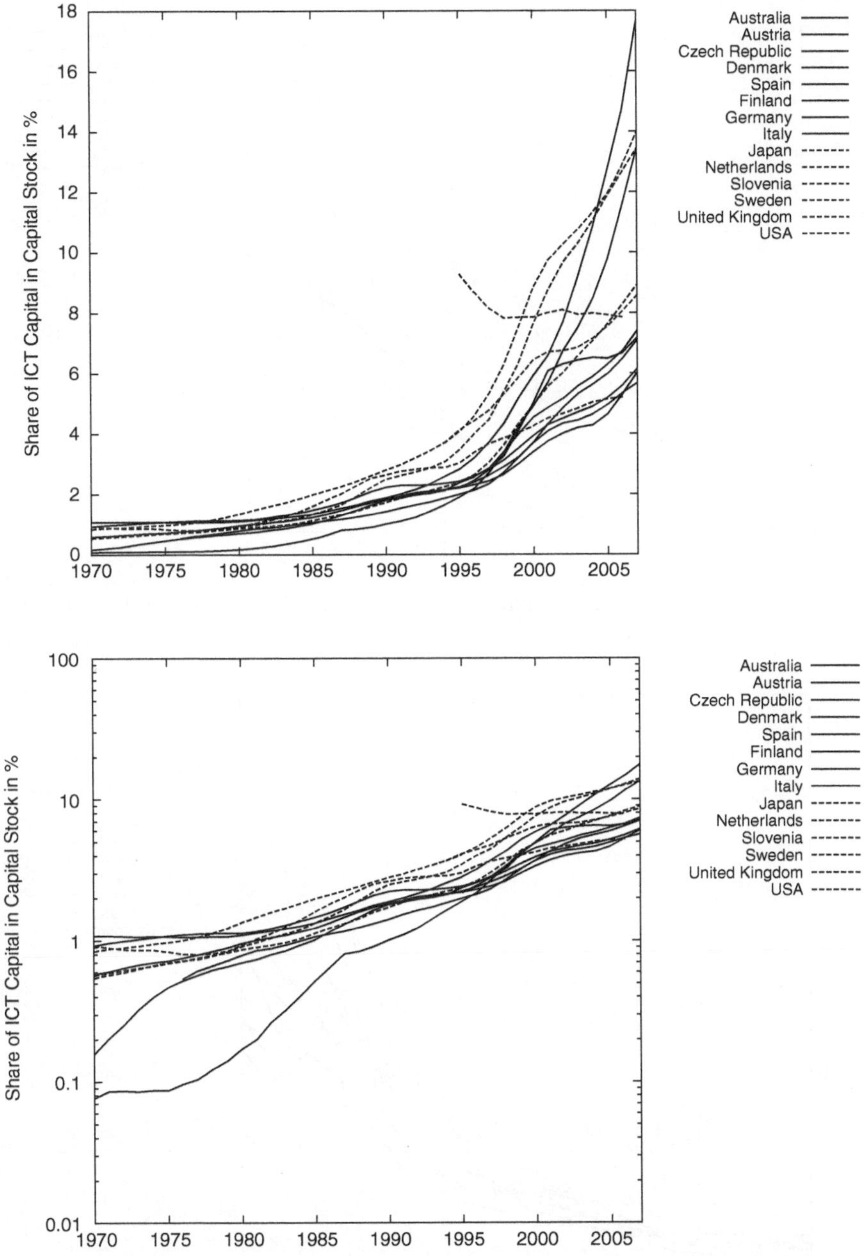

Figure 6.6 Share of ICT in capital stock. Real 1995 prices, data from EU KLEMS Project (2011).

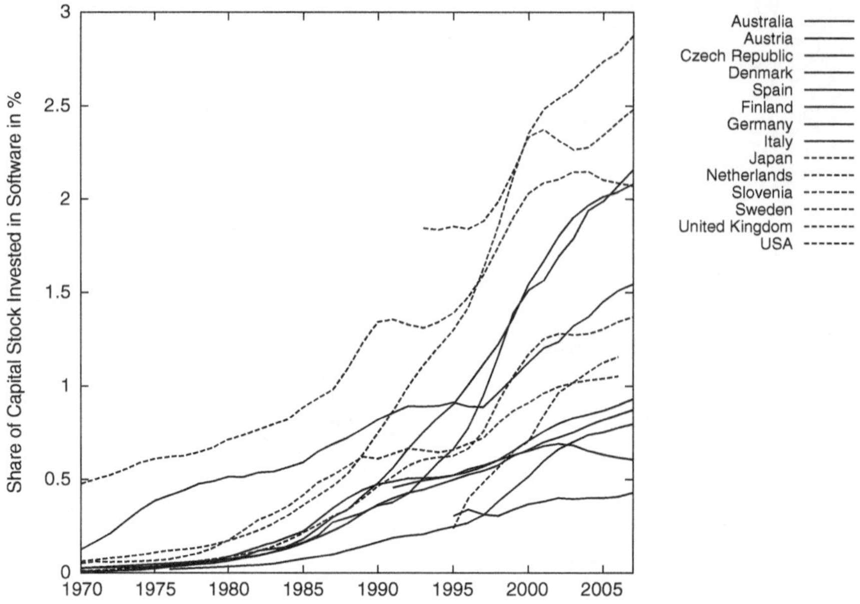

Figure 6.7 Share of software in capital stock. Real 1995 prices, data from EU KLEMS Project (2011).

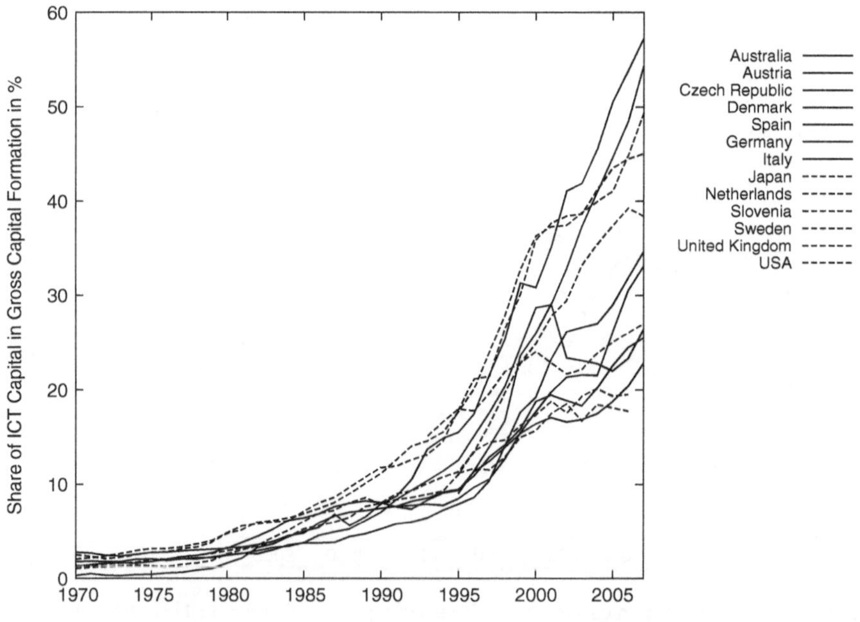

Figure 6.8 Share of ICT in gross capital formation. Real 1995 prices, data from EU KLEMS Project (2011).

serving the public representation of firms or direct sales and advertisement purposes. Furthermore, the 2000s saw for the first time a direct impact of obviously ICT-related developments on general economic growth with the 'dot-com bubble' (see Figure 5.1).

The patterns seen in quantitative data confirm that ICT as an industry sector (1) is rapidly growing, thus displacing shares of other sectors (in investments, capital stock and other variables), (2) develops in waves and (3) this development is internationally correlated. It is very likely, that the observed waves do not constitute proper cycles but rather repeating patterns of occasional significant technological changes and subsequent growth processes, a very similar pattern was observed in the simulations in Chapter 5. These periods of change and more rapid growth are in the mid 1980s, the late 1990s after 2005. It may be suspected that the growth periods are the era of the rise of the personal computer in the 1980s, the internet and 'new economy' boom in the late 1990s and finally the rise of social networks, the so-called 'Web 2.0' and the response of commercial vendors and advertisers to this shift of significant parts of the life of people into online communities in the 2000s. All these developments are subject to their own standardization processes and lock-ins.

Having considered the macro-development of the ICT sector in the current section, the following sections will focus on specific examples and the micro-level patterns of lock-ins and their effects on the industry structure in the ICT sector.

6.2 Standard wars, lock-ins and oligopolistic structures

6.2.1 Cases of lock-ins in information and communication technology

As discussed above, lock-ins occur almost naturally with standardization but they may be especially strong with network technologies. Devices must be able to work with the technology of the network and to communicate with other devices on the network or they are otherwise useless. The numerous standards of modern information and communication technology mentioned above provide a wide variety of examples of various cases and various ways to deal with the problem of incompatibility and lock-in.[15]

- Some technologies such as e-mail were developed as open and public pragmatic solutions for the developer's immediate problems or the immediate requirements of their work or research (Ziewitz and Brown, 2011).
- Others, such as the lowest layers of computer network protocols, were developed by the operator company; the protocols, were proprietary and incompatible. Instead of trying to implement a common standard, it was decided to develop a third layer protocol (after the physical layer and the lowest network protocol layer) that is able to integrate all proprietary designs

and to provide a unified mechanism of packet layout, addressing, etc. (TCP) (Ziewitz and Brown, 2011).

- On the other hand, the POSIX standard as a common basic definition of application programming interfaces (APIs) for the Unix operating system was commonly agreed on in a committee created by the IEEE for exactly this purpose (West and Dedrick, 2001).
- Many processes of standardization, however, have not taken place as explicit standard setting but rather by designing an at the time completely new product for a completely new sector and subsequently driving all competing products out of competition, at which point the technical properties of the product at hand become industry standard. This happened in most cases of standard wars; examples include the personal computer operating system[16] (West and Dedrick, 2001; West, 2005; Diker and Scholl, 2001) and web browsers (Windrum, 2004).

Further variants and mixtures of these approaches may have been employed for the creation of other standards. There are plenty of obvious cases of standard wars with different outcomes, but usually ending in the domination (but not complete monopolizion) of the sector in question. Gallagher and West (2009) have compiled an extensive list. As the most popular and well-understood examples, consider the following four cases:

- The competition between personal computer operating systems. The industry took off in the late 1970s with Apple as the at first uncontested industry leader. This was essentially because Apple's competitors including large, well-established IT firms (i.e. IBM, Altair and others) did not recognize ease of use as a crucial factor in general, and the benefits of supplying an operating system along with the machine in particular. Later, Microsoft succeeded in gradually pushing Apple out of the industry sector by working with hardware vendors (IBM and later Intel) (Bresnahan and Greenstein, 1999; West and Dedrick, 2001; Diker and Scholl, 2001; West, 2005). Crucial aspects of Microsoft's success strategy were reportedly the focus on being the first mover rather than quality,[17] tolerating product piracy,[18] and its strategic alliances with established IT firms.[19] The exact extent of Microsoft's success, due to the lack of reliable data, is difficult to assess; it appears, however, that Microsoft's dominance of the sector was already substantial in the late 1990s. Kretschmer (2004) presents data gathered in telephone interviews with industry firms in the United Kingdom in 1998 – almost 75% of the firms were exclusively using Microsoft operating systems for their personal computers, about 90% used mostly (more than 50%) Microsoft operating systems. In 2009 (after dominating the sector for at least 15 years), Microsoft Windows had a usage share of about 90%. (See Figure 6.9.) Apple, however, was never completely driven out by Microsoft (unlike a few minor competitors such as BeOS); with greater ease of use, better reliability and better integration of hard- and software, they were able to create a niche

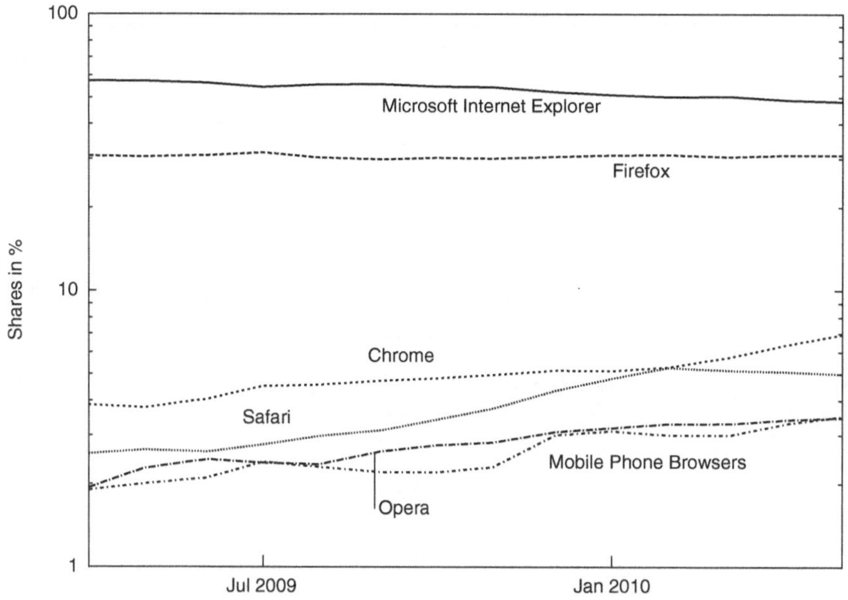

Figure 6.9 Usage shares of web browsers 2010–11, logarithmic scale, data from Wikimedia Foundation (2011).

which they – contrary to the expectations of some experts at the time (West, 2005) were able to defend. Recently, Microsoft also faces another growing competitor, the free and open source operating systems GNU/Linux and BSD,[20] systems that were originally more often used for servers and less for personal workstations. As such, however, they were superior to both Microsoft Windows and Apple/Mac in terms of network handling and network security[21] (Anderson, 2001, 2002), which slowly became more important in the personal computer operating system sector (or, at least, users become more aware of this fact). This is probably also why Microsoft products never became as dominant in the server operating system sector as they are in personal computer operating systems – according to data from the Netcraft SSL survey 2009 (Netcraft Ltd., 2009), Microsoft's server operating systems were loosing usage shares in most countries between 2006 and 2009, as shown in Figure 6.11.[22]

- Web browsers. With the rise of the http protocol and wide popular usage of the internet as a medium for distributing and acquiring information – preferably with sophisticated graphics – a new class of programs was needed; a sector with a huge potential as the growth of the internet during the 1990s (Figure 6.1) shows. After the initial success of Netscape, Microsoft entered this sector rather late but – mainly by packaging its browser Internet Explorer with its operating system Windows – eventually succeeded in taking over almost

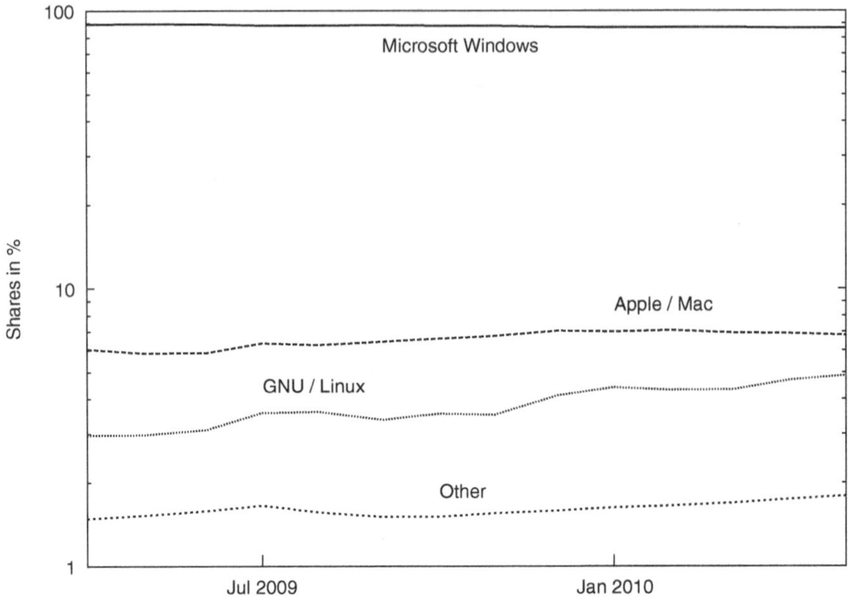

Figure 6.10 Usage shares of personal computer operating systems 2010–11, logarithmic scale, data from Wikimedia Foundation (2011).

all of Netscape's usage share. Netscape was subsequently abandoned as a commercial project, however releasing the software as free and open source (Windrum, 2004). The project (later called Mozilla Firefox) succeeded in becoming a major competitor to Microsoft's Internet Explorer again, although as of 2010 Internet Explorer is still dominant, as shown in Figure 6.10. Several smaller competitors have also recently entered the field, namely Google's Chrome browser and Apple's Safari which are now ported to the Windows operating system – an obvious attack on the Internet Explorer's and the Firefox's usage shares. What is remarkable about the web browser sector is again the existence of small niches. Not only was Firefox able to survive a long period of strong dominance by Internet Explorer, there is also the Opera browser, which has existed since the mid 1990s, since the very beginning of http; Opera seems to maintain a rather constant share of just above 1% of internet users.

- The Apache webserver. Maintaining the server side of http and the graphical internet (i.e. providing websites to be rendered by web browsers and viewed by internet users), Apache is one of the great and early success stories of the open source communities. Commercial vendors, primarily again Microsoft, try to compete with Apache (in Microsoft's case with the webserver Microsoft IIS), but remain largely unsuccessful[23] (West and Dedrick, 2001). Like Apache, many server-side applications are dominated by free and open source software

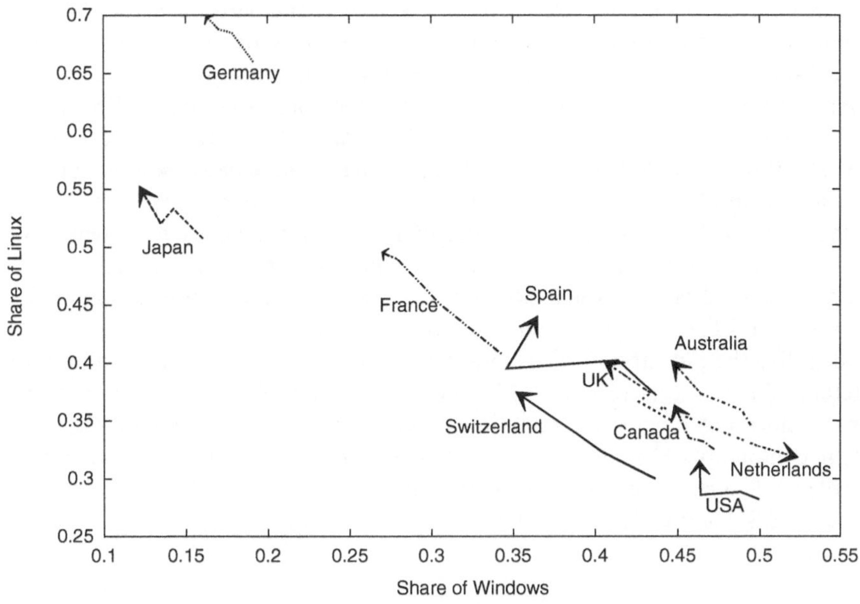

Figure 6.11 Development of usage shares of server operating systems 2006–9 for various countries, data from Netcraft Ltd. (2009).

(including e-mail servers, vpn servers, dns servers, etc.), Apache shall suffice as an example in this section.

- The Programming Language Java. Java entered the stage in the mid 1990s and was almost instantly an enormous success – even though it was widely thought to lag behind other comparable languages in performance, security properties and memory management (Prechelt, 2000; Oancea *et al.*, 2011). Java was distributed as a free and partly open platform with commercial backing and support (as well as ownership) by Sun Microsystems. Sun's strategy has been analyzed as to have used the openness and the emerging Java community in order to successfully contest with established competitors (Fortran, C++, Pascal, Python, Ruby, PHP, Perl, Delphi, .NET), both free and commercial (e.g. firms like Borland and Microsoft) (Luksha, 2008). Luksha called the strategy 'niche creation' (Luksha, 2008) – however, it seems more like an attempt to break into occupied niches by using the concept of free and open source software and the benefits of community style development. It appears that Sun has also employed the very same strategy in other cases, for instance by developing OpenOffice[24] as an alternative to Microsoft Office (with a set of OpenOffice native file formats, etc.) and by acquiring MySQL as an open competitor to Oracle's and other vendors proprietary database software.

Though these examples are merely qualitative, there are several remarkable aspects that are immediately obvious about the competition of standards in the 'new digital economy'. Most importantly, of course, there is the presence and prevalence of lock-in effects. Many scholars have discussed this issue from an economic perspective and also in the context of models of network externalities (Bresnahan and Greenstein, 1999; Anderson, 2001; Windrum, 2004; Kretschmer, 2004; Stango, 2004; West, 2005). It has also been noted that these network externalities, contrary to simple models, do not lead to complete monopolization (West, 2005); in the light of the above consideration of the cases of operating systems and web browsers this may be seen as a result of niche construction. In section 4.2.3 it was discussed how product differentiation can lead to a segmentation of the user base and hence higher switching costs, which may make it impossible to monopolize the industry. Furthermore, the ICT economy is a complex system of interconnected sectors (Figure 6.12); niches may be broken off certain sectors, other sectors may be merged, sometimes even deliberately as a strategic action by strong competitors;[25] as discussed in Chapters 4 and 5, this may lead to complex interdependencies. The diffusion of free and open source Unix-type operating systems (BSD and GNU/Linux) into the personal computer sector may serve as an example.

6.2.2 Some empirics

Though again quantitative data is difficult to obtain, data on personal computer operating systems and web browsers from the Wikimedia project (Wikimedia Foundation, 2011) suggests that software usage in these two sectors is closely linked. Correlation coefficients between the usage shares of operating systems

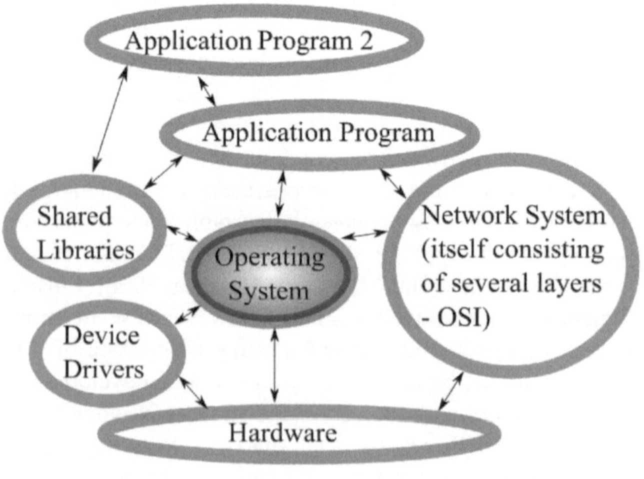

Figure 6.12 Complex interconnected system of personal computer operating systems (very simplified scheme).

Table 6.1 Correlation coefficients between usage of personal computer operating systems and web browsers, monthly data from Wikimedia Foundation (2011), data listed in Table A-2

	Windows	*Apple/Mac*	*GNU/Linux*	*Other operating systems*
Microsoft Internet Explorer	**0.912**	0.264	0.030	−0.497
Firefox	**0.956**	**0.750**	0.599	0.104
Chrome	−0.006	**0.650**	**0.824**	**0.930**
Safari	−0.097	**0.683**	**0.700**	**0.948**
Opera	0.158	**0.805**	**0.880**	**0.939**
Other non-mobile-phone browsers	0.380	**0.775**	**0.696**	**0.655**
Mobile phone browsers	−0.194	0.594	**0.723**	**0.981**

and web browsers in a sample of the monthly numbers of hits using that software between April 2009 and May 2010 are shown in Table 6.1. Recognizing that internet usage differs from month to month, the correlation coefficients should be high and significant between the numbers of users of different software. However this is only partly the case (t-test, confidence level 99%), in particular pairwise between (unsurprisingly) Microsoft Windows and Microsoft Internet Explorer, Microsoft Windows and Firefox, Apple/Mac and the browsers Firefox, Chrome, Safari and Opera (and miscellaneous browsers), as well as both GNU/Linux and other operating systems and several browsers except Firefox and the Internet Explorer. This suggests that there is a significant difference between the user groups of Microsoft and all browsers except Internet Explorer and Firefox, between the user groups of Apple/Mac and Internet Explorer as well as GNU/Linux and others and Internet Explorer.[26] This indicates that the development of software usage in the two sectors is indeed tied. Note that this only indicates statistical relations – the development of the usage shares of the software in question is correlated; it does not allow any conclusions about how particular users of a specific software would choose in the other sector (though this seems rather obvious from unrelated qualitative evidence). Another, similar, example is the cross country data on the usage of server operating systems (Microsoft Windows and Linux in particular) and web server software (Microsoft IIS and Apache) in a sample of 23 countries in 2009 as given in the Netcraft SSL survey (Netcraft Ltd., 2009). The correlation coefficients given in Table 6.2 indicate that a strong relation exists between Microsoft Windows as an operating system and Microsoft IIS as a webserver as well as between Linux as an operating system and Apache as a webserver.

Of course, the datasets analyzed so far focus on relations that are already more or less obvious from qualitative accounts. There are, however, also hints that suggest a connection between less obviously related sectors: Table 6.3 contrasts data on *server* operating system usage (Windows and Linux) and web browser usage (Firefox and Safari) on personal computers in a set of seven countries

Table 6.2 Correlation coefficients between usage of server operating systems and web server software, 2009 data for 23 countries from Netcraft Ltd. (2009), data listed in Table A-2

	Windows	*Linux*	*Microsoft IIS*	*Apache*
Windows	1.000	−0.944	1.000	−0.936
Linux		1.000	−0.943	0.998
Microsoft IIS			1.000	−0.935
Apache				1.000

Table 6.3 Server operating system usage shares and usage of Firefox and Safari as web browsers in seven countries, data from Netcraft Ltd. (2009) and Wikimedia Foundation (2011). Correlation coefficients between Windows and Firefox (relative), –0.79; between Linux and Firefox (relative), 0.79; between Windows and Safari (relative), 0.76; and between Linux and Safari (relative), –0.76

Country	*Server OS 2009*		*Web browser 2010 (just Firefox and Safari)*		*Relative values*	
	Windows	*Linux*	*Firefox*	*Safari*	*Firefox*	*Safari*
USA	**46.3**	31.5	13947011	5625562	0.713	**0.287**
Germany	16.3	70.0	4888089	693083	0.876	0.124
Spain	36.5	44.0	2407787	353650	0.872	0.128
France	27.0	49.5	2323592	595152	0.796	0.204
Italy	43.8	38.0	1279229	310977	0.804	0.196
Netherlands	**52.4**	31.8	346902	122845	0.738	**0.262**
Sweden	**56.5**	22.6	287327	97107	0.747	**0.253**

in 2009.[27] Countries with a smaller share of Microsoft Windows server operating systems (and a larger share of Linux servers) seem to have a larger share of Firefox usage compared to Safari. This must of course be treated as merely a hint not hard statistical evidence since the sample is very small and the data quality is quite bad as well. Another interesting fact (though also relying on very weak evidence) is the relative shares of Windows and Linux as a server operating system in different countries in the Netcraft SSL survey (Netcraft Ltd., 2009); as Figure 6.11 shows, English speaking countries (Australia, Canada, the UK and the US) form, together with the Netherlands, a cluster in the lower right (relatively high Windows, relatively low Linux usage) while the server operating systems sector obviously took a very different path of development in countries such as Japan and Germany (both in the upper left) and also different in France, Spain and Switzerland (middle part of the graph).

6.2.3 The weightless economy of ICT

Essentially, apart from, the compared to other sectors, a notably more obvious presence of lock-ins in information and communication technology, nine specific aspects can be identified in relation to lock-ins in the 'digital economy':

1. The ICT sectors in general are developing and growing rapidly (see Figures 6.1 to 6.5), hence losing usage shares may not always indicate absolute shrinking of the standard's network. The fast growth also allows for diversity and the continuing and rapid creation of new industries and niches.
2. Often, even after fierce 'standard wars' no complete monopolization takes place; rather weaker standards survive in niche industries (Java).
3. Open source communities are so successful in the ICT sectors that they evidently form an integral part of the strategies of commercial firms. Indeed, the presence and strength of open source communities may be linked to the success of the sector. In the modelling chapter (Chapter 5) it was found that openness of research and innovation is probably one of the strongest determinants of growth and technological change. The evolutionary economic literature describes this mostly as technology diffusion, and hence an indirect kind of openness. The very direct approach of the open source communities may have facilitated the rapid development of the ICT economy.
4. Industry structures are frequently strongly asymmetric in ICT sectors, which corresponds to the presence of one large competitor and several smaller niche vendors. It is also not always clear whether, in which way and to which extent the niche sectors are still a part of the greater industry sector they originated from. This may relate to the power law distribution of the size of the standard's networks, the importance of standards and as a consequence also the sector shares of the standard's commercial vendor as analyzed in section 5.2.4 above.
5. Tying of standards is evident and frequent, both directly as a deliberate strategic action[28] and indirectly as a consequence of competitors' actions[29] or user choices that may develop with path dependence (e.g. as a result of peer group effects).
6. Strong advantages for first movers. The first vendor with a sufficiently large network may, due to a rapid lock-in in this rapidly growing sector, dominate the sector for a long time.
7. Fierce 'standard wars' that may include tolerating software piracy or directly distributing the product in question for free. This may be especially important in the initial stages of the development of a sector in order to rapidly gain an installed base and possibly dominate the sector.
8. Many 'ad-hoc technologies' that were invented and developed as needed because the required infrastructure happened to be around and the technology was felt to be beneficial (e.g. e-mail, the RFC-standardization system (Ziewitz and Brown, 2011)).
9. Very open and fault-tolerant designs that evolved as a necessary response to too many lock-ins into proprietary closed standards in connected industries that needed a way to introduce compatibility. This is especially the case for crucial layers of the internet, i.e. the TCP/IP standard and the related DNS system. This design includes three aspects in particular: (1) it is open, in the sense that it is possible to connect and integrate one or many devices into the network without further changes to the remainder of the network and that the required technical information is openly available, (2) it is decentralized

and redundant, and hence robust against random malfunctions, and (3) it is designed as an end-to-end network, i.e. connections are made directly to another network node not requiring excessive intermediation.[30] This open design was most likely crucial in the great success of the internet. It is also a promising approach for overcoming at least some negative aspects of lock-ins; section 6.4 will go into more detail on this.

6.3 Technical consequences of lock-ins

Aside from the usual consequences of lock-ins, i.e. the failure to maintain a variety of alternative standards and the subsequent difficulties of applying possible improved designs, other severe results can be observed in the 'digital economy'. As this industry has been developing rapidly over the recent decades, consequences of the lock-ins may have become more apparent and much more quickly than in other industry sectors. One of these consequences was already discussed in the previous section: with too many proprietary designs of network connection systems at the physical and the data link (second) layer, it would have been very difficult to interconnect these networks. The solution implemented with TCP/IP at the network and transport layers (third and fourth layers) was a very ingenious one, in that it left the lower layers unaffected and just added open designs to run on top of the lower layers. While an adequate solution, it instantly became locked-in and very difficult ever to be replaced since that would again require the compatibility of all lower layer designs and eventually also higher layer applications without offering large incentives (such as general connectivity when there was none before as in the case of TCP/IP). There are many more examples, some of which will be discussed in detail in the current section showing that consequences of lock-ins may indeed be both complex (thus probably unexpected ex ante) and very undesirable. It should, however, also be pointed out that lock-ins are not an isolated inconvenient result of network effects. The increasing returns resulting from standardization or universal connectivity may greatly increase production in the sector or utility to the consumers (or both). Lock-ins are the side-effect that is generated by the very same mechanism (increasing returns to the number of users) and that may or may not consume a part or all or even more than the benefits resulting from this.

6.3.1 Emerging problems

The emergence of information and communication technology, especially the internet, is largely a great success story that has inspired scholars and writers around the world and surprised many people. However, while the technological framework was able to support the network even under such fundamentally changed circumstances and under conditions of exponential growth, the standards had to meet new challenges that no one could have anticipated.

As Evron puts it, 'Because the protocols used to run the Internet were originally designed in an open, all-inclusive environment, information streams can easily be falsified' (Evron, 2008: p. 123) Moreover, the underlying technology of the internet is not at all prepared for attacks from within the system, but was designed to counter external (military) attacks or otherwise catastrophic incidents by the distribution of databases as well as of communication systems itself on a network.[31]

Hacking as a phenomenon surfaced in the 1980s. And though hackers usually do not have a malicious intent,[32] hacking was correctly perceived as a serious threat to the network structure as well es the integrity of network-connected computers in terms of data security. Apart from the targeted activity of individual programmers to hack into remote systems, other even more dangerous threats were soon at hand. First there are, self-replicating programs, viruses and worms, that may or may not do any real damage to an infected system. They do, however, spread through the network in an uncontrollable manner. Second, as an improved version of the first category, there are cooperative self-replicating programs which may be able to communicate and act purposefully, they can be pre-programmed or their objectives can be changed later remotely. Such programs may download other malicious software or provide illegitimate access to the infected system (trojan horses) and may even group together to form so-called 'botnets' to jointly engage in much more powerful attacks. Third, the target of the attack might be the network protocols itself. For an overview see Asosheh and Ramezani (2008), for more detailed accounts of particular vulnerabilities see Kaminsky and co-workers (Kaminsky, 2006; Kaminsky *et al.*, 2010) and Vaughn and Evron (2006). For a history including an economic analysis of computer hacking see Coyne and Leeson (2006).

The very fact that a relatively great variety of these attacks do work, is stunning – even more so, since critical infrastructure for energy supply, government institutions, financial markets, logistics systems and even military systems depend on the internet. In fact, the internet was largely developed by a military organization (the DARPA) and was designed to serve at least partly military purposes. The reason is of course that an attack was assumed to originate from outside the information infrastructure. The networks are indeed largely invulnerable, even to weapons of mass destruction, since the data is distributed and if parts of the network break down, the remaining nodes will mostly still be connected and will continue to work. An attack from within the network, however, was not anticipated (Evron, 2008).

Of course, that these attacks are possible is due to identifiable and mostly known shortcomings in the computer operating system architecture and the network functioning.

6.3.2 Vulnerabilities in operating systems

Basically, a computer executes exactly and only the programs it is ordered to execute by the operators (the people working with that computer). In theory, if a potential attacker has no physical access to the computer, she has no chance to 'hack' into it. If, however, she can communicate with the computer via a remote interface or can get her files smuggled into the computer and executed on that computer, this is not necessarily true any more. The most common reasons for intrusions are (Tanenbaum, 2001; Coyne and Leeson, 2006):

1. permissions for remote interfaces not appropriately restricted
2. no file protection system (an attacker's files are executed or the attacker can write into system executables)
3. human failure (an operator accidentally executes the attacker's code; mostly due to insufficient knowledge and training)
4. vulnerabilities of the operating system and programs itself (a vulnerability to an attack method introduced by poor programming)

Of course, these reasons may be interconnected or combined to cause an exploitable vulnerability.

Again, it might be asked, why this can happen and, if it does, why it is not quickly corrected? Two answers to this question are important. (1) It may not even be detected; it requires a lot of knowledge and resources to scan a complex network of programs or of computers for intrusions. And it gets more complex, the more complex the system is, while an intrusion is made more easy at the same time.[33] (2) It might cause problems to switch to a more secure system in terms of compatibility with past data as well as in terms of effectiveness due to lack of experience, and lack of connectivity with other systems.

6.3.3 Vulnerabilities in the network infrastructre

To study the shortcomings of the network itself, it is important to review the working of the underlying protocols of the internet. First, a computer is addressed using its IP number, in standard IPv4 (Internet Protocol version 4) number, a quadruplet of 8-bit numbers (numbers between 0 and 255), which has to be unique on the network. TCP/IP (and other) packets are addressed to this number and are routed from node to node on the network until the packet reaches its destination.[34] For this purpose, each computer maintains a routing table that defines to which neighbouring host to pass packets bounded for certain destinations. To keep these routes consistent, the routers communicate using several protocols (most importantly BGP) announcing their routes to their neighbours. This structure depends crucially on correct detection of forged announcements, i.e. that the network 'believes' only route announcements from the servers that are authoritative for that route.

Second there is the DNS (Domain Name System): each computer on a network may have a URL (Uniform Resource Locator), a memorable name consisting of several strings (usually three) of characters separated by dots. The last of the characters denotes the top level domain. Top level domains (country domains such as .it or .fr as well as special domains like .org, .com, .gov) are managed by a domain name authority and recognized (or distributed) by the IETF. The DNS relies on a network of domain name servers that may be queried by other computers for the translation of URLs into global IP numbers; if a DN server does not know the answer, it will query another DN server. Again, this system relies on the observation of the sections of authority of the respective domain name servers.

Third, TCP (Transmission Control Protocol)/IP (Internet Protocol)[35] specifies the layout of packages and communications: IP specifies, among other things, that each packet is to contain source and destination IP and port.[36] TCP includes that a communication starts with a SYN package, replied to by a SYN ACK package, that packages are to be acknowledged by ACK, and that the connection may be reset by RST and ended by a FIN package.[37] Of course a source IP and port can be forged and the TCP communication may deviate from the specified order. Both will render effective communication dysfunctional and can be used for attacks.[38]

There are plenty of known attacks targeting the network structure. To mention a few:

- *DoS* (*denial of service*): 'Flooding' a host with malformed (or not even malformed) TCP connections. At some point it will exceed the target's capacity (in terms of network bandwidth or in terms of the number of open connections – TCP stack exhaustion) and it will subsequently 'deny' legitimate requests.
- *DDoS* (*distributed denial of service*): Like DoS but using a multitude of source computers for the attack, thus being more powerful and harder to detect.
- *DNS amplification attack*: A DoS attack is run not directly against the target, but DNS queries with a forged source IP address (the IP of the target) are sent to lots of DNS servers; as DNS responses can be very large compared to the query and as certain (malformed) queries can be predicted to evoke very large responses, the target will get a huge amount of incoming traffic consisting of unwanted DNS responses, while, for the target, there is no possible way to detect the attacker's real IP address[39] (Vaughn and Evron, 2006).
- *DNS cache poisoning by unauthoritative announcements*: DNS servers add additional sections to answers to queries from other DNS servers. These additional sections contain wrong information causing wrong domain name resolutions. The original servers are therefore unreachable while the connection is established to another host, possibly under the attacker's control and possibly used to 'spoof' passwords, access personal information, access bank account information, etc (Kaminsky, 2006).[40]
- *DNS cache poisoning by stealing authority*: Until 2008, DNS queries and answers were identified only by a 16-bit number (between 0 and 2^{16}). An attacker can trigger a query and send lots of answers with random 16-bit numbers as identifiers. She will succeed within finite time. In 2008, source port randomization, a method already used by several DNS servers,[41] was widely introduced, substantially increasing the difficulty of this form of attack from $1 : 2^{16}$ to $1 : 2^{32}$ (Perdisci *et al.*, 2009).

6.3.4 Empirical evidence

These vulnerabilities, both those in the individual computer system and those in the network structure have caused serious damage the whole time the internet has existed and will probably continue to do even greater damage. The US CERT provides statistics about catalogued vulnerabilities (Figure 6.13) as well as received incident reports (Figure 6.14). The incident report statistic, however, was stopped in 2003, when botnet attacks became more abundant and therefore one single attack could cause lots of incidents, rendering the statistic meaningless (US CERT, 2009). Nevertheless, it is clear that this does not make an incident less dangerous; with the rise of distributed attacks, the history of the internet has entered a new age in terms of security considerations.

Kaminsky and co-workers (Kaminsky, 2006; Kaminsky *et al.*, 2010), while analyzing the DNS vulnerabilities discussed above, stresses, that the vulnerabilities are temporarily fixed, but not entirely resolved. The security measures are insufficient given enough computing power on the attacker's part, or a more technically connected vulnerability. Cryptographic authentication methods could be available, but the system would have to be changed fundamentally. Furthermore, a wrong DNS resolution could be detected by cryptographic authentication (e.g. the use of SSL/https for http connections) but these are used very rarely. The same is true for fraud checking of downloaded software packages,

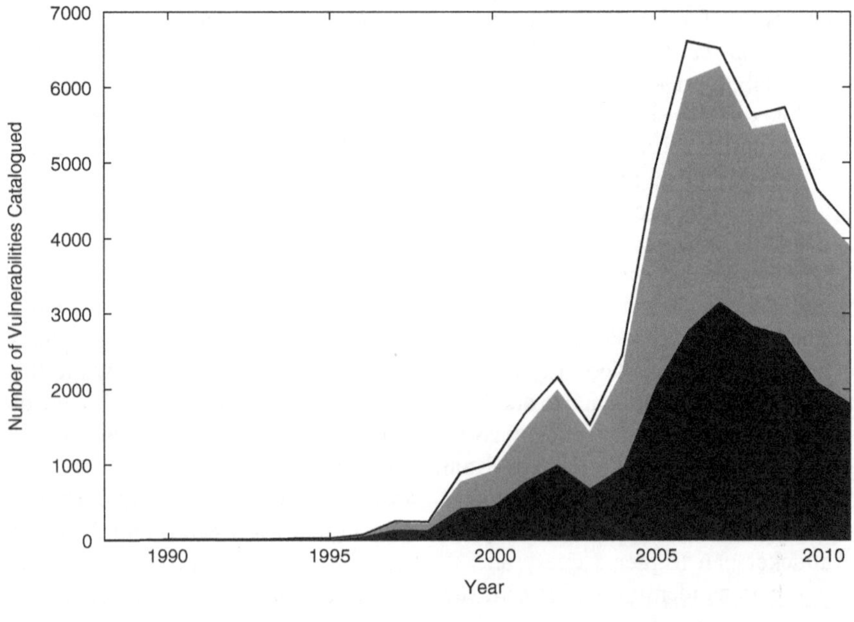

Figure 6.13 IT vulnerabilities catalogued in the US national vulnerabilities database; high severity (CVSS classification ≥ 7) solid area, high and medium severity (CVSS classification ≥ 4) shaded, all catalogued vulnerabilities as a line. Data from the US National Vulnerabilities database (US NIST, 2011).

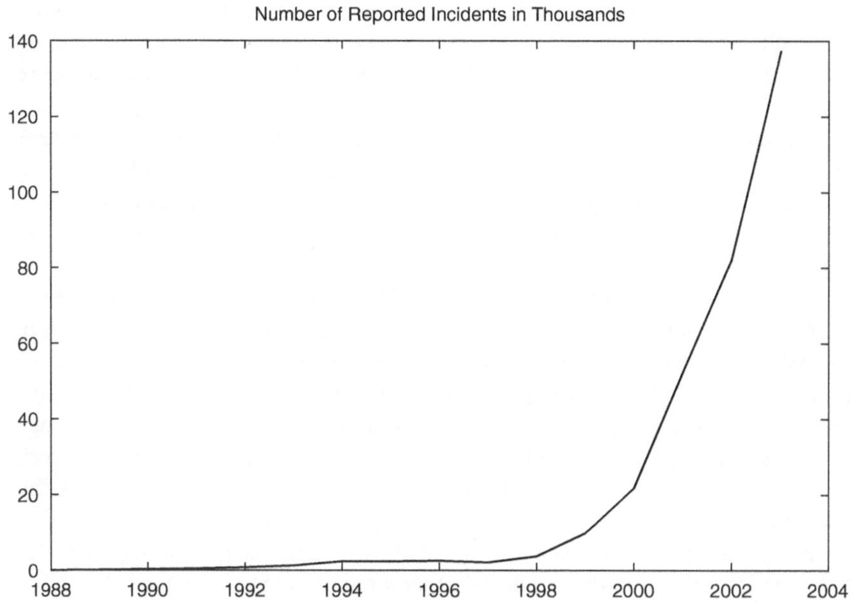

Figure 6.14 Incidents of attacks against IT systems reported to the US CERT; data from the US CERT (2009).

which should use sufficiently secure cryptographic algorithms like SHA256, but if used at all relies on the MD5 algorithm which is quite old; it was developed in the 1993 and was finally broken in 2004 (Kaminsky, 2004).

John *et al.* (2008) studied the composition of internet backbone traffic over several month and found that about 20% to 30% of the traffic is malformed and likely malicious, while another 40% to 50% is used by P2P (file sharing) applications. This means that vast amounts of internet traffic is at least unnecessary if not worse. Another interesting fact is that IP ranges that have never been used by real servers receive a substantial amount of traffic (Vrancken *et al.*, 2008).[42]

Finally, Evron (2008) provides a detailed account of distributed DoS attacks on Estonian servers in late 2006, apparently originating from Russia and motivated by nationalist attitudes. The damage was serious, NATO and other international and government institutions examined the incident (NATO, 2009). This exemplifies the vulnerability of more complex internet-dependent infrastructure (and therefore virtually all sections of modern economies and societies).

As illustrated in the case of DNS vulnerabilities, cryptographic hashing and insecure http-connections[43] above, the problem is not that there are no better alternatives. There are even very functional alternatives that have been used in practice for many years and that could be used by everyone (djbdns, SHA256 and SSL); however, there are still some remaining compatibility issues: A webserver administrator may fear that her clients are not able to use SSL or that

SSL-incompatible browsers are still in use and that people would rather stop using her web service than update their software. Therefore, she may decide to work without SSL on the server side,[44] thereby preventing other users from connecting with SSL and implicitly suggesting that unencrypted http connections were secure enough.

Of course there are also less security-critical failures to switch to better standards for incompatibility reasons. For example, the IPv4 address space will be depleted by the end of this decade; IPv6 is, however, virtually unknown, although it was standardized as the successor of IPv4 more than a decade ago (Vrancken *et al.*, 2008).

In short, the history of the internet is not only a history of creative adaptation of an infrastructure built for a completely different reason (e.g. telephone wire) but also a history of obstacles to the success of fundamental innovations and lock-ins with severe technological consequences. Moreover, the gap between knowledge and necessity of knowledge has largely been closed not by increasing knowledge but by lowering the necessity to know;[45] the congestion of capacity of the network by malicious or at least irrelevant traffic has largely been resolved by increasing the capacity.

Nevertheless, there are also a few positive examples of well-functioning security-critical IT standards, e.g. the BSD server operating system class (West and Dedrick, 2001) and stream ciphers (Bjørstad, 2008) involving creative discussion in an open community.

6.4 Open standards and other strategies: implications and attempts to solve the problem of lock-ins in ICT

The case of the TCP/IP protocol as discussed above in section 6.2 is instructive for analyzing lock-ins especially in the ICT sectors. When the protocol (or rather, the two combined protocols, TCP and IP) slowly emerged in the 1970s and 1980s,[46] there were a lot of incompatible protocols running in different networks and it had been recognized that this posed a problem. The chosen solution was to add gateways for direct transmission of information between the networks and, more importantly to install a generalized mode of addressing and packet handling and transmission.

GATEWAYS

If several incompatible systems are competing in an industry, causing all the negative consequences, especially that the benefits resulting from network effects cannot fully be enjoyed (because parts of the network are incompatible with others) and that the variety of technologies may decline with the ensuing standard war, a prudent solution, if applicable, is just to remove the incompatibility. For technological networks that means adding a device that is properly connected to both networks and 'translates' for communication between the networks. The concept is well understood (David and Steinmueller, 1994) and is used for problems of physical incompatibility, but also for translating between incompatible network protocols,[47] for connecting physical devices with different interfaces (by using a converter between the two), running software on another operating system than it was compiled for (using emulator software). From the economic point of view, the solution is sound: with the incompatibility removed, the whole extent of the positive network effects becomes immediately available for all users, the competition through network size (instead of intrinsic properties of the product or price) is removed and it becomes unlikely that one competitor would drive another one out of the industry.[48] The problem, however, is that this solution is often impractical. Additional hard- and/or software or at least additional configuration is needed, and, more importantly, in the case of gateways between networks, the entire inter-network traffic and communication relies on the gateway. That is, if the gateway breaks down, if only because it cannot handle the amount of traffic, communication between the networks ceases. The gateway will have to cater for translating each packet (possibly a great effort) and the users would nevertheless have to cater for using proper addressing for all the networks the packet is intended to pass. (The great strength of TCP/IP protocol was to add another layer with generalized and open specifications that made direct handling of the lower layer network by the users unnecessary.)

STATE INTERVENTION

Another way to deal with lock-ins is regulation. This is not a satisfactory solution, but rather a strategy to shape the race for the industry standard in favour of the standard candidate that is expected to be the most desirable. Furthermore, and at least as important, the race is brought to an end, hence generating certainty for the actors that would have hesitated to join one of the networks out of fear of choosing the wrong one.[49] Regulation may also succeed in bringing incompatible networks together by just prescribing a standard, and it may help to facilitate the eventual break-out from the lock-in as soon as a promising better technology is available. It does, however, not offer a solution to the lock-in as such, since this will nevertheless ensue, and even faster than it would have without intervention.

In terms of slow and fast growth rates that were analyzed to be the result of lock-ins in Chapter 5, state intervention is limited to keeping the time lag between the invention[50] of a promising technology and its successful standardized implementation as small as possible (hence shortening the periods of slow growth) and ensuring that the best possible technology becomes the next industry standard. Still, it may be questionable how the government (or any standard setting body) can always know in advance which standard would turn out to be the best choice. It may also be necessary for a successful intervention to happen at the right time, the window of opportunity (David, 2005) before and after which the system's dynamics may be too strong to respond to simple endorsements or specific legislation. In addition, commercial vendors may take the chance to try to shift the system in their favour by issuing an endorsement of some standard.

Official intervention is most likely an important tool, but that alone in itself is not a satisfactory solution.

OPEN STANDARDS

The arguably most striking argument in favour of TCP/IP was its openness. The specification was publicly available and even open to discussion, the design was open, fault-tolerant and decentralized without any possibility for any actor to exercise asymmetric control over the standard or the network as such. Complete openness (as discussed in section 5.3.4, a crucial aspect in technological progress) makes it possible to anticipate problems in advance, to reuse the technology in suggestions for technological improvements, and also for the actors to trust the network, to trust its technological performance, and to trust the other participants. Apart from the conceptual openness of the standard (i.e. the public availability), the technological openness convinced participants that the network would run at almost any size,[51] whether only two or three commercial networks were linked using the TCP/IP protocol or millions and billions of computers around the world were linked; redundant network layout makes the network also robust against the failure of individual nodes. Conceptual openness proved crucial not only to that, but also to many of the later developments which eventually gave rise to the open source communities. Section 5.3.4 emphasized that all the models under consideration rely crucially on the openness of technological knowledge, which triggers a network effect of the research side. The body of preliminary research one can rely on in the development of a technology is much greater if all or some of the technological knowledge is public. It induces a positive feedback as the original developer may in turn employ the further developments based on her technology. The non-open part of the industry, reduced to relatively slow rates of progress but not excluded from using the public and open knowledge is pulled along. A growing body of research discusses open standards as a concept both for resolving negative effects of lock-ins and for improving technological change in general.[52] Yet, openness does not prevent lock-ins; it just prevents negative effects such as the failure to recognize and resolve shortcomings (as discussed above), obstacles to further development brought about by the proprietarity of the technology, and unresolved incompatibility caused by non-public definition of the standards.[53]

It may be concluded that none of the solutions considered is a universal one but they (especially open standards) constitute promising strategies and may – as in the case of TCP/IP be combined. There are a number of related issues recently discussed both among economists and in the wider public, all related to openness. The remainder of this section will discuss them briefly.

OPEN SOURCE COMMUNITIES

Open source communities are a phenomenon largely unique to the development of digital products (though attempts to employ user led innovation and open source in other areas have been made (von Hippel, 2001)). What is striking for standard economics about open source communities is the simple fact that developers work for free and provide sophisticated and well-maintained products without charging fees to compensate themselves for the work put into development of the products at hand. Open source developers do not appear to be motivated by monetary incentives[54] but rather by social recognition, reputation, self-fulfillment (i.e. intrinsic motivation) and practically solving problems (Raymond, 2000; Lakhani and von Hippel, 2003). Open source as a concept has roots in how programmers worked before the take-off of the ICT industry[55] and finally materialized in the 1980s with the founding of the free software foundation (FSF) by Richard Stallman and others. Subsequently, a wide variety of open source software was produced (or commercial software made open source) including operating systems (GNU/Linux, BSD), server-side software (Apache, Fetchmail and others), code libraries for the implementation of specific things (notably OpenSSL for strong encryption, and many more) and a huge number of user applications for all major operating systems. Open source projects soon received substantial funding and support from commercial firms who recognized that open source development was in some cases simply more successful than closed source approaches.[56] Among the benefits is the efficient integration of potential user's suggestions and wishes into the development, but also the involvement of others in bug-fixing (reporting and solving flaws in the program) and development by openly discussing the source code.[57] Arguably, open source and the technical impossibility of preventing diffusion of knowledge contributed a great deal to the rapid development of ICT during the past decades, and may be one practical example of the importance of openness for strong and rapid technological change and growth. It has been suggested that the rise of digital goods in general and open source in particular has made digital information non-excludable, and hence turned knowledge generation and essentially research (at least as far as communities on electronic networks are involved) into a public good (Elsner *et al.*, 2010).

INTELLECTUAL PROPERTY RIGHTS

A related issue is the debate around intellectual property rights. Following the Coasian assumption that property rights must be complete for a 'market' to function properly (Coase, 1960) and also the rise of the 'new digital economy' there is a movement directed towards establishing stronger intellectual property rights, allowing the patenting of increasingly trivial things like pieces of code, story lines and sports moves but also genomic products, life forms, etc. (Boldrin and Levine, 2009). It is easy to see, and has often been argued, that this is not only ethically questionable but also has a negative effect on technological progress and economic growth in the sector (Adkisson, 2004; Simcoe, 2006; Boldrin and Levine, 2009). Reasons for this include: portfolio-only patents, i.e. patenting just for the value of the patent without intention to use the patented technology; strategic patenting, i.e. patents aiming at preventing others from using or charging others for using a technology; but also other patent-related obstacles to growth without deliberate strategic intent from the part of the patent holder. For instance, with a large number of patents on basic findings and mechanisms, it may be increasingly difficult and costly to conduct further research or use such technologies in innovations (as an increasing number of licenses would have to be obtained first). Essentially, intellectual property rights threaten to destroy the technological openness that was found to be integral for technological progress in section 5.3.4. Furthermore, empirical studies were also unable to detect any supposedly negative effects of 'digital piracy' (i.e. illegal file sharing) on the performance or the revenue of the holders of the intellectual property rights in question (Oberholzer-Gee and Strumpf, 2010).

NET NEUTRALITY

Finally, a debate has taken place about the neutrality of information transmission on the internet (Economides and Tåg, 2011; Ganley and Allgrove, 2006; Choi and Kim, 2010), that is, whether or not service providers would be allowed to treat packets differently depending on their nature, content, origin, provider of the origin's network, destination, or network provider of the destination node. Most of the debate developed in public between politicians, representatives of commercial interests and social groups, but eventually it hit economics – though most economists agree that deregulation (abolishing the requirement to honour net neutrality) would most likely be harmful. This is true for both mathematical model approaches (an overview of which is given by Choi and Kim (2010)) and evidence from attempts by private companies[58] to regulate the internet. Needless to say, allowing internet service providers to interfere in packet transmission and priority would (or at least could easily) disrupt and destroy the very foundation of the open, decentralized, flexible, end-to-end network. Ganley and Allgrove (2006) offer some rather stark examples of what might follow – indeed what has already happened – if net neutrality is not honoured, including blocking e-mails with specific content (criticizing the provider in question), blocking competitors VoIP services or blocking access to trade union websites. Though these examples provide an overly pessimistic view, the argument makes clear that the issue includes a political and strategic element that would prove dificult to control, not just questions of regulation or deregulation in an otherwise 'perfect market'. In all, net neutrality is one of the central elements of the internet as a system of – mostly – open standards that was crucial in facilitating the later phases of the rapid technological development of the ICT sectors.

6.5 Lessons from the 'new digital economy'

Reiterating the conclusion of the model in Chapter 5, in the light of the current case study it has to be stated again that network effects, lock-ins and standardization are not primarily an obstacle to rapid growth. Rather, they are the very structure that by means of positive feedback loops makes rapid growth possible. It is a side effect that it has also severe and very likely socially harmful consequences for the industry structure and will, at the same time as they facilitate the current wave of rapid growth, retard or even prevent the next one, by making breakout of the technological lock-in very difficult. Paradoxically, it would be most desirable to be locked-in and not locked-in at the same time – which, of course, is absurd and impossible. Nevertheless, ICT teaches some important lessons about how to at least approach this unattainable goal – most importantly by enforcing technological openness and if possible using open standards (alongside, of course, technological gateways as needed and a sound technology regulation policy if possible).

It has been established that the ICT sectors are – unsurprisingly – subject to network externalities and lock-ins – however, with a number of specific phenomena. These include the very complex industry structure with a complex pattern of interlinked and tied standards, some of which are open standards while others are not. Open standards as well as, arguably, the remarkably strong open source communities in this sector played a crucial role in facilitating the rapid growth of ICT over the recent decades. The complex interlinked structure alongside rapid growth, in turn, facilitates strategic defences against stronger competitors in 'standard wars' by creating niche sectors. Less desirable phenomena that have been observed include the shipment of immature products (to secure first-mover advantages) as well as the use of infrastructure for purposes which it was never intended for, which can both lead to substantial shortcomings.

Alongside the crucial result of the importance of openness for technological change and progress, which could be confirmed, a wave-like pattern has been found in macro-data for ICT (with phases of rapid growth in the 1980s, the 1990s and since 2005). This may support another central result of the model in Chapter 5, namely the oscillating pattern of slow and rapid growth that accompanies the development of a complex structure of industry sectors with strong network effects and lock-ins that change, not so much by internal developments, but rather by developing derivative industries that feed back to the original industry as well as other co-existing sectors that work to mitigate lock-ins. One prime example is the diffusion of open Unix-like operating systems (i.e. GNU/Linux and BSD) from the server operating system sector into the sector of personal computer operating systems. This may not only support the finding that growth cycles (or waves) are related to network effects, but also that waves in different sectors may couple, in turn creating resonance effects and hence wave-like patterns at the macro-level.

7 Conclusions

The technologies, the networks and the next wave of growth

7.1 Economic growth, technology and network effects

Two things are remarkable in general about the mainstream theories of economic growth judged from a more heterodox perspective. First, there is very little that is established and undisputed about economic growth. As certain literature traditions of Marxian and ecological economics (Daly, 1968; Georgescu-Roegen, 1975a,b; Minsky, 1980; Smith, 2010) show, it is even disputed if economic growth (1) really exists and (2) is a good thing at all. However, even what the different approaches to economic growth *claim* or *imply* to know is still rather meager: for more mainstream economics this seems to be limited to statistical relations of certain country-level macroeconomic data, hence the conclusion that different macroeconomic data series relate and correlate in certain ways. This is what is both the theoretical core and the focus of empirical justifications from[1] Solow (1956, 1957) to the theories of endogenous growth (Cass, 1965, 1966; Uzawa, 1965; Lucas, 1988). Some newer and more sophisticated approaches add aspects of simple microfoundations like perturbations as the origin of cycles (Lucas, 1972, 1975, 1981), non-linearities (Romer, 1986), and the effect of technology and technological change (Aghion and Howitt, 1992). Evolutionary growth theory (Nelson and Winter, 1974, 1982a,b; Nelson *et al.*, 1976; Conlisk, 1989; Silverberg and Lehnert, 1993, 1994) and other alternative approaches[2] are more serious in including microfoundations, thereby emphasizing the importance of direct interdependence, innovation, technological change and technological diffusion. This results in more robust agent-based models that also imply certain simultaneous developments on the micro-level, such as the industry structure, research investment, etc. (which indeed have been extensively studied by Nelson and Winter (1982a,b)). Still, the real microstructure of firms, entrepreneurs, employees, etc., as well as the microfoundations for many macro-level phenomena are somewhat obscure.

Second, on the other hand, given the complexity of the system, the detail to which the available growth models illustrate how the process supposedly works and how the (rational) reasoning of the agents supposedly functions, is astonishing. Studying such models, the feeling imposes itself that fitting the models into certain commonly used narratives is more important than living up to economic reality.

This, in turn, seems to happen at the expense of the simplicity (in the sense of making as few assumptions as possible) of the model. To explain constant growth rates, i.e. exponential growth, a simple exponential function suffices. Nevertheless, in growth models that try to reflect an empirically observed growth with an on average approximately stable[3] growth rate, a large number of variables with complicated interrelations, including capital, labour, output, costs, price and the market are usually represented. Endogenous growth theory also extends the models to force a rational optimization device in. Though this helps to generate a comprehensive neoclassical theory that integrates with other neoclassical models, whether such a construction and the necessary assumptions are justified is highly doubtful (Kaldor, 1972; Chen, 2002).

Still, the many models of economic growth offer a vast repertoire of tools that may play a part in real-world growth, though the focus on the tool, the crucial mechanism of the model that generates the desired behaviour of the model is often lost. The careful observation of such mechanisms may – given sucient empirical data and more sophisticated theories of how these mechanisms may work in reality, how they may or may not interact and with which results, this rich collection of powerful tools may in the future help to form a more satisfying theory of economic growth. For now, however, it is unlikely that the arrangement of the variables and interpretations considered there – usually in great detail – holds for the socio-economic reality. After all, these models – no matter how complex – neglect several important aspects. Among these aspects are widely prevalent positive feedback mechanisms and especially network externalities that lead to 'markets' with increasing returns and all other sorts of – from a mainstream point of view – irregularities. The current analysis has tried to incorporate into a theory of economic growth and technological change the aspect of network effects and resulting externalities, lock-in and positive feedback. There may nevertheless be plenty of further phenomena contributing to a completely 'irregular', non-neoclassic structure of economic reality, as many analyzes by Albin (1982, 1987), Mirowski (1989), Arthur (1989), Albin and Foley (1992), Foley (1998), Keen (2001, 2009b), Chen (2002, 2005, 2010b) and others indicate.

Indeed, growth models are on the one hand not very successful in explaining the socio-economic reality, and on the other hand rather heroic assumptions and detailed claims concerning the results are made. This is to some extent also true for the model of the current thesis, though care has been taken to keep the models as simple as possible for the mechanisms under investigation to work. The analysis of interaction structures with network externalities in Chapter 4 suggested that economic systems do almost generally include some form of network effect, and will thus result in non-linearities, path-dependence, lock-ins, etc. Path dependence is simple to model and will almost certainly emerge if the system at hand is complex enough. An example of a path dependent growth model consisting only of an output, a capital and a technology term[4] was given in the last section (before the chapter conclusion) of Chapter 4. In Chapter 5, this model was extended to an agent-based version and eventually a simulation model. Finally, it has been attempted in section 5.2 to bridge the micro-to-macro gap that poses a problem for so many of the growth models, and that only during the last few decades both neoclassical and heterodox scholars have tried to close.

7.2 Finding ways to unleash the next wave of growth

Starting with a relatively simple agent-based vintage production model with cumulatively improving technology and scarce labour, it has been determined that with the addition of network effects quite realistic growth data patterns are generated. The model accounts for superlinear growth in output and capital stock, for waves of growth, for occasional technological paradigm changes, as well as for successions of oligopolistic regimes that remain each stable for some time. Of course, the extent to which the model could be used for forecasting has yet to be examined by matching the model with empirical growth data – which is, however, beyond the scope of the current theoretical analysis. Further improvements and calibrations to the model will also be necessary. The model notably lacks a demand side[5] as well as a financial sector;[6] research-side network effects have not been investigated in detail in the current study, and the various macro-level effects as discussed in section 5.3, including scale-free distributions in a number of variables and resonance between sector-level cycles, also warrant further analysis.

To the knowledge of the author, it has never before been attempted to include network effects of technologies into a theoretical model of economic growth and technological change. As discussed, network effects may, contrary to what is usually assumed, be a very common phenomenon. They change the basic dynamics of economic systems substantially, and even to an extent that does not allow these systems to be modelled realistically with usual standard assumptions such as decreasing returns. The model, as developed in the current study, may be used to explain persistent (but probably irregular) fluctuations in the growth rate and also the prevalence of oligopolistic industry structures, especially in new, dynamic and recently developing industries.

Among the results are two further important insights. First, the process of lock-in to one technology standard, the related efficiency gains, and the subsequent expansion of the respective industries is the part where the largest growth effects are realized – although this very lock-in becomes an obstacle to further progress later on. This process could, given the knowledge and the power, theoretically be guided. Second, most obvious in this model, but not limited to it, is that the openness of technological knowledge, and thus the diffusion of innovative insights, is what drives the long-term speed of economic growth. Recent technological developments (the entire ICT economy) may be used to facilitate this productive diffusion of knowledge, an opportunity that may be taken or may be missed, something for which our time will be remembered in history.

In fact, the core of growth policy may not lie in either taxation or monetary policy but rather in the policies of technological development. It is probably not so much a problem of how to divide the cake, but rather one of how to make the cake bigger, a problem of standardization, public knowledge as well as expertise (human capital) and industry structure instead of an allocation problem; it may include aspects of public good problems, coordination problems and social dilemmas. Most crucial, however, is to recognize the time to break-out of a technology regime and to open the field to new innovations. Being aware of the fact that this induces friction and inefficiencies, economic actors could strive to accomplish such transitions in a cooperative, sustainable (for instance, using open standards) and least harmful way. Open standards, open source communities, net neutrality, intellectual property rights, software piracy and a number of court cases involving monopoly and market power are among the crucial factors in this field that are currently about to be decided, and that will shape not only future economic and technological development but also how contemporary society will be judged, and what it will be remembered for.

Returning once more to the case of the invention of the printing press independently in China and Europe, the only possible conclusion is that history may have turned out very differently, either with China being the cradle of modern industrialism or – had the crucial technologies failed to take off in both Europe and China – with the world still locked-in to technologies of the late middle ages or the early modern period.

Appendix Data on information and communication technology

A-1 Data on personal computer operating systems and web browsers

The data are from the squid (logging demon) logs of every 1000th server request to one of the servers of the Wikimedia Wikis (including Wikipedia, Wikimedia and others). It should be noted that the browser and operating system identity is recognized from the self-identification of the web browser in the user agent string. This value may be changed by the user; however, it is unlikely that a significant portion of users do this.

Further, the data are biased compared to the actual values of operating system and web browser usage since only a minority of internet users use Wikipedia. It is conceivable that the sample is thus biased towards younger and better-educated internet users from developed countries, especially English speakers. It is also conceivable that more rare (more experimental) software is overrepresented in the sample. However, there are no reliable data about this bias, its extent, and its direction.

A-2 Data on webservers and server operating systems

The data used for this analysis are from the 2009 Netcraft SSL survey. Such data are collected by sending requests to a precompiled list of webservers and analyzing the servers' responses. Usually, webservers indicate the type and version of both the webserver and the operating system, though this may be changed by the server's operator (however, it is unlikely that many operators make use of this option). Using the SSL certificate, the servers can further be categorized by country (SSL certificates contain a country code); however, since according to the survey a substantial proportion of the servers used self-signed certificates, the country code may be misleading. Further, Netcraft's precompiled list is probably heavily biased for servers in developed countries, and especially English-speaking countries.

Table A-1 Personal computer operating system and web browser usage data from Wikimedia squid logs (sample relates 1:1000 to real numbers)

Date	*Personal computer operating systems*				*Web browsers*						
	Microsoft Windows	*Apple/ Mac*	*GNU/ Linux*	*Other operating systems*	*Microsoft Internet Explorer*	*Mozilla Firefox*	*Google Chrome*	*Apple Safari*	*Opera*	*Other non-mobile-phone browsers*	*Mobile-phone browsers*
Apr 2009	3367634	227797	55999	111433	2158580	1155484	145132	96663	72767	61549	71471
May 2009	3505608	229137	59680	116102	2240575	1189987	147175	103688	88967	62823	78229
Jun 2009	3554225	233443	62961	123028	2239193	1224598	160680	103711	97320	65279	83459
Jul 2009	3290290	236765	61687	132681	2029995	1173085	167987	103137	88692	70735	89352
Aug 2009	3301567	233805	58446	134068	2072733	1127819	170188	110787	87575	71932	85715
Sep 2009	3743852	272173	63773	142054	2354608	1257440	199400	131794	110723	78003	92948
Oct 2009	3766132	281934	64484	150826	2332891	1286074	205509	145306	117470	80595	93755
Nov 2009	3933522	301282	69354	156526	2427861	1336341	221108	166596	125788	77908	102537
Dec 2009	3657893	295537	66652	172678	2190294	1276886	215607	181822	129035	72357	125753
Jan 2010	3345202	269143	62689	169498	1962035	1186681	197344	185008	122413	73026	119217
Feb 2010	3439876	280005	65374	170799	1989459	1224686	209007	208304	130051	75110	118658
Mar 2010	3233440	257361	62879	160897	1873439	1130595	213476	191005	122432	71122	111404
Apr 2010	3228623	256788	65080	174511	1823648	1148812	237120	189425	126747	75310	122890
May 2010	3233130	254100	67070	181917	1802607	1153437	258951	186404	129228	73091	130705

Table A-2 Data on the numbers of Microsoft Windows and Linux server operating systems and Apache and Microsoft IIS webservers from the Netcraft SSL Survey

Windows	*Server operating systems*		*Webservers*		*Servers*
	Linux	*MS IIS*	*Apache*	*Total*	
USA	166282	113118	174646	133117	358840
Japan	7452	33870	7365	46714	61294
UK	22827	22636	24903	23737	56323
Germany	7461	32076	6372	34491	45843
Canada	13620	11038	12523	13694	30401
Australia	9649	8682	9332	9331	21611
Netherlands	9796	5953	8843	6142	18706
France	2886	5296	2484	6058	10692
Spain	2865	3452	2659	3600	7846
Switzerland	2676	2850	2576	3644	7607
Sweden	4089	1640	3782	1944	7241
Denmark	3938	1040	3484	1190	5770
Italy	2475	2146	2145	2295	5648
Korea	2587	1874	2399	1978	5348
Brazil	2582	1564	2366	1493	4672
Turkey	2441	1384	2577	1445	4305
New Zealand	1978	1523	1869	1703	4215
Norway	2321	1215	1844	1202	4136
Austria	1237	2225	1144	2408	4092
Finland	1193	1799	1039	1954	3689
Poland	549	1791	711	2033	3333
Ireland	1333	1241	1369	1361	3052
Belgium	1199	973	1090	927	2736

Notes

1 Introduction: the poor craftsman and the wealthy inventor

1 This book will follow much of the literature in using the term 'cycle' in spite of the fact that it is highly doubtful whether these cycles are regular and thus fit the name 'cycle'.
2 It will be argued in Chapter 5 that the 'openness' of knowledge, standards and technology (the term 'openness' here being taken from the open source communities and the open standards approach (West, 2007; Simcoe, 2006)) is crucial for rapid technological change and growth. For the system, it does not matter whether the knowledge is made open (public) intentionally or not; nor is there a major difference between complete openness (everyone who wants to can access the information) and probabilistic access (anyone willing to make certain efforts or take certain risks has a good chance of gaining access to the knowledge). The latter is identical to technological diffusion in evolutionary growth theory. It should also be noted that access to technologies is generally non-competitive, though, as shall be shown in Chapters 4 and 5, first-mover advantages play a crucial role with network effects, which makes the profitable employment of a technology competitive.

2 Methods for an analysis of economic growth and network effects

1 Some textbooks add a mathematical appendix where the optimization method is derived from fixed point theorems and some mathematical properties are discussed. However, it is treated like a necessary evil, not as an integral part of the theory. Some few textbooks on mathematical economics go more into detail (Baldani *et al.*, 2005) but it is still almost impossible to understand *why* the Lagrangian method solves this kind of problem. The scholars who actually explain the theory seem to be only those who are critical of it (Mirowski, 1989; Hands, 2006; Smith and Foley, 2008; Keen, 2009a; Lehmann-Waffenschmidt, 2007).
2 For a detailed history, see Mirowski (1989).
3 That is transportation and administration costs, bargaining time, etc.
4 For an assessment of how 'markets' without a central auctioneer work and are indeed different from the perfect neoclassical market in their outcome, see Albin and Foley (1992)
5 That is, aside from completeness (for every possible bundle of goods there is a utility), reflexivity (if for two bundles of goods $A \succ B$, then $B \prec A$), transitivity (if $A \succ B_1$ and $A \succ B_2$ then $A \succ B_2$) and continuity.
6 Either $pA + (1-p)B \succ A$ or $pA + (1-p)B \succ B$ or both.
7 This includes most importantly also production with a concave production function such as a Cobb–Douglas function.
8 Expenditure by consumption and investment must not exceed the income. See the restrictions in the section on Romer's (1986) model.

9 For example, $\sum_i$ Pric $e_i x_i - Budget \leq 0$

10 This is not necessarily the case, since μ must fulfill both conditions $\mu = \frac{\frac{\partial F(x_1, x_2)}{\partial x_1}}{\frac{\partial G(x_1, x_2)}{\partial x_1}}$ and $\mu = \frac{\frac{\partial F(x_1, x_2)}{\partial x_2}}{\frac{\partial G(x_1, x_2)}{\partial x_2}}$.

11 The following formulas closely follow the argument and formal mathematical considerations of Sato (1981).

12 ϱ_j is thus the inverse demand function of the *j*th good $\frac{\frac{\partial F(x)}{\partial x_j}}{\frac{\partial F(x)}{\partial x_n}} = \varrho_j(x)$.

13 $MRS(x_j, x_n)$ denotes the marginal rate of substitution between goods *j* and *n* at quantities x_j and x_n.

14 Ordinality again requires the assumptions of completeness (for every possible bundle of goods there is a utility), reflexivity (if for two bundles of goods $A \succ B$, then $B \prec A$) and transitivity (if $A \succ B_1$ and $B_1 \succ B_2$ then $A \succ B_2$).

15 While the idea was known long before the advent of game theory (for instance, in the form of Cournot's duopoly market outcome Cournot (1897 [1838])), it was formalized by von Neumann and Morgenstern (1944) and its range as a concept was shown by Nash (after who it is named) who proved that for any normal form game at least one such equilibrium exists in mixed strategies (Nash, 1950).

16 σ^*_{-i} denotes the strategy of all other agents leading to the Nash equilibrium except agent i.

17 Pareto optimality of an outcome (a combination of the strategies of all agents) $\sigma_i^P \forall i$ requires that there no alternative outcome $\sigma_i \forall i$ exists, such that at least one agent is better off while no agent attains a worse outcome than in $\sigma_i^P \forall i$.

18 Normal form games are non-repeated games where all agents choose their strategies simultaneously, and are perfectly informed about both the rules of the game and the payoffs resulting from all possible combinations of strategies.

19 Since $1 + 3\varepsilon > 0 + 2\varepsilon$.

20 This requires *common knowledge of rationality* as a further assumption: not only are all agents rational with a well-defined preference ordering, they also know that all (other) agents are rational and that all agents are informed about this common knowledge, etc.

21 As the game is symmetric, the game matrix is reduced to a matrix

$$\mathcal{A} = \begin{pmatrix} \Pi_1(\Gamma, \sigma_{1,1}, s_{2,1}) & \Pi_1(\Gamma, \sigma_{1,1}, s_{2,2}) \\ \Pi_1(\Gamma, \sigma_{1,2}, s_{2,1}) & \Pi_1(\Gamma, \sigma_{1,2}, s_{2,2}) \end{pmatrix}$$

containing only the payoffs of the first player. This is convenient, as this matrix multiplied by the vectors of the mixed strategy (the σ_i) of the agent and her opponent gives the payoff of this interaction.

$$\Pi_1(\Gamma, \sigma_1, \sigma_2) = \sigma_1 \mathcal{A} \sigma_2.$$

Note that since the games are symmetric the strategy sets of the agents are identical. Hence, σ_i and σ_i^* stand for universal strategies in the game (the index i is retained to avoid confusion with the combinations of strategies σ and σ^*). A strategy σ_i^* is evolutionary stable if it performs better than any other strategy σ_i against a population composed mainly (share $1 - \varepsilon$) of agents playing σ_i^* and including a small part ε of agents playing σ_i, hence

$$\sigma_i^* \mathcal{A}(1-\varepsilon)\sigma_i^* + \sigma_i^* \mathcal{A}\varepsilon\sigma_i > \sigma_i \mathcal{A}(1-\varepsilon)\sigma_i^* + \sigma_i \mathcal{A}\varepsilon\sigma_i.$$

As ε may be arbitrarily close to 0, the left hand side of the inequality must for $\varepsilon = 0$ be at least equal to the right hand side, hence (first condition of evolutionary stability)

$$\sigma_i^* \mathcal{A} \sigma_i^* \geq \sigma_i \mathcal{A} \sigma_i^*.$$

If the left hand side is larger than the right hand side, the condition above is fulfilled and σ_i^* is evolutionary stable. If otherwise both sides are equal (and thus the condition still holds) and since both expressions are present in the general condition, and one may be substituted for the other, yielding (second condition of evolutionary stability)

$$\sigma_i^* \mathcal{A} \sigma_i \geq \sigma_i \mathcal{A} \sigma_i.$$

22 The more colourful but needlessly violent story of reproductive success of successful agents with simultaneous extinction of less successful ones yields equivalent results. For a discussion of this issue, especially the fact that the 'survival of the fittest' model is neither realistic nor necessary, see for instance Kirman (1997) or Nowak (2006).

23 This is required by and mathematically identical to the first condition of evolutionary stability in footnote 21.

24 A replicator dynamics system can always by constructed out of the above setting by computing each agent type's payoffs against a mixed strategy representative of the composition of the population. The resulting payoffs may than be used as the individual fitness of a common replicator equation.

25 For details on how to derive the equilibrium and its stability, see section 2.4.

26 This is the mixed strategy Nash equilibrium of that game.

27 See, for instance, Hicks (1935), Harrod (1939), Samuelson (1939a,b), Kaldor (1940) and (later) Goodwin (1967).

28 In turn, for a repeller there thus exists an environment $\mathcal{B} : x \pm \varepsilon$ (where ε is a n-dimensional vector of arbitrary small length $|\varepsilon|$) such that $lim_{T\to\infty} x_{t+T}(x_t) \neq x^* \; \forall x_t \in \mathcal{B}$.

29 This is basically because eigenvalues λ relate to the general solutions systems of difference equations in a specific way. For systems of linear difference equations, the general solution is of the form

$$x_t = \mathcal{C}\mathcal{J}^t$$

where $\mathcal{C}$ is a vector of constants (the initial conditions); note that $\mathcal{J}$ is in this case a simple transformation matrix of scalar elements. The behaviour of the system in this transformation is governed by the eigenvalues of the transformation matrix (the Jacobian) $\mathcal{J}$. Systems of difference equations of lower order can be transformed into lower-dimensional (e.g. one-dimensional) systems of difference equations of higher order. For the, in this respect easier, case of a single homogeneous difference equation (one-dimensional system) of high order with n distinct and simple roots of the characteristic equation (corresponding to eigenvalues of the Jacobian in first-order difference equation systems) $\lambda_1, \lambda_2, ..., \lambda_n$ the equation has a general solution

$$x_t = \mathcal{C}_1\lambda_1^t + \mathcal{C}_2\lambda_2^t + ... + \mathcal{C}_n\lambda_n^t.$$

where the $\mathcal{C}_1, ..., \mathcal{C}_n$ are arbitrary constants. (For multiple eigenvalues, the solution would contain expressions of the form $t\lambda_i^t$ that do, however, have the same dynamic properties.) Consequently, each term of the equation with $|\lambda_i| > 1$ follows the positive part of an exponential function $F_t = \lambda_i^t$ (growing, expanding, diverging to $+\infty$ for

$t \to \infty$). The resulting motion is henceforth not converging. For $|\lambda_i| = |\lambda_i^t| = 1$ (the limit cycle case) the motion is neither contracting nor expanding but preserves its extent. For $|\lambda_i| = |\lambda_i^t| < 1$, the terms of the general solution follow the negative part of an exponential function $F_t = \lambda_i^t$, which converges to zero and translates into damped motion of the dynamic system.

30 Eigenvalues λ relate to the general solutions of differential equations in a specific way. For systems of linear differential equations, the general solution is of the form

$$x_t = \mathcal{C}e^{\mathcal{J}t}$$

where $\mathcal{C}$ is a vector of constants (the initial conditions); note that $\mathcal{J}$ is in this case a simple transformation matrix of scalar elements. The behaviour of the system in this transformation is governed by the eigenvalues of the transformation matrix (the Jacobian) $\mathcal{J}$. Systems of differential equations of lower order can be transformed into lower-dimensional (e.g. one-dimensional) systems of differential equations of higher order. For the, in this respect easier, case of a single homogeneous differential equation (one-dimensional system) of high order with n distinct and simple eigenvalues $\lambda_1, \lambda_2, ..., \lambda_n$ of the corresponding Jacobian as

$$x_t = C_1e^{\lambda_1 t} + C_2e^{\lambda_2 t} + ... + C_n^{\lambda_n}t.$$

where the $C_1, ..., C_n$ are arbitrary constants. (Multiple eigenvalues would lead to terms of the form $te^{\lambda t}$; the dynamic properties remain the same.) Consequently, each term of the equation with $\lambda_i > 0$ follows the positive part of an exponential function $e^{\lambda_i t}$ (growing, expanding, diverging to $+\infty$ for $t \to \infty$). The resulting motion is henceforth not converging. For $\lambda_i = 0$, $|e^{\lambda_i t} = 1|$ (the limit cycle case) the motion is neither contracting nor expanding but preserves its extent. For $\lambda_i < 0$, the terms of the general solution follow the negative part of an exponential equation $e^{\lambda_i t}$, which converges to zero and translates into damped motion of the dynamic system.

31 If the divergence or convergence effects are too fast, the cycles may not be visible in the trajectories.

32 For details on complex and chaotic systems, see Prigogine and Stengers (1984), Bar-Yam (1997), Mandelbrot and Hudson (2004) or Nowak (2006).

33 Everything that is not a part of this system is either left out of the simulation or fed to the simulation model and program as an exogenous parameter.

34 Unlike highly specialized scripting languages such as MATLAB, Mathematica (both for mathematical analysis) and R (for statistics), Python is more shapeable for the simulations needs, though it lacks some of the specialized functions that are predefined in Mathematica, R, etc.

35 This was found by Prechelt (2000) in a study comparing seven scripting (TCL, Rexx, Perl, Python) and compiled (C++, C, Java) languages. Compiled (except for Java) languages performed generally a little better in runtime, scripting languages needed generally fewer lines of code to accomplish the same tasks.

36 An interesting model of the self-organization of groups (e.g. groups of speakers of certain languages) in space is given by Yegorov (2003).

37 This is most obvious when it comes to explaining empirical patterns observed at the macro-level that do not fit the mainstream view of the micro-level. An example of this (growth cycles) as well as various mainstream and heterodox attempts to solve it are discussed in more detail in section 3.5.

38 This may include both natural resources or commonly produced resources and adherence to informal institutions (which is, provided that the informal institution in question is somehow economically or socially beneficial, similar to a resource in the strict sense of the term).

39 For a detailed analysis and real-world examples of common pool resource games, see Ostrom (2000).

3 A literature review

1 Though Adam Smith's work bears strong theological traits and is thus very different from modern economic thought. For an extensive discussion of the theological nature of Smith's writings, see Waterman (2002). Such traits were also much less present in the later classic economic scholars mentioned above.

2 The seminal works were of course Smith's *The Wealth of Nations* (1776), Ricardo's *Principles of Political Economy and Taxation* (1817), Cournot's *Researches into the Mathematical Principles of the Theory of Wealth* (1897 [1838]), Marx's *The Capital* (1964 [1867], 1963 [1885], 1894), Marshall's *Principles of Economics* (1990 [1890]) and Veblen's *Theory of the Leisure Class* (2008 [1899]). For an overview see Kurz and Salvadori (2003).

3 See Marshall's chapter 'Growth of Wealth' in his *Principles of Economics* (Marshall, 1990 [1890]).

4 This was certainly the case for economics in the nineteenth century. As Marshall put it in an appendix ('Note on the Statistics of the Growth of Wealth') to the chapter 'Growth of Wealth' in his *Principles of Economics* (Marshall, 1990 [1890]), 'The statistical history of growth of wealth is singularly poor and misleading'.

5 Among others, the models of Harrod (1939), Samuelson (1939a) and Kalecki (1962) use this approach.

6 Examples include the models of Domar (1946), Arrow (1962) and Goodwin (1967).

7 For an elaboration of Schumpeterian concepts, technological change and economic growth, see Nelson and Winter (1982a).

8 A simplification of the Solow–Swan Cobb–Douglas production function $Y = A(K,L)K^{\alpha} + L^{1-\alpha}$ with $A(K,L) = AK^{(}1-\alpha)$ and normalizing $L = 1$, thus $Y = AK$. (The variables denote output (Y), capital (K) and the technology level (A), respectively; α is a parameter adjusting the relation of the marginal factor productivities of Y and K.) This model is due to Solow (1956) himself (who also named it the 'Harrod–Domar case') and other growth theorists such as Frankel (1962); for a discussion of the name and history of this class of models (AK models), see Mulder *et al.* (2001)

9 The variables denoting output (Y), capital (K) and the technology level (A), respectively; s is a parameter for the (exogenous) savings rate, in essence in this model the constant capital-to-output ratio.

10 See for example the recent book by Aghion and Howitt (2009). Harrod and Domar are cited – among others – but their models are barely recognizable from the depiction in this textbook.

11 Such restrictions are discussed more extensively in the context of Marx's and Georgescu-Roegen's theories in section 3.3.2.

12 Basically the argument is that the development function of the growth rate $g = \frac{I'}{K'}$ assumes the form $g_{t+1} = \alpha g_t + \beta g_t^2$ (with $\alpha > 0$, $\beta > 0$), which is increasing, crosses the origin and has leftward curvature in its entire domain. In order to have a stable fixed point – the only case for which a meaningful equilibrium growth rate exists – the model needs to cross the 45°-line from left to right, which does not happen for a function of the described shape. This shape, however, is Kalecki's own interpretation of the Harrodian formulation $\Delta Y_{t+1} = C(\Delta Y_t) + F(K_t(Y_t))$, approximating the first difference by squaring. Further, Kaleckis solution of the problem – a positive axis intercept – is not excluded in Harrod's model and might have a different meaning than the 'intensity of innovations'.

13 Note the similarity to the first equation quoted from Harrod in section 3.1.1.

14 Kalecki defines $\Delta Y = h\Delta I$. The equation is thus equal to $I_{t+1} = (1-\lambda)I_t + \lambda Ch\Delta I$.

15 This is the paper in which the Solow–Swan model was first presented. Solow starts with some general considerations and proceeds to give three specific examples of growth models, one of which is what later became known as the Solow–Swan model, another one (which Solow calls the Harrod–Domar case) is what Aghion and Howitt present as a general example of the Harrod–Domar model.
16 Labour may be neglected or capital may be modelled as per capita capital $k = \frac{K}{L}$ in more recent descriptions of the model (Aghion and Howitt, 2009; McGrattan, 1998).
17 The figure was generated by using a simple computer simulation script; the Python code is available from the author on request.
18 In AK models, minimal shocks in capital or output would alter the growth trajectory forever.
19 The title of Harrod's seminal paper 'An essay on dynamic economic theory' as well as the developed argument are clear about this (Harrod, 1939).
20 Though his papers develop the model in more general terms and add several examples, the most well-known and influential of which is that with the Cobb–Douglas production function. Another is what Solow called the Harrod–Domar case – which continues to shape the way Harrod's and Domar's models are perceived today.
21 In this case, the production function is just stretched vertically shifting the equilibrium to a higher or lower level of equilibrium capital endowment since the production function changes while the depreciation remains the same linear function.
22 See, for example, Nelson and Winter (1974).
23 For example, Barro and Sala-i-Martin (2003) but also – less orthodoxly – Aghion and Howitt (2009).
24 However, other theorists later proposed other models with better empirical performance for the same dataset, most notably Nelson and Winter (1974, 1982a) and Mulder *et al.* (2001).
25 Solow was very careful to provide more general formulas, or at least to give several different examples, before assuming particular alternatives. The more general formulation would therefore be $sF(K,L) = sf(k,L)$.
26 Again, Solow also includes other examples with several stable or unstable fixed points as well as systems without stable equilibrium.
27 Solow gave this equation without a depreciation term; however, the depreciation behaves just like the population growth rate n, which was also discussed by Solow. δ is included here since it is usually given as a central part of the Solow model in recent textbook accounts. See, e.g. Aghion and Howitt (2009).
28 Assuming full employment, that equals per capita terms.
29 At the microeconomic level, the modified system is defined as $y = A(t)f(k)$.
30 His even earlier lecture notes were published recently (Swan, 2002 [1956]). For historical details, see Pitchford (2002).
31 'The inclusion of an element of technical progress … [with one factor] … is equivalent to a reduction in the other [factor], at least from the point of view of the production though not necessarily from the point of view of the distribution of incomes between factors.' (2002 Swan, 2002 [1956]: p. 380).
32 See subsection 3.3.3.
33 The second constraint is equivalent to equation (3.9). For the formal solution of the corresponding differential equation see section 3.1.4.
34 δ is the discount rate, σ a measure of risk aversion. $L(t)$ is, as above, the infinite sequence of labour employed in production and $c(t)$ is the infinite sequence of period consumptions.
35 Or also with diminishing returns to the factor inputs $F(\psi k_i, K, \psi x) \leq \psi F(k_i, K, x)$.
36 Here $\overline{e}$ is the production in the first period, that is either consumed in c_1 or invested in knowledge k.
37 Romer uses λ as the Lagrangian multiplier instead of μ as used in the current volume to avoid confusion with the eigenvalue.

38 Note that for this optimization problem, the results are an interior solution, thus $X > 0$ and $\mu \geq 0$ since the Kuhn–Tucker conditions $X\frac{\partial \mathcal{L}}{\partial X} = 0$ with $X \in c_1, c_2, k, x$ with both factors ≥ 0 and $\mu\frac{\partial \mathcal{L}}{\partial \mu} = 0 \forall \mu$ with both factors ≥ 0 are satisfied.
39 Note that $\frac{\partial K}{\partial k}$ is a constant.
40 See section 3.4.
41 For technical reasons the variable labels used here differ from those originally used by Aghion and Howitt (1992).
42 Aghion and Howitt discuss such a case (Aghion and Howitt, 1992).
43 As shown for the different models of this class, this is usually either capital or labour but might in theory also be any other feasible factor.
44 δ_t denotes a time preference parameter with which future payoffs are discounted.
45 Mathematically, meaning that the optimization would only find interior minimum points, yielding the least possible utility, which are not valid solutions; the corresponding shadow prices μ_h would be negative. The true solutions would be boundary solutions such as consuming everything in one period.
46 In this case $e^{\delta}\frac{\frac{\partial U(C_t)}{\partial C_t}}{\frac{\partial U(C_{t+1})}{\partial C_{t+1}}} = e^{\delta}$ and $\frac{\partial F(I_t)}{\partial I_t} = e^{\delta}$ does generally not hold.
47 Note that this is the general form, not the current value Hamiltonian for time t.
48 As Verdoorn's paper is difficult to obtain (and especially in a language other than Italian) this consideration follows Thirlwall (1980) who was familiar with the original article as well as much of the debate on the matter over the following decades.
49 Thirlwall's law has also been published in a magazine that is very difficult to obtain (the Italian *Banca Nazionale del Lavoro Quarterly Review* in 1979), but was extensively discussed by Thirlwall and others (Thirlwall, 1986).
50 Demographic transition refers to the era of huge population growth after death rates dropped when modern medical methods became available but before the subsequent decline of birth rates.
51 In the more easy discrete case of the Gibbs entropy, this is $S = -k_b \sum_i \rho_i ln \rho_i$ where k_b is the Boltzmann constant and ρ_i is the probability of the microstate i to occur.
52 See e.g. Solow (1957), Frankel (1962); one notable exception being Kaldor (1957)
53 Arrow notes that the use of the capital goods is optimal in this model at any time given that decisions are rational (and information is perfect). Capital goods are ordered in the sequence of their productivity and the lowest G' are out of use.
54 Evolutionary growth theories drawing on Schumpeterian ideas including neo-Schumpeterian approaches, such as the theories by Freeman and Perez, will be discussed in detail in the next two sections, 3.4 and 3.5.
55 Or output growth, which – given enough bargaining power of the producers and sufficient demand – reflects the amount of investment and the return on capital expectations.
56 It predicts convergence towards a stable equilibrium which is not generally supported by empirical evidence (Knack, 1996; Pritchett, 1997) and was famously attacked later for introducing results as exogenous (technology) factors (Nelson and Winter, 1974).
57 Without heterogeneity, entry and exit were nothing more than replacing something with itself.
58 To say it using Dawkins' metaphor: 'A chicken is just an egg's way to create another egg.' (Dawkins, 1976).
59 As the probably most important aspect, Nelson and Winter (1974) cite empirical evidence, that competitive advantage and the resulting advantage and rents arising from monopoly conditions or simply from lower costs are the main stimulus for innovation.
60 For a detailed summary of the history of evolutionary economic thought, see Antonelli and de Liso (1997) and Antonelli (2009). For an analysis of evolutionary economic ideas in the work of Veblen, see Berger (2009a).

61 The latter, entrepreneurship, might as a slightly different concept already be found in Veblen's writings. For a more detailed account of this context, see Berger (2009a).
62 For an historical perspective on game theory, see for example Hargreaves Heap and Varoufakis (1995).
63 First and foremost the co-evolution of the industry structure.
64 For example, Dixit and Stiglitz (1977).
65 However, the heterogenity and the specific qualities of the firms are produced only by their technology. Thus two firms with the same technology are, from an evolutionary perspective, essentially the same.
66 The current section attempts to use the same notation as the 1976 paper of the developers of the model; this is, however, not always possible since Nelson *et al.* (1976) describe some features more extensively and represent several variables not as mathematical variables but rather in verbal descriptions (e.g. the unit price of output, here named p) or apply their numerical values (e.g. 1.16 for the threshold value here denoted as $1+\tau$).
67 Note that the following equations are not given explicitly in either one of the papers Nelson, Winter and Schütte wrote about the model; the relations are however explained in Nelson *et al.* (1976: pp. 94–98).
68 p stands for the – fixed – unit output price.
69 Nelson *et al.* used $\delta = 0.04$ (Nelson *et al.*, 1976).
70 $w(\sum_i \frac{a_{L,i,t}}{a_{K,i,t}} K_{i,t}) \frac{a_{L,i,t}}{a_{K,i,t}} K_{i,t}$ where $w(\cdot)$ is the wage depending on total labour demand $\sum_i \frac{a_{L,i,t}}{a_{K,i,t}} K_{i,t}$, which is multiplied by the firms, labour demand $\frac{a_{L,i,t}}{a_{K,i,t}} K_{i,t}$.
71 The capital costs, being revenues to the capital owners, are assumed to be an increasing linear function of the capital stock.
72 This is defined as a stochastic function $r(\cdot)$, providing the firm with a technology slightly better than its current one $(a_{K,i,t}, a_{L,i,t})$, or, in the case of imitation, with the technology currently being used by one of the other $N-1$ firms of the industry. Finally, the firm will only adopt the newly available technology if it is an improvement over the currently used one under the ceteris paribus assumption.
73 In Nelson *et al.*'s original computation, τ was set to $\tau = 0.16$ (Nelson *et al.*, 1976).
74 In their later 1982 model concerning evolutionary R&D as a process, Nelson and Winter (1982a) add some considerations about the model's behaviour in special cases. Further simplifying assumptions may render the model mostly or completely predictable and lead to a stable, unique and computable steady state equilibrium. The same is true for the 1973 model.
75 At later stages, however, the sector will be in a monopolistic state and growth will thus occur suddenly (when the monopolist detects a new technology).
76 For example, the GDP per capita growth data displayed in Figure 1.2.
77 The inclusion of both labour supply and demand for goods would also result in two limiting effects that would be very similar in nature; hence, it does not add any important aspect to the model and would only complicate analysis of the cause of observed results. This point will be argued in more detail in section 5.3.2.
78 The labour productivity of the individual enterprise $a_{L,i,t}$, since it determines the extent of labour required by the enterprise as well as that of the rest of the industry, since it affects the wages.
79 c are the production costs per unit of capital and $r_{i,t}$ denotes the investment in research and development.
80 For their simulations, Nelson and Winter employ function

$$I = max\left(0, min\left(1+\delta - \frac{2-\frac{Q_{i,t}}{Q_t}}{\frac{P_t A_{i,t}}{c}(2-2\frac{Q_{i,t}}{Q_t})}\right)\right).$$

(Nelson and Winter, 1982a: p. 303) General additional requirements assumed by Nelson and Winter (1982a: p. 286) (and satisfied by the function above) are $I > 0$, $\frac{\partial I}{\partial \frac{P_t A_{i,t}}{c}} \geq 0$, $\frac{\partial I}{\partial \frac{Q_{i,t}}{Q_t}} \leq 0$, and $lim_{\frac{Q_{i,t}}{Q_t} \to 0} I\left(1, \frac{Q_{i,t}}{Q_t}, \pi_{i,t}, \delta\right) = \delta$.

81 The equeation in Nelson and Winter is included in the previous equation, substituting I by this functional form.

82 Should several firms possess the best-practice technology, their industry shares develop according to their initial industry shares.

83 Nelson and Winter consider two cases, one in which the firms can only reinvest their profits apart from some investment to recover depreciation, and one in which this investment may reach 2.5 times the profit of the firm. The latter regime obviously requires lending and thus introduces a financial sector.

84 Where m_t is derived from the number of plants in operation during the last period, N_t, multiplied with a reinvestment parameter $0 < s < 1$, thus $m_t = sN_t$.

85 It decreases in k_t and increases in σ and m_t, $lim_{k_t \to \infty}(E\Delta\mu) = 0$, $lim_{m_t \to 0}(E\Delta\mu) = 0$.

86 For a more detailed discussion of the Slutzky–Yule effect, see section 3.5.1.

87 This mechanism is generally named a Phillips curve mechanism (Goodwin, 1967; Silverberg and Lehnert, 1993, 1994). It is not immediately obvious how this relates to the general notion of the Phillips curve, i.e. the inverse relation of price level (inflation) and employment (unemployment). The line of reasoning for this labour limiting model feature is that with high employment rates the bargaining power of the labour side (employees) increases, hence increasing wages, which without additional productivity leads to a rising price level.

88 Here Silverberg and Lehnert's replicator for cross technology investment $s(r_i - \bar{r})$ is neglected in (this) equation (4) in Silverberg and Lehnert (1993), it would constitute an unnecessary complication.

89 It is important that the evolving entity biologically is not (necessarily) the organism, an animal, a human, etc. Rather it is the individual gene (Dawkins, 1976) that has no separate phenotypic representation. The organism in turn is the result of the combination of a multitude of genes. Still, the genes themselves are not entirely separable from each other, as their specific combination in the genome (the entirety of the genes of one individual organism) of an organism contributes to the fitness and performance of that organism. Still, genes are partly separable as they may appear – in different combinations – in multiple organisms, which also contributes to the fitness and performance of these organisms (a frequently quoted example being mothers defending their children). In the earlier evolutionary economic theories, the distinction of genotype and phenotype was reduced to the strategy or technology (genotype) employed by a firm (phenotype), this (represented as a single value) did not possess the scope and flexibility of a DNA sequence.

90 In biological systems these are DNAs (deoxyribonucleic acids), long sequences of which form an allele of a chromosome. Two alleles together form one chromosome, which may differ although the information serves the same purpose and is active at the same time. Thus the system also involves a considerable amount of redundancy that has not been transferred to evolutionary economics.

91 For recent discussions of this debate see Bormotov (2009) or de Groot (2006)

92 See, e.g. Nelson and Winter (1974).

93 Among the explanations are notably the acquisition of new resources (according to de Groot, Kondratieff's original argument (de Groot, 2006: p 98)), change of technology regimes (Perez, 1983), the length of a working life or a generation, the cycle of capital construction and crisis (de Groot, 2006), the entropy in the economic system (Devezas and Corredine, 2002), and many more.

94 Using a Poisson regression model, de Groot and Franses actually find five cycles of respectively 5, 13, 24, 34 and 61 years length in the frequency of basic innovations, four of which match the suggested lengths of economic cycles (de Groot and Franses, 2009).

95 For an overview, see de Groot (2006).

96 According to de Groot, Kuznets cycles are for instance said to be specific to data from the USA, while Kitchin cycles are suspected to have changed over time, recently more closely resembling the US electoral cycle (de Groot, 2006).

97 Positive feedbacks have been identified for the returns on human capital (Arrow, 1962; Romer, 1986), institutions (circular cumulative causation) (Berger and Elsner, 2007; Berger, 2009a) and network goods (Arthur, 1989; David, 1985; Katz and Shapiro, 1985; Farrell and Saloner, 1986). But it may also be applied for much more basic relations. Macroeconomically, a shock in income will alter demand, which influences the growth of savings, investment, output production and in turn again income. At a micro-level, supposed that the prices in a more or less neoclassical 'market' only react with a certain time lag, demand shocks lead to shortage (translated into higher demand in the next period), rising price and rising production, which will not go back to normal immediately but only after a certain time, a much longer period than the above mentioned lag in the price reaction. Compare Wheat (2009).

98 Shorter cycles might exist as well but, if they do, are dominated by the strong seasonal variations and thus impossible to detect.

99 Stationary meaning $E(X(t)) = 0$ and $E(X(t)^2) = \sigma^2$.

100 Kondratieff used a 9 year moving average filter on growth rates of the USA (de Groot, 2006).

101 Slutzky used a Gaussian curve to approximate the probabilities for numbers of successive values to be consistently below (or above) the mean of the distribution (Slutzky, 1937 [1927]).

102 In order to derive the corresponding homogeneous system, the exogenous part is left out, thus

$$\begin{pmatrix} Y \\ D \end{pmatrix} = \begin{pmatrix} a & \beta \\ a^2 - a & a\beta \end{pmatrix} \begin{pmatrix} Y \\ D \end{pmatrix}.$$

103 The eigenvalues also govern whether the motion is contracting (stable fixed point), which is the case if the absolute value of both eigenvalues is < 1 (case A and B) or expanding towards infinity (the other cases).

104 Cyclic or non-cyclic, expanding or contracting behaviour may occur, but may be limited to a certain region of phase space. Suppose, the system expands beyond the area of phase space where the eigenvalues are positive, the eigenvalue – depending on the position in phase space – may become zero (a limit cycle) or negative again. For a more detailed explanation of the concepts of complex dynamic theory, see Chapter 2.

105 Lotka (1925) and Volterra independently developed a model of population dynamics on successive steps of the food chain using a system of differential equations derived from a modified Verhulst equation. As the population development of the predator depends positively on its food source, and that of the food animal (or plant) is however negatively dependent on the predator population, the dynamics are not trivial and lead to circular motions of different kinds, that is, more or less stable limit cycles. Lotka considers a number of examples including ecosystems with predator and prey animals, but also epidemics and examples from chemistry to which the same dynamics apply.

106 This is the Phillips curve assumption of increasing market power as the workers share of profit crucially depends on the wage rate *w* (increasing nominal or even real wage) of the work force, possibly mediated by increased bargaining power of the trade unions.

107 Due to the square root being a monotonic transformation function and the radicand being by $4BD$ larger than the value that would yield $D-B$.

108 Fixed points are hyperbolic if none of the eigenvalues of the system at this point are purely imaginary. The orbits in the vicinity of the hyperbolic fixed point are hyperbolic, hence the name. For details on dynamic theory see Chapter 2.

109 Keen (1995) did exactly this using Goodwin's model as a framework to apply Minsky's (1980) ideas (section 3.5.5). For other more complex models based on the 'Marx–Keynes–Goodwin' system see, e.g. Silverberg and Lehnert's (1993) model (section 3.4.3).

110 Both functions are assumed to be continuous.

111 More accurately, the boundary applies to the relation of investment and capital.

112 For details on the method, see section 2.4.

113 If $det(\mathcal{J}) < 0$, the square root becomes positive and larger than the rest of the equation

$$\frac{tr(\mathcal{J})}{2} < \sqrt{-det(\mathcal{J}) + \left(\frac{tr(\mathcal{J})}{2}\right)^2},$$

consequently, one of the eigenvalues will be positive while the other one will be negative (except for the special case in which the trace of the Jacobian matrix is $tr(\mathcal{J}) = 0$, which will be detailed below). The negative determinant requires

$$\frac{\partial S}{\partial Y}\left(\delta - -\frac{\partial I}{\partial K}\right) < \frac{\partial I}{\partial Y}\delta.$$

114 This is valid only for the case depicted in Figure 3.12, i.e. $\frac{\partial I(Y^*,K^*)}{\partial Y} > \frac{\partial S(Y^*,K^*)}{\partial Y}$. For $\frac{\partial I(Y^*,K^*)}{\partial Y} < \frac{\partial S(Y^*,K^*)}{\partial Y}$ the fixed point is stable for $\alpha > \alpha_{Hopf}$, and unstable for $\alpha < \alpha_{Hopf}$. Everything else including the bifurcation at the critical value and its nature (as a Hopf bifurcation) remains.

115 This may be different for other values of the net investment function.

116 There were some early approaches that made heavy use of political and historical facts and were less concerned with the economic system, notably Kondratieff's original hypothesis of war and peace in capitalist societies and Kuznets' observation of migration into the USA. For a detailed account, see de Groot (2006).

117 This is true, notwithstanding the fact that the dynamics of the recent cycle, especially the IT industry, tends to interfere with the known product and technology life cycle itself. This in turn raises the question whether the description as a continuation of a regular cycle pattern is accurate. This argument will be detailed in Chapter 6.

118 In order to experience the full scale of a technology's possibilities, including spillover effects, decreasing cost of the input factors and modular compatibility of multiple products of different sectors, it must be employed by a number of different sectors. In order to accomplish that, these sectors must be coordinated on one and the same technology even in the presence of the older incumbent technology (of the previous cycle) and possible other alternatives.

119 Chen, however, argues that Lucas' theory does not withstand its own critique (Chen, 2002, 2005).

120 Lucas explores different kinds of real and monetary cycles using a similar model in another paper (Lucas, 1975).

121 A simple Markov process assigning probabilities to each size the 'population' may assume (basically any positive integer) that consists of a probability for the growth of a smaller population in the period before, the probability of population decline in a larger population (in the previous period) and the probability of a population of the same size (in the previous period) to remain constant.

122 Specifically, Samuelson does not go into detail on how to work with and how to analyze difference equation systems in his papers on the oscillator model. In effect, he shows that his model is capable of diverse dynamics depending on the parameters, he discusses which exact configurations of parameter values lead to this behaviour, but he does not discuss what causes this, specifically not the eigenvalues of the system.

123 In Lucas' case, the observation of the micro-level as a basis for modelling has been explicitly emphasized; still the simplifications in other parts of the model (for details see section 3.5.7 and Chen (2002, 2005)) reduce it to another instance of the general equilibrium theory – which doubtlessly had been anticipated as a desired result.

124 In Kaufmann *et al.* (2000) this is a mere side note: they argue that technological progress is faster at the start of the lifecycle of a new technology. Given that the invention of new technologies is a rather rare event, such a pattern would be visible in the form of cycles.

125 At least for the last 70 years, shrinking economies have been exceptions notably occurring in cases of disasters, wars and severe economic crises like the Great Depression following the crash of 1929 and the current crisis since 2008.

126 See the section on the Marx–Keynes–Goodwin model, 3.5.3.

127 See section 3.4.1, especially Figures 5.12 and 5.13. This question will be discussed in more detail in the modelling Chapter (5) including a possible solution.

4 Network economies

1 Depending on wheather Arthur and his collaborators, from mathematical and evolutionary economic backgrounds, and David and his group, of more historically oriented economists, are seen as one and the same group. They came to virtually the same conclusion, however, employing very different approaches.

2 Arthur and David were both at Stanford University in the 1980s; Arthur is also affiliated to the Santa Fe Institute of Complex Systems and was – before becoming a Stanford professor – in the late 1970s and early 1980s also a reasearcher at the International Institute for Applied Systems Analysis at Laxenburg, Austria, an institution of scientific cooperation by US and Soviet reaearchers (as well as scientists from other eastern and western bloc countries) during the time of the Cold War. This is where Arthur worked with Kaniovski and Ermoliev; Kaniovski in turn was still a researcher in Laxenburg when Dosi led an international research project there in the 1990s. Kaniovski was also a visiting researcher at the Santa Fe institute in the late 1980s and in the University La Sapienza in Rome in the early 1990s, where he may also have worked with Dosi.

3 At the time, Katz and Shapiro were both at Princeton University; Farrell and Saloner were both at MIT. Three of the four, Katz, Shapiro and Farrell later moved to the Haas School of Business of the University of California at Berkeley.

4 The idea of external effects was already discussed by Marshall (1990 [1890]) but discussed more thoroughly by Coase. Coase introduced negative external effects (which he called 'harmful effects' (Coase, 1960)) as a criticism of the standard neoclassical theory and provided an analysis on how to deal with them (Coase, 1960). Positive externalities are principally the same type of phenomenon: an influence on the utility of payoff of another agent that is not subject to optimization by the market (usually because the external effect is unilaterally caused).

5 See, e.g. Farrell and Klemperer (2007), Arthur (1990a) and Silverberg *et al.* (1988).

6 For a discussion of the history of economic thought on network effects see Farrell and Klemperer (2007).

7 For an overview see Berger and Elsner (2007) and Berger (2009a).

8 Standard textbook maximization problems, for instance, use linear isoclines of the cost function and convex indifference curves.

9 For convenience, this consideration will focus on marginal utility as the device responsible for increasing returns. This is more close to the form of network

effects discussed in the following sections. However, this is equivalent to decreasing marginal cost.

10 The exception of course being natural resources, the supply of which is exclusively controlled by the monopolist. This is also the traditional example given by Cournot (1897 [1838]) and Marshall (1990 [1890]).

11 In fact every Cournot oligopoly with $n-1$ oligopolists (and with constant marginal costs as usual) would produce a quantity of $\frac{n-1}{n}$ times the welfare optimizing perfect competition quantity. Every further (nth) newcomer would be able to make gains by producing a further $\frac{1}{2}(1-\frac{n-1}{n})$ (half of the difference of the current oligopoly quantity and the perfect market quantity) thus appropriating a monopoly rent on the uncleared part of this neoclassical market. Subsequently the competitors would reduce their quantity and the newcomer would expand till all (now n) competitors produce n^*, such that (the quantity per competitor being x, the demand function being $\eta_1 - \eta_2 x$ and the costs being set to zero without loss of generality)

$$max_x \Pi_{all} = max_x x(\eta_1 - \eta_2 x - (n-1)\eta_2 x_{others})$$
$$= max_x(\eta_1 x - \eta_2 x^2 - (n-1)\eta_2 x_{others}),$$

hence,

$$\frac{\partial \Pi_{all}}{\partial x} = \eta_1 - 2\eta_2 x - (n-1)\eta_2 x_{others} = !0.$$

After applying equal market power of all n competitors, hence $x = x_{others} = x^*$, this yields

$$x^* = \frac{\eta_1}{(n+1)\eta_2} = \frac{1}{n+1}\frac{\eta_1}{\eta_2} = \frac{1}{n+1}x_{perfect\ competition}$$

and in total

$$nx = \frac{n}{n+1}x_{perfect\ competition}.$$

For a detailed description of the model, see Cournot (1897 [1838]), for a more detailed explanation of the behaviour of the model with several firms, see, e.g. Varian (1992).

12 Marshall (1990 [1890]) also includes licenses owned by the monopolist, transportation using infrastructure controlled by the monopolist, etc. In essence, however, this is the same situation: the monopolist is effectively the only one able to provide the good.

13 For an overview of the debate on Marshallian monopoly in the 1920s/1930s, see Hicks (1935).

14 In fact, most modern neoclassical textbooks add a section on game theory and apply game theory concepts extensively, especially when it comes to industrial economics, oligopoly dynamics, etc. See, e.g. Varian (1992) or Mas-Colell *et al.* (1995).

15 For example, personal computer operating systems (quasi-monopolist Microsoft), office software – (quasi-monopolist Microsoft), personal computer CPUs/chipsets (quasi-monopolist for some time Intel), data bases (quasi-monopolist Oracle), webserver software (quasi-monopolist Apache Foundation), web browsers (quasi-monopolist Microsoft), web search engines (quasi-monopolist Google), social networks (quasi-monopolist Facebook) and many other examples. See Chapter 6.

16 To be exact, the physical utility should be written as $r + v(0)$ which guarantees that this is equal to $u(r, 0)$. As $v(n)$ is monotonic it will, however, be assumed for convenience that $v(0) = 0$.

17 Note that this is a simplified model, as it disregards market power and divisibility of production charges, etc.

18 The term *intrinsic utility* was used by Arthur (1989); Katz and Shapiro used the less illustrative wording 'basic willingness to pay' (Katz and Shapiro, 1986).

19 Or, rather the same or any compatible standard. Compatibility issues are specifically analyzed by Katz and Shapiro (1985, 1986). This requires a slightly different specification of the model and will be discussed below in section 4.2.3.

20 The limit case, both sides of the inequality being equal, would mean indifference between the two choice options, *A* and *B*.

21 Testing the conditions (4.5), (4.6) and (4.7) for the other standard (*B*) results in the network utilities remaining unchanged in these inequalities, while the differences of the net instrinsic utilities (Δ_1 and Δ_2) have to be multiplied by -1 (hence the $-\Delta_1$ instead of Δ_1 and $-\Delta_2$ instead of Δ_2). From case 2 (according to Table 4.1), it follows from $v(N_1+N_2)-v(N_2) \ngeq \Delta_2$ that this condition holds for *B* ($v(N_1+N_2)-v(N_2) \geq -\Delta_2$) as the left side of the inequality is always positive (from the definition of v) while either Δ_2 or $-\Delta_2$ must be negative (or zero), such that only one, either Δ_2 or $-\Delta_2$ can be greater than the left side of the inequality. Furthermore, the second condition (inequality (4.7)) was not fulfilled for *A* ($v(N_1+N_2)-v(N_1) \nleq -\Delta_1$). It follows directly that condition 2 (inequality (4.6) since condition 1 holds for *B*, hence $v(N_1+N_2)-v(N_1) \geq -\Delta_1$) must hold for *B*. Along similar lines it can be shown that for case 4 of Table 4.1 the equilibrium considering the other standard as the choice of period 1 consumers always fulfills this expectation. In this case condition 1 for *A* proved true, condition 2, however, did not. This does not allow any conclusion about whether condition 1 holds for *B*. However, as inequality (4.6) did not hold for *A* ($v(N_1+N_2)-v(N_1) \ngeq \Delta_1$), it follows that for *B* both inequality (4.6) ($v(N_1+N_2)-v(N_1) \geq -\Delta_1$) and inequality (4.7) ($v(N_1+N_2)-v(N_1) \leq \Delta_1$) must always hold true. In other words, period 1 consumers will in this case always choose standard *B* no matter what they expect period 2 consumers to do, in turn justifying period 2 consumers assumption.

22 The question as to how individual utilities of different agents may be compared (the relation of 'welfare' and Pareto optimality) will be neglected at this point.

23 This includes, for instance, Farrell and Saloner (1985, 1986) and Klemperer (1987) to mention a few early theorists.

24 This application for a very similar model was made by Arthur in one of his later papers (Arthur, 1989).

25 In Arthur's, model (Arthur, 1989), the difference of the net intrinsic utilities could take only two possible values, a positive one indicating preference for *B* and a negative one indicating preference for *A*. In this case, these two values are identical to the absorbing barriers. In the current model, the variable is continuous, therefore, the tipping point from which agents decide to forego their net intrinsic utilities for the higher network utility may differ from agent to agent. The argument, however, remains applicable since the difference of the net intrinsic utility is bounded to both sides.

26 Though Farrell and Saloner also used a notation slightly different from the one considered above and employed by both Katz and Shapiro and Arthur, it may be captured in terms of the above notation. They used two very different notations in their fairly consecutive papers in 1985 (Farrell and Saloner, 1985) and 1986(Farrell and Saloner, 1986), the first one just considered the preferences of strategies (to adopt or not to adopt a particular standard at a given time) while the second added a discounted utility function over an infinite horizon, and consequently also beliefs about the behaviour of the rest of the system (which standard will or will not prevail). This might have been an extension in order to fit the approach with the contemporary neoclassical theory (like for example the theories of endogenous growth by Lucas, Romer and others, as discussed in Chapter 3). The results, however, are fairly similar.

27 If the agents are considered homogeneous with respect to their intrinsic utilities, equations (4.15) and (4.16) are changed further (using $\Delta_\kappa = \Delta \forall \kappa$) to

$$v(N) - v(g(\kappa)+1) \leq r_B - r_A \equiv \Delta$$

and

$$v(N - ch_t) - v(ch_t + g(\kappa) + 1) \leq r_B - r_A \equiv \Delta.$$

The dynamics of the system, however, remain unaffected by this.

28 At least for true network technologies, this assumption is not realistic. The voice over IP telephone technology for instance may have great advantages over the ordinary telephone networks, but it is unlikely that a telephone user would want to be the only one to switch to Skype or the SIP VoIP network, knowing that for some time (or possibly forever) there will be no gateway technology to the regular wired telephone and cellphone network.

29 Competing technologies, competing standards and competing products employing different technologies behave similarly.

30 However, it is of course subject to the result of the first period.

31 Repeatedness is in this case essentially the possibility of communication and signaling in order to agree on one Nash equilibrium, preferably the Pareto optimal one. The phenomenon could also be modelled as a repeated game.

32 In addition, if an agent anticipates that it may come down to her own choice, the decision to join the risk-takers would also be a dangerous one, as the situation shows that the success of the establishment of the new network is threatened and the agent in question would have to bear the full risk.

33 Note that the vertically arranged terms in brackets in the above equation do not denote column vectors but binomial coefficients.

34 The number of specific sequences of i adopters among $N-2$ actors is $\binom{N-2}{i}$.

35 For a more detailed discussion of power laws in the context of network structure, see section 4.3.6 below.

36 It has been shown that humans are mostly risk averse, although the extent of this risk aversion has been found to vary greatly (Smith, 1989).

37 An example for this would be industry-wide standard setting cooperation as has been the case in the definition of the UNIX/POSIX standards (Axelrod *et al.*, 1995; West and Dedrick, 2001; West, 2003, 2005; Tanenbaum, 2001).

38 Note, however, that in this case, $\tilde{g}$ is beyond κ's control and does not stand for a bandwagon effect exercised by κ's choice, risk averse agents will therefore revert to the risk-free wait-and-see strategy.

39 For details on these systems, see, e.g. Kunert and Kuhfeld (2007) and Butler and Neuhoff (2008).

40 Symbols are used as above.

41 However, it is difficult to estimate marginal costs in neoclassical markets with increasing returns. (The term 'market' is appropriate here as marginal cost itself is an idea from neoclassical theory and related to the idea of the market.) See section 4.1.1.

42 Note that this logical step is common for the neoclassical perfect market considerations; it is the step from a market demand schedule to a smooth demand function.

43 Otherwise, the firm would not be able to recover the initial investments and at a macro-scale the development of this industry would not even be socially desirable.

44 $c_{prod,var}$ are the constant marginal production costs or variable unit production costs, which may also be ≈ 0, for instance for digital goods.

45 This is the easy case of a standard without competitors. Note that $I_{network}$ changes according to the investment necessary to drive out competing technologies, the price war discussed earlier.

46 Note that there are a number of possible reasons for such heterogenity in real-world standards, including consumption habits, skills, earlier investments, geographical location, etc.

47 It may be argued that this strategy was successfully employed by Apple/MacIntosh in order not to be routed by Microsoft in the intense competition for dominance in the personal computer operating system sector in the late 1980s and early 1990s.

48 Also note that the overall social benefit (for a moment making the heroic assumption that such a value is measurable as the combined utilities of all agents including the standard vendors) increases considerably.

49 This might be a counter strategy to niche construction by minor competitors, as considered in the previous section.

50 This could be accomplished, e.g. by supplying frequent revisions of the technology, the subsequent versions changing to incompatibility, then back to compatibility, etc., or by modifications to the infrastructure. With interference from the user base, through changes to the competing technology, or through gateway products the second step may also follow without any efforts on the part of the major competitor.

51 In the case of a communication technology, e.g. telephone, either two agents can communicate or they cannot. One-way compatibility does, however, work for other standards, say file formats: software A may be able to read and write the native file format of software B while software B cannot use software A's native file format. A good example for this is word-processing software: the open source software LibreOffice Write is able to work with its native file type (.odt) but also with Microsoft Word's file type (.doc) while no version of Microsoft Word (as of 2011) can read or write .odt files.

52 Returning to the example from footnote 51, users of word-processing software do not only want to read files constructed by users of other standards, they also want their own texts to be readable by other people. LibreOffice users can of course save their texts in Microsoft's .doc format – however, this does not always work very well. Though this is probably not intended by the programmers of LibreOffice, if it did work perfectly, they would possibly find users exclusively using .doc, and consequently their software becoming a part of their competitor's network.

53 However, this finding is valid just for the case that the producer cannot exercise price discrimination; otherwise she could simply provide the new version at a lower and, if necessary, zero cost to established customers. Except for the welfare effect of using the new version, this case would be equivalent to designing the new standard with backward compatibility (since both types of consumers would be part of the same network and obtain the same network utility).

54 Consider, e.g. the competition of several standards for personal computer operating system. While this competition is one of the most well-known examples for network externalities and standard wars (David and Steinmueller, 1994; Katz and Shapiro, 1994; Bresnahan and Greenstein, 1999; Stango, 2004; Stango and Greenstein, 2008; West and Dedrick, 2001; West, 2003, 2005; Gallagher and West, 2009), the personal computer as such is a standard that had to prevail against older technologies that had been serving partially the same purpose for many decades before the rise of personal computers (such as typewriters or mainframe terminals), and that more recently faces competition from newer technologies like laptop computers, tablet computers, mobile phones, etc.

55 With the evolutionary game theory approach followed here, the standards are treated equivalently to pure strategies played by a specific agent type in a population, hence the symbol σ that is otherwise used for strategies.

56 Note that $N^{max}_{ch(\kappa)} - N^{Im}_{ch(\kappa)} > N_{Im}$, thus the third term remains positive.

57 For convenience, presuming the risk to be proportional to the produced quantity of plagiates.

58 Note that the terms 'legalist' and 'pirate' are just one of several possible interpretations; in this case implying the incorruptible legalism of the members of the first group

as their motivation to allow themselves to be exploited by the monopolist. Another interpretation would be that the 'legalist' group just lacks the technical knowledge or the connections and the social integration to get in touch with the imitating supplier.

59 That is, the utility of and thus the possible revenue from these two groups can be determined mathematically by simply splitting the utility function $v(n+1)+r_K(n)$

$$
\begin{aligned}
v(n+1)+r_K(n) &= v_{pirate}(n+1)+r_{K,pirate}(n)+v_{legalist}(n+1)+r_{K,legalist}(n) \\
&= \frac{N_{legalist}}{N_{legalist}+N_{pirate}}(v(n+1)+r_K(n)) \\
&+ \frac{N_{legalist}}{N_{legalist}+N_{pirate}}(v(n+1)+r_K(n)).
\end{aligned}
$$

60 This result is consistent with King and Lampe (2003) (who are rather critical of 'profitable piracy').

61 Though this may be rather similar to the strategies discussed in the above section on private sponsoring of technology standards that would better be described as giving the product away for free. It is of course equivalent to letting someone 'pirate' it.

62 Darmon *et al.* (2009) quote Jeff Raikes, the former head of the Microsoft business group saying 'If they're going to 'pirate' somebody, we want it to be us rather than somebody else' (Darmon *et al.*, 2009: p. 449). Peitz and Waelbroeck (2006) mention somewhat obscurely that 'some analysts of the software industry have argued that the weak intellectual protection of Microsoft products together with the increasing lobbying of Microsoft for increased future intellectual property in China is based on a strategy that exploits network effects' (Peitz and Waelbroeck, 2006: p. 469). They do not cite any of those analysts; the text probably refers to analyses related to if not based on statements by Microsoft officials like the one quoted above, or those made by Bill Gates during the decade before, such as '... [some of the users in Asia] don't pay for the software. Someday they will, though. And as long as they're going to steal it, we want them to steal ours. They'll get sort of addicted, and then we'll somehow figure out how to collect sometime in the next decade.' (Bill Gates as quoted in Odlyzko (2003: p. 81).)

63 To this group belongs first and foremost Liebowitz (2005) who operates on the basis of his earlier research together with Margolis (Liebowitz and Margolis, 1990, 1994, 1995), which led them to believe that network externalities to not exist at all.

64 The issue will be discussed in more detail in Chapter 6. For an overview of the relevant literature, see Peitz and Waelbroeck (2006); however, for a more formal, and also neoclassical, discussion of the effects identified in game theory and market models as well as empirical case studies until 2004, see Farrell and Shapiro (2004), for more recent empirics, see Oberholzer-Gee and Strumpf (2010).

65 Most importantly, in 'the weightless economy' (as Quah describes the information, software, etc., sectors (Quah, 1999b)), variable production costs are indeed zero, which means that investments necessary in order to set up a product 'piracy' business, are negligible. In fact, imitating the monopolist's product does in this case come down to copying the software (or information good other than software), or providing a version of the software for copying via filesharing. Removing any copy protection mechanisms may be required, but given the imitator has the necessary programming expertise at her disposal, this is not overly costly. Consequently, almost everyone can become a software (or even more easily music/video) 'pirate', law enforcement is correspondingly difficult and this kind of 'piracy' is consequently rather frequent, leading to a fierce public debate centring around intellectual property rights and possible punishment for copyright infringement. For economic aspects of this debate and empirical findings

see Peitz and Waelbroeck (2006), Farrell and Shapiro (2004) and Oberholzer-Gee and Strumpf (2010).

66 In the terminology above that is a sufficiently high expected network utility $v(N_{ch(\kappa),A}+1)$.

67 Studies on the effect of monopolistic industry structures on the intensity and effectiveness of R&D find ambiguous evidence (Gilbert and Newbery, 1982; Reinganum, 1983; Winter, 2006). Schumpeter, as one of the earliest scholars to have discussed innovation and technological progress in detail, was under the impression that only monopoly rents allow extensive research (Winter, 2006). On the other hand, monopolies do not face competition and may not have sufficient incentive to innovate. This is especially true if the monopolist is able to establish high entry barriers for access to competitors in the industry. Network externalities as well as several similar mechanisms (for instance, preemptive patenting) certainly encourage and facilitate the development of entry barriers (Gilbert and Newbery, 1982). Neoclassical as well as heterodox (neo-Schumpeterian) economists are divided on this issue; recently, with the debate on intellectual property rights and the corresponding legislation, the question has gained new attention (Winter, 2006).

68 If utilities in the system cannot be expressed in the same system of utility using well-defined prices in a common notation (common to all commodities and the utility of all agents) the aggregation of welfare is not possible, and the concept is thus not defined.

69 It is also common to simply say 'Pólya urn scheme'; however, it is more appropriate to use the term Eggemberger–Pólya urn scheme as the original paper (Eggenberger and Pólya, 1923) was the joint work of George Pólya with his student Florian Eggenberger. (The same goes for the similar terms related to it.)

70 Note, however, that the standard deviation, following the sum of the standard deviations of the Bernoulli distributed $\zeta_{t+1}\left(\frac{n_{A,t}}{n_t}\right)$, rises in t. The variance of $\zeta_{t+1}\left(\frac{n_{A,t}}{n_t}\right)$ computes as

$$Var\left(\zeta_{t+1}\left(\frac{n_{A,t}}{n_t}\right)\right)=\frac{n_{A,t}}{n_t}\left(1-\frac{n_{A,t}}{n_t}\right),$$

hence the variance of the Eggenberger–Pólya process is

$$Var\left(\frac{n_{A,t+T}}{n_{t+T}}\right)=\sum_{\tau=t}^{T-1}\frac{\frac{n_{A,\tau}}{n_\tau}\left(1-\frac{n_{A,\tau}}{n_\tau}\right)}{(\tau+3)^2}$$

yielding the standard errors

$$s.e.\left(\frac{n_{A,t+T}}{n_{t+T}}\right)=\sqrt{\sum_{\tau=t}^{T-1}\frac{\frac{n_{A,\tau}}{n_\tau}\left(1-\frac{n_{A,\tau}}{n_\tau}\right)}{(\tau+3)^2}}.$$

This function increases the number of time steps: with each time step, a positive term $\frac{n_{A,\tau}}{n_\tau}\left(1-\frac{n_{A,\tau}}{n_\tau}\right)/(\tau+3)^2$ is added to the square root. (This term cannot be negative or zero, as the divisor $(\tau+3)^2$ is positive and both colours of balls are present in the urn; hence both shares are >0 and <1.)

71 In Arthur *et al.* (1982), this is done for all three rules the authors proposed for the one-dimensional ($d=1$) case. For cases of higher dimensionality, it is slightly more

complex, as with different combinations of shares the number of balls of a type that is necessary for this type to acquire the largest share in this draw changes between at least $\frac{L}{d}$ and at most $\frac{L}{2}$.

72 Since $\sum_{i=1}^{d+1} x_{i,t} = 1 \forall i \forall t$, it does not matter which one is left out, the system remains the same.

73 In order to discuss the model as a replicator model, the basic unit has to be shifted from 'agents' (the balls) to types of 'agents' (the colours). Note that virtually all scholars working on generalized urn schemes and their application to technological change (including Arthur, Ermoliev and Kaniovski, the authors of the original 1983 paper (Arthur *et al.*, 1982)) view the model as an agent-based model of decisions of the individual agents, not as a replicator dynamic on shares of technologies.

74 $[0,1]_{d+1}$ denotes a $d+1$-dimensional constrained space between 0 and 1 in each dimension.

75 A more detailed discussion of methods for the evaluation of stability in dynamic systems is given in section 2.4.

76 Likely, not certain, because the system is stochastic, not deterministic.

77 In order to have a global attractor this attractor would have to be the single element of $\mathcal{B}$.

78 Note that the transition probability function of R4

$$p(x_{i,t}) = \frac{3x_i^2 - 2x_i^3}{\sum_{j=1}^{d+1} 3x_j^2 - 2x_j^3}$$

reduces to

$$p(x_{i,t}) = 3x^2 - 2x^3$$

in the one-dimensional case.

79 Note that, like in case of R4, the transition probability function of rule R5

$$p(x_{i,t}) = \frac{1 - 3x_i^2 + 2x_i^3}{\sum_{j=1}^{d+1} 1 - 3x_j^2 + 2x_j^3}$$

reduces to

$$p(x_{i,t}) = 1 - 3x^2 + 2x^3$$

in the one-dimensional case.

80 For a more detailed account, see section 2.4.

81 One notable and recent example is monopolistic competition, as investigated by Dixit and Stiglitz (1977).

82 For R4, the complete standardizations are the only attractors, interior solutions are unstable. This turns the process into a positive feedback loop.

83 In real-world systems that are generally more complex than symmetric urn schemes, the size is not the only important small effect. Another one is the compatibility with other elements in the complex network of interlinked technologies, yet another the availability of the required infrastructure. The internet as a new technology, for instance, relied in its early days on the preexisting infrastructure of telephone communication technology, more specifically wires and technologies to use them. See Chapter 6.

84 The stochastic system does not have clear trajectories since the path the system takes is subject to random influences; hence, there would be multiple, possible infinitely many trajectories a system could take from any point in the phase space.

85 This follows immediately from Arthur *et al.*'s (1982) theorem concerning the convergence to attractors in generalized urn schemes as discussed above.

86 For instance, an infinite and symmetric system with a single global attractor (such as the urn scheme with rule R5 as discussed above) is doubtlessly non-ergodic but may hardly be called complex.

87 This example is to be seen as a purely theoretical experiment. The evolution of humanity and its complex technology implies that there was a time at which the ancestors of humans and primates did not possess sufficient cognitive ability for the conscious development of the use of tools and techniques. The properties of the technological progress before and after this transition are not difficult to derive; the consideration does, however, not claim detailed historical or anthropological accuracy.

88 The specific polynomials are

$$\frac{\partial^2 p_1}{\partial t \partial p_1} = -3p_1^2(q_1(\chi_{11} - \chi_{12} + \chi_{21} - \chi_{22}) + \chi_{12} + \chi_{22})$$
$$+ 2p_1(q_1(\chi_{11} - \chi_{12} + 2\chi_{21} - 2\chi_{22}) + \chi_{12} + 2\chi_{22})$$
$$- q_1(\chi_{21} - \chi_{22}) - \chi_{22}$$

$$\frac{\partial^2 p_1}{\partial t \partial q_1} = -p^3(\chi_{11} - \chi_{12} + \chi_{21} - \chi_{22}) + p_1^2(\chi_{11} - \chi_{12} + \chi_{21} - \chi_{22})$$
$$- p_1(\chi_{21} - \chi_{22})$$

$$\frac{\partial^2 q_1}{\partial t \partial q_1} = -3q_1^2(p_1(\chi_{11} - \chi_{21} + \chi_{12} - \chi_{22}) + \chi_{21} + \chi_{22})$$
$$+ 2q_1(p_1(\chi_{11} - \chi_{21} + 2\chi_{12} - 2\chi_{22}) + \chi_{21} + 2\chi_{22})$$
$$- p_1(\chi_{12} - \chi_{22}) - \chi_{22}$$

$$\frac{\partial^2 q_1}{\partial t \partial p_1} = -q^3(\chi_{11} - \chi_{21} + \chi_{12} - \chi_{22}) + q_1^2(\chi_{11} - \chi_{21} + \chi_{12} - \chi_{22})$$
$$- q_1(\chi_{12} - \chi_{22}).$$

89 Any agent added to the neighbourhood must have five other contacts to the same peer group, therefore there are 5×10 open links from the 11 agents of an agent's distance-1 (direct) neighbourhood, 5 have to connect to the same exterior agent, hence all in all there are $\frac{10 \times 5}{5}$ new agents in any larger distance neighbourhood group.

90 This is still not realistic when it comes to model real-world systems of how agents interact. Boccaletti *et al.* (2006) discuss a number of systems to construct network topologies that resemble real-world networks more or less closely.

91 From their paper (Lee *et al.*, 2006) it is not entirely clear how they conducted this simulation. They operate with an abstract value for the 'importance of complements'. However, they are more precise about the result, which is also plausible regarding the findings of the section on standard tying (section 4.3.5); Lee *et al.* also give an illustrative example. The successful VHS videotape standard was primarily used for prerecorded tapes. Though most customers were reportedly more interested in being able to make own recordings, this did, according to Lee *et al.*, at some point shift the standard war towards VHS and away from Betamax.

92 However, Delre *et al.* (2007) use a more complex setup, considering also distance-2 neighbours as direct influences with respect to the local network externalities, hence increasing the number of relevant neighbours to up to $4 + 4 \times (4 - 1) = 16$ (which still is a rather small number).

93 Dunbar operated with cross-species statistical relations of the largest observed group size and the brain weight. His exact estimate for the cognitive limit to peer group size for humans was 147.8 but with large confidence intervals (Dunbar, 1993). There are some difficulties with this methodology as it is not sufficiently established that brain weight relates to cognitive abilities with sufficient reliability.

94 The diffusion speed of a technology A with $n_A(t)$ adopters at time t is usually measured as (see Delre *et al.*, 2007)

$$\zeta(\tau) = \frac{1}{\tau} \frac{\sum_{t=0}^{\tau} n_A(t)}{\sum_{t=0}^{\tau} n_A(t) - n_A(t-1)}$$

which yields values $\zeta(\tau) = [\frac{1}{\tau}, 1]$, small values denoting a very slow adoption, values close to 1 stand for a rather speedy adoption.

95 For details on power law distributions, see e.g. Newman (2005), Clauset *et al.* (2009), or Andriani and McKelvey (2007).

96 Distribution (4.59) may also be slightly different, such as including a slowly varying multiplicative term.

97 Substituting $p(n_K)$ in equation (4.60) for the probability density function (4.59) and integrating yields

$$\frac{\Upsilon}{1-\alpha} \left[n_K^{-\alpha+1} \right]_{min(n_K)}^{\infty} = 0 - \frac{\Upsilon}{1-\alpha} (min(n_K))^{-\alpha+1} = 1$$

Resolving for Υ yields the required value of the normalization parameter

$$\Upsilon = (\alpha - 1)(min(n_K))^{\alpha-1}$$

98 For this method, nodes are selected at random, and all nodes up to a specified distance from this node are assigned to a new cluster. This is repeated with the next randomly selected node, now only assigning unassigned nodes to the newly formed clusters. The process continues until all nodes are assigned to a cluster; empty clusters are discarded (Kim *et al.*, 2007).

99 Mandelbrot's investigation of the development of cotton prices played a crucial role in the discovery of power laws (Mandelbrot and Hudson, 2004).

100 This type of graph is constructed using an iterative mechanism which connects already 'near' (e.g. just distance-2) nodes with greater probability than random other nodes.

101 The network externalities used in these models has mostly been defined as linear for the individual user, below called *Metcalfe's Law*.

102 Shy uses averages, i.e. $\frac{total\ Revenue}{\sharp\ Subscribers}$ instead of an explicit price.

103 For instance, David (1985) reports results of a US Navy experiment that allegedly showed that the costs of retraining QWERTY-keyboard typists with a Dvorak-keyboard would amortize within ten days of full employment. Later investigations critical of David claimed that it was most likely actually August Dvorak himself (the patent holder of the Dvorak-keyboard) who filed the report for the US Navy (Liebowitz and Margolis, 1990).

104 That is, there is for perfectly rational optimizers or for a perfect social optimizer no way to foresee the ex-post situation; hence given that information set, the outcome would have to be considered optimal.

105 However, Mokyr seems to rely on a very narrow definition of network effects that involve the physical presence of a network (such as the railway, gaslighting and telegraph networks) (Mokyr, 1993).

106 It is resentful in the way that it answers defection with defection. Although as Axelrod pointed out in his comprehensive analysis of the surprisingly good performance of *tit-for-tat* in a repeated prisoners' dilemma, it must not be too resentful (Axelrod, 1984).

107 It does not have to be infinitely repeated, but to avoid end-game-effects, the agents must not be aware of which repetition will be the last.

108 This strategy is generally cooperative without any intention to exploit other agents but answers defection with defection to avoid being exploited itself, and most importantly to ensure punishment for the earlier defection (which is ascertained by the payoff structure, particularly $\Pi_{11} > \Pi_{21}$).

109 This solution is discussed in detail by Axelrod (1984) and by Elsner and Heinrich (2009, 2011).

110 Even for an unrealistic social situation with a social dilemma and just one possible social norm the process would be path dependent, since it is by no means clear whether the norm can successfully be established or whether the trivial (defecting) handling of the social dilemma prevails.

111 Specifically, Schmookler shows that innovation depends on demand side effects. See section 3.3.4. Innovation in turn is undisputedly one of the main sources of economic growth, especially in Schumpeterian evolutionary growth theory (Nelson and Winter, 1982a) but also in the mainstream neoclassical Solow–Swan model of the time, where it takes the form of the technology level A (Solow, 1956, 1957).

112 For detailed discussions of several examples, see section 3.2.

113 See section 3.1.

114 Though some Kaldorian models, specifically by León-Ledesma (2000) include effects that work to reinforce international convergence in spite of Kaldor's feedback mechanism. León-Ledesma's argument, technology catch-up by imitative research conducted in technologically more backward regions, is, however, empirically questionable. In fact cases of smooth technological catch-up by less-developed countries are extremely rare, which is one of the most striking criticisms of the convergence-hypothesis of neoclassical models and in turn one of the strengths of Kaldor's original idea, which rules out the necessary existence of general long-term convergence.

115 As the system is a difference equation system, stability is indicated by modulus smaller than 1. See section 2.4.

116 For example, consider the system for $\gamma = 2.0$ and 3 units of initial capital. An equal allocation of (1.5; 1.5) would lead both systems to collapse, allocating more than 2 to one type, however, would lead the respective part of the production to succeed and flourish.

117 Details of this will be discussed in section 5.2.2 of the following chapter.

5 A complex model of growth, technological change and network effects

1 The few non-evolutionary Schumpeterian theories, e.g. in Aghion and Howitt (1992, 2009), use a very similar theoretical fundament. (For details, see section 3.2.4.)

2 Other different (or even similar) ideas for such a device have been developed in the tradition of the neo-Schumpeterian theory of innovation, among others a process of regular fluctuation through the periodic creation of new sectors and the temporal monopoly position of the first-mover, as developed by Saviotti and Pyka (2008). When moulded into a model of macro-level growth, Saviotti and Pyka's device will most likely generate effects very similar to those resulting from Nelson and Winter's device, as the simulation results reported by Saviotti and Pyka suggest.

3 In linear systems, i.e. $H(x) = hx$, the growth rate declines in x

$$\frac{H(x+1) - H(x)}{H(x)} = \frac{h(x+1) - hx}{hx} = \frac{1}{x}.$$

4 The model developed in this section follows inspirations taken from Nelson and Winter's model (Nelson and Winter, 1974, 1982a), Conlisk's model (Conlisk, 1989), as well as various considerations on network externalities discussed in the previous chapter.

5 Equivalently, the model could use a slightly increasing total available labour defined as a function of time $L = L(L_0, t)$. The result and the basic mechanisms, however, would remain stable.

6 This implies an efficient allocation of labour. In real economic systems this generally does not hold; however, it is safe to assume that scarce input factors (labour in this case) are generally allocated to those with high reservation prices in capitalist economies – which gives a near-efficient allocation. Structural inefficiencies are of course also present in real economic systems. This would, however, require a more complex model and generate a number of special cases that cloud the dynamics of the model that is currently under investigation.

7 This is the case if several plants have the same labour–output ratio.

8 In effect, standardization in these examples reduces the need for in-house production or additional procurement of non-standard-conforming special equipment. It might also reduce the administration costs caused by this.

9 This would still hold for much weaker forms of network externalities, e.g. by applying a higher order root, thus $\sqrt[E]{\frac{1}{S_{TEC,t-1}}}$ with E being a large positive number.

10 In the current form of the network effect, the labour requirements for all technologies of the type would become zero as soon as no other technology is in use any more. This is of course not realistic and can also be easily avoided while still fulfilling property (5.11) with the network externality. See the simulation model in section 5.3.4.

11 If network effects for certain technologies lead to a faster technological development of only these technologies, any other alternatives will inevitably fall back technologically, leaving one type of technologies developing far ahead of all others and therefore the only viable alternative for efficient production in this system. With the variety reduced to just one type of technologies (the system locked-in) the network effect would – apart from a different speed of technological improvement – cease to be relevant for further technological development.

12 Examples include basically any major technological change, e.g. from medieval human power and (at most) wind power to steam engines; from steam power to gasoline for transportation; from mechanical to electronic and computer-based technology.

13 If the success of research is heavily linked to the available capital, e.g. as in Nelson and Winter (1974, 1982a,b) as well as Conlisk (1989), this does of course not hold, as it will almost always be the same firm that innovates – at least as soon as this firm gains a significant share of the available capital.

14 Note that the rather complicated term for the case $y_t > \sqrt[\gamma_t - 1]{\gamma_t}$

$$\gamma_{t+1} = \frac{\gamma_t}{2} + \frac{y_t - \sqrt[(\gamma_t - 1)]{\gamma_t}}{y_t} \frac{\gamma_t}{2}$$

uses the relation of by how much y_t exceeds the equilibrium value $\sqrt[\gamma_t - 1]{\gamma_t}$ to the value of y_t, that is $\frac{y_t - \sqrt[(\gamma_t - 1)]{\gamma_t}}{y_t}$, applies this relation – which is < 1 – as a reducing factor to γ_t and averages the reduced value with the current value of γ_t to avoid a too abrupt reduction of γ_t in this area.

15 This means, the technology level increases as well as the scope of production and thus the total output.

16 In fact, trajectories of continuous systems must not intersect; however, the discrete system analyzed here is to be seen as just an approximation. The intersections are

rare and the context allows extrapolation of how the dynamics of the approximated continuous system (in which the trajectories do not intersect, of course) behave.

17 For details on mechanisms leading to the emergence of power law distributions, see Newman (2005).

18 For details, see Newman (2005).

19 It does of course depend on the comparative superiority of this technology but it does not matter which technology it is.

20 In turn, in conservative systems, the trajectories at least do not converge. Systems for which the trajectories diverge in an extreme way, such that the development of a trajectory allows almost no conclusions to be made about adjacent trajectories (i.e. trajectories that are very close initially), are termed chaotic.

21 According to the theoretical model considered above, the technology regime changes occur irregularly. Note, however, that the regime lengths are not power law distributed as discussed above, but follow a binomial distribution with a radical innovation either being made in the current period or not being made, with constant probabilities for each agent and (as the number of agents per sector is kept constant (100) in this simulation model) also per sector.

22 Note though that the right end of the spectrum is questionable in its significance with, e.g. a sample size of merely 20 values left for a lag of 80 periods.

23 More sophisticated sector-level dynamics are to be studied in section 5.3.4.

24 For this argument, the computer and microchip industry must again be given as the primary example. This industry gave rise to different kinds of software production industries, a computer and software related services sector, internet based technologies and enterprises, etc.

25 Application software could be ported to a new operating system if the currently dominant one should lose this position. The operating system's industry would have to make changes if the industry standards for computer CPU architecture should change (e.g. recently, 32-bit to 64-bit), but the consequences are not catastrophic. There are plenty of other examples.

26 This ensures that b is never 0 which allows the mathematical considerations below to be written more conveniently. (In a philosophical way a technology may also generally be considered derivative to itself.)

27 That is, technologies with h (e.g. twice) times the impact on the economic system of a benchmark technology size are h times less often (e.g. half as often) as technologies of the benchmark size.

28 These theories are discussed in detail in two sections of Chapter 3, section 3.1 and section 3.2.

29 For details, see section 3.5.7.

30 See section 2.4.

31 The characteristic equation of the corresponding homogeneous system (or of course the system itself if it is homogeneous) which is used to compute the eigenvalues must be a second degree equation. Complex eigenvalues λ would have to be multiplied to give a real result that fulfills the equation. This does not happen for purely linear – first degree and one-dimensional – equations. First degree equations are possible if the model contains at least two equations (a Jacobian matrix of the size 2×2 or greater) such that the characteristic equation may result to be a second degree equation, i.e.

$$0 = det(\mathcal{J} - \lambda\mathcal{I})$$

where $\mathcal{J}$ is the Jacobian matrix of the system and $\mathcal{I}$ denotes the unit matrix for both matrices being two-dimensional or larger. (However, second-order differential or difference equations may usually conveniently be transformed into first-order systems of two equations and vice versa.)

32 That is, all eigenvalues being $|\lambda| \geq 1$ for difference equations or having a real part $Re(\lambda) > 0$ for difference equations.
33 For details on Samuelson's model, see section 3.5.2.
34 Without scarcity, every firm would produce an infinite amount of output regardless of capital endowment and technology level. The mechanism would thus not work.
35 This possibility has not been explored by Nelson and Winter but would yield very similar results.
36 Nelson and Winter also investigate the possibilities for firms in their more complex simulation model to focus on imitation of innovation, etc. (Nelson and Winter, 1982a,b). This is, of course, not possible with the simplified version. However, it is not the focus of the current considerations.
37 The Herfindahl index is a common measure of the skewness of the size of entities, especially of firm size, shares of an industry belonging to different firms hence provide an indicator of the monopolization of this industry but also (when applied to technologies rather than firms) of the degree to which an industry is locked to a dominant technology. The normalized Herfindahl index gives the Euclidian distance of the distribution among n entities from the perfectly equal distribution (share $s_i = \frac{1}{n}$ for each entity) in an n-dimensional space where every one of the n dimensions is used to indicate the size of one of the n entities. Hence, the normalized index is computed as

$$HI = \frac{\sum_i s_i^2 - \frac{1}{n}}{1 - \frac{1}{n}}.$$

38 Discussed in more detail in Chapter 6. For open standards and open source as a concept, see also West *et al.* (West, 2007; West and Dedrick, 2001; Gallagher and West, 2009) Raymond (2000), as well as Lakhani and von Hippel (2003); for a theoretical assessment of motivation and incentive for the provision of open knowledge (specifically open source), see Lerner and Tirole (2002) as well as (sharply dissenting with Lerner and Tirole) Bitzer and Geishecker (2010) and Greiff (2011).
39 Some policy implications resulting from this consideration will be discussed in the context of the the case study in Chapter 6.
40 For the following simulation runs this number is set to 128.
41 For the following simulation runs, $c_{Plant} = 1.6$ and hence, $k_{i,0} = 1.6 \forall i$
42 $t_{Plant\ TTL}$ is assumed to be the same for every plant no matter when it is constructed. TTL stands for time-to-live. For the following simulation runs, $t_{Plant\ TTL} = 20$
43 For the following simulations, $L_0 = 32$
44 The revenue for the firms is equal to $k_{i,t} = \frac{1}{4} n_{i,t,active}$ since the plant revenues per timestep are normalized to 0.25 by definition, as mentioned above. Consequently, the industry-level production follows as the sum over all the production for all firms, which is the equation above.
45 Combining the innovation with two or more previously known technologies would not be appropriate since these technologies may actually have been part of the same technology line, and hence one of them might have previously been enhanced with an innovation to generate the other one. In that case, combining the innovation with both of them would include the previous innovation two times.
46 Such that after about 150 iterations the program will need more than 24 hours to complete another iteration on a 2 GHz processor.
47 This variable (measuring the industry structure) considers only firms with at least one active plant. The variable behaves, however, very similarly to the normalized Herfindahl index, counting all 128 firms in each period.
48 Though it is possible that firms will be driven out of the industry (spending capital to construct plants that do not return any revenues if another competitor has simply

developed a better technology in the meantime). The entity does, however, continue to exist.

49 For excessive system requirements, the runs for a system with all technologies being open have not been extended beyond 52 periods.

50 In this case, only the innovations of one firm can be put into practice. Even with some (10%) open technologies, this causes very slow progress.

51 For details on public good problems and prisoners' dilemma population settings, see Ostrom (2000), Axelrod (1984), Lindgren (1997) and Elsner and Heinrich (2009, 2011).

52 For example, for $s_{TEC} = 1$ (the industry is already completely standardized and the technology type is exclusively used), the labour–output ratio would compute as $\widetilde{a}(\theta) = \frac{(5-1)a(\theta)}{4} = a(\theta)$, while for the other extreme $s_{TEC} = 0$ (the technology type is not used at all), $\widetilde{a}(\theta) = \frac{(5-0)a(\theta)}{4} = 1.25a(\theta)$.

53 With 20 researchable technology types, these are approximately 5% of the new technologies.

54 This may seem strange, as the competitors do not collect any revenues without active production and nevertheless recover their competitiveness. In fact, they do, however, still own production facilities and as the labour–output ratio is declining with the network externality and increasing technology level, the labour demand of the proto-monopolist is at some point lower than the labour supply, allowing the competitors to operate their plants again. Of course, they will need to improve their technological knowledge for this to be a lasting recovery. This effect is discussed in more detail below.

55 The t-value is denoted as $\overline{t}$ in order not to be confused with the time indices t, ν indicates the correlation coefficients, and $d.f.$ denotes the degrees of freedom.

6 Case study

1 Parts of the arguments and analyses presented in this chapter have also been presented at various conferences in 2009 and 2010. The author is grateful to the various audiences and his co-author of two of the conference papers, Miriam de Blasi, for helpful comments.

2 This is especially true for the samples available to the Wikimedia Foundation and the Netcraft Corporation.

3 This refers to those given in Tanenbaum (2001), a recent information science textbook with a comprehensive review on the history of software and operating systems. Tanenbaum is one of the leading developers of operating systems (MINIX microkernel). Leiner *et al.* (2004) reflects on the development of the modern internet; Vint Cerf, one of the so called 'founding fathers' of the modern internet is one of the co-authors. Finally, Evron (2008) details the distributed DoS attack on the Estonian IT infrastructure in 2006 from the point of view of the IT security specialist, Gadi Evron, who participated in the subsequent investigation.

4 That is, papers by cryptography researcher Dan Kaminsky (Kaminsky, 2006; Kaminsky *et al.*, 2010) and IT security specialists Randal Vaughn and Gadi Evron (Vaughn and Evron, 2006) on particular structural vulnerabilities of the internet on a technical level as well as the NATO report on the Estonian crisis in 2006 (NATO, 2009).

5 The entire section about the history of the information and communication sector is largely based on several comprehensive reviews of different aspects of this history, Tanenbaum (2001), Glowniak (1998) and Bresnahan and Greenstein (1999); on technical development of hard- and software (especially before the rise of the personal computer industry), Glowniak (1998), Leiner *et al.* (2004) and Ziewitz and Brown (2011) on the history of the internet and West and collabor ators (West and Dedrick, 2001, 2004, 2006; West, 2003; Gallagher and West, 2009), Diker and Scholl (2001),

McKelvey (2001), Anderson (2001, 2002), Kretschmer (2004) and Windrum (2004) on the competition between different computer operating systems and between different web browsers.

6 The ARPANET, sometimes refered to as the predecessor of the internet, was the network of the DARPA (Defense Advanced Research Projects Agency), an organization sponsored by the US Department of Defense.

7 Later to be replaced by *BGP, the Border Gateway Protocol.*

8 The 'standard wars' fought by Microsoft and Apple over personal computer operating systems standards, between several Unix-vendors – notably Sun and AT&T – over the standards for Unix-systems, as well as IBM, Intel and several other companies in the (micro)processor and PC-hardware sector show this very clearly (West and Dedrick, 2001).

9 Of course, the vendor will need some way to decide when to start charging money again, this may, however, arguably be more accurately done using market research on consumers' willingness to pay for the product.

10 The vendor could eschew filing lawsuits (direct tolerance) or just not implement copy protection.

11 For details on the dataset, see O'Mahony and Timmer (2009).

12 Consider, for example, that in 1970 Spain, of all countries in figures 6.2 to 6.4, had the highest share of capital stock in ICT. Spain at the time was, by Western European standards, a backward economy in the final years of the Franquist era. Another obvious inconsistency is the accounting of communication technology in Slovenia. Contrary to all other countries, the share of the capital stock in communication technologies does not develop proportionally but falls sharply in the late 1990s.

13 It even appears exponential with respect to the shares of the total capital stock (i.e. linear with logarithmic axis) in 6.7.

14 That Slovenia and Finland do not conform to the general pattern may be caused by specific patterns in the country's ICT sectors (e.g. a strong dominance of a communication technology firm) or also by differences in accounting.

15 In fact, ICT standards were the prime examples that fueled the early research on network effects by David (1985), Arthur (Arthur *et al.*, 1987; Arthur, 1989), Katz and Shapiro (1985) and Farrell and Saloner (1985).

16 Technical specifications of peripheral devices did for a long time (and mostly do even today) only specify one or several versions of Microsoft Windows which they were guaranteed to be compatible with; vendors did (and do) not bother with support for other platforms such as Apple/Mac and GNU/Linux, even if the product is as simple as an external hard drive with a USB interface that will support all three platforms without problems.

17 Anderson calls this strategy illustratively 'we'll ship it on Tuesday and get it right by version 3' (Anderson, 2001).

18 Compare section 4.2.3.

19 That guarantees also the advantage of pre-existing infrastructure that may be used for the system. The phenomenon is similar to the use of telephone wires for early computer networks because they happened to be there.

20 BSD and GNU/Linux are technically similar; BSD is essentially an open variant of Unix while both GNU and Linux (the operating system commonly known as Linux is composed of a Linux kernel and GNU periphery) are designed after Unix. Their expansion into the sector of personal computer operating systems has been facilitated by the development of graphics environments for Unix-type operating systems, the X Window System (since the 1980s) and appropriate toolkits, and window managers in the 1990s. Without today's sophisticated graphics environments and the huge variety of application software compatible to that system, it would have been considerably more difficult for GNU/Linux (and BSD) to gain a foothold with personal computers.

21 Apple eventually decided to use a BSD kernel (Darwin) as the basis for Mac OS X.
22 With the introduction of a new version of Windows Server (2008), this trend may or may not have stopped.
23 Microsoft IIS only runs on Microsoft Servers while Apache is largely platform independent.
24 Particularly in the case of OpenOffice, Sun (and nowadays the Oracle corporation which acquired Sun Microsystems) would use the free and open source product (OpenOffice) to gain a foothold in the industry sector followed by the introduction of a non-free enhanced version of the product (StarOffice). However, as a consequence of Oracle's less favourable policy regarding open source projects, the OpenOffice community abandoned the Oracle-related to create a community owned project, LibreOffice.
25 This is done by making a product compatible with the environment to which the firm intends the product to expand. As an example, consider Apple's recent advance in the web browser sector by making its browser Safari available for Windows platforms after having been a Mac-only browser for almost a decade.
26 The correlation between GNU/Linux and the Firefox is significant with a confidence level of 95%.
27 This set is very small: only countries for which reliable data on server operating system usage exists from the Netcraft SSL survey (Netcraft Ltd., 2009) and that are also readily identifiable from the user agent strings in the Wikimedia squid logs (Wikimedia Foundation, 2011). The latter is unreliable for some countries – most English language browsers for instance identify their country as the USA which leaves very little data about other English-speaking countries. The seven countries are the USA, Germany, Spain, France, Italy, the Netherlands, and Sweden. Furthermore, the two browsers indicated in Table 6.3 are Firefox and Safari, those with the largest usage shares in 2009 after Internet Explorer, which generally does not indicate a country code in its user agent string.
28 The most famous case is the lawsuit against Microsoft in the late 1990s when Microsoft was charged for bundling its operating system, Windows, with its web browser, Internet Explorer.
29 This may, for example, be the case for Unix-type operating systems and software such as Apache that Microsoft tried to drive out by introducing a Windows-native application, Microsoft IIS. As a consequence, Apache is much more prevalent in the Unix-sector than on Windows machines.
30 Many applications do, however, today rely on central servers, which has the advantage that users do not have to be online and reachable at all times. Other reasons include that a centralized design requires less technical expertise from the individual users and enables a central provider to charge fees for access to some network or common resource.
31 This is the case for at least some of the early developments, reportedly for instance in the case of ARPANET developer Paul Baran and his colleagues (Ziewitz and Brown, 2011).
32 Sometimes different terms are used for righteous technology-interested *hackers* and shady, mostly rent-seeking, *crackers* especially by self-identified hackers themselves. 'Whitehat' and 'blackhat' are used equivalently. In reality, however, there is no such sharp difference; 'crackers' may perceive what they do as a good or necessary thing and therefore refuse to identify themselves as 'crackers'.
33 A system is as secure as is its weakest component.
34 Or is finally dropped after a certain number of hops to avoid crowding the network – e.g. if the specified destination does not exist.
35 Actually two protocols, TCP, more sophisticated than IP but relying on IP.
36 The port, a number between 1 and 65536, 2^0 and 2^{16} is used to identify which application on a computer a protocol is destined for or coming from.

37 Actually SYN, ACK, RST and FIN are flags in the headers of the packages, bits at the start of the package that are set to 1 if the flag is set, or to 0 if the flag is not set. For example, in a SYN package, the SYN flag is set to 1, ACK, RST and FIN are set to 0.
38 For example, a forged source IP pretends that the package stems from another host and may introduce wrong data pretending that it is coming from a trusted source; 'malformed' TCP communication can be used to 'scan' whether a computer or a certain application on a computer is online.
39 The proposed countermeasures focus on requiring DNS servers to answer only queries they are authoritative for (therefore the authoritative server would send the answer only once in a given time to the target and the attack would fail).
40 This attack was famously used by Eugene Kashpureff in 1999 who was unhappy with the IETF controlling the Top Level Domains, which he countered by announcing his own Top Level Domains and wrong DNS resolutions of IETF websites (Kaminsky, 2006). Since 1999, the vulnerability has been fixed. (Kaminsky, 2006)
41 For example, the open source DNS server djbdns (Perdisci *et al.*, 2009).
42 These statistics refer to the number of connections; P2P accounts for as much as 90% of transferred data. The question whether or not file sharing is harmful to certain sectors of the economy will not be addressed in the current thesis but it should be pointed out that file sharing is not necessarily illegal, depending on the concept, the content and the legal context of the transactions in question. Even in the case of illegal file sharing, it is highly doubtful, that it is actually harmful to the economy in general and to specific firms in particular (Shy, 2001; Peitz and Waelbroeck, 2006; Darmon *et al.*, 2009; Le Texier and Gordah, 2011). For recent empiric work, see Oberholzer-Gee and Strumpf (2010); as both file sharing and the very sharp response of the software and entertainment industry are very recent phenomena, empirical data is limited, especially with respect to long-term consequences.
43 http is actually not a vulnerability in this sense, but nevertheless a security leak if used to transfer confidential data such as passwords.
44 Today most webservers are able to simultaneously allow http and https connections.
45 The most prominent example for that is, of course, the Windows and to a lesser extent the Apple PC operating systems.
46 TCP/IP was proposed in 1974, three major networks were successfully linked operating with this protocol in 1977, the ARPA officially endorsed the protocol as a standard in 1983 and the vendors of the major operating systems followed in adding standard support for the protocol to their systems. For a more detailed account of the history, see Ziewitz and Brown (2011).
47 For example, different chat protocols of which there are plenty; ICQ, MSN, QQ, XMPP, Skype, and many others.
48 With network externalities, this happens largely through the increasing difference in utility as one network grows larger at the expense of the other, thus inducing a positive feedback loop. For details, see sections 4.2 and 4.3.
49 The resulting effect is described in section 4.2.2.
50 Or possibly, if funding of research is counted, starting with the conception of the idea of the technology.
51 The final limit is of course the exhaustion of the IPv4 address space.
52 See, e.g. West and Dedrick (2001), Simcoe (2006) and West (2007). Simcoe in particular discusses an extensive number of examples.
53 Usually, at least some programming interfaces (APIs) of a software including the internal or public standards it uses are made public; however, the vendor of proprietary software will keep the core of software and standards to herself in order to prevent competitors from developing alternative products that are able to work with the same standard.

54 Though attempts have been made to explain open source as a signaling strategy guaranteeing the programmers better jobs in the future (Lerner and Tirole, 2002). This could, however, not be documented with empirical evidence (Bitzer and Geishecker, 2010).
55 Consider, e.g. the development of email as a simple and adequate solution to a practical problem without commercial purposes (Ziewitz and Brown, 2011).
56 Examples include the various open source projects of Sun Microsystems (Luksha, 2008) but also the funding of the development of Linux by large firms (Campbell-Kelly and Garcia-Swartz, 2008).
57 This especially makes security shortcomings more unlikely in open source software than in closed software that enjoys the benefit of 'security by obscurity' but can only rely on in-house programmers for finding and resolving bugs (not mentioning the effect of the lack of incentive to the firm to quickly resolve an unpublished flaw) (Anderson, 2001, 2002).
58 It should be mentioned that governmental regulation, as is usual practice in many countries, is by no means better and is in some cases, in dictatorial regimes, much more harmful.

7 Conclusions

1 Or even earlier with the Keynesian dynamic approaches of Harrod (1939), Samuelson (Samuelson, 1939a,b), Kaldor (1940) and Domar (1946)
2 Such as Kaldorian (Lorenz, 1987) and Marxian (Minsky, 1980; Keen, 1995) models.
3 Small changes over time that may be explained by changes in the organizational environment, changing preferences, etc., do not fundamentally change this growth pattern.
4 This term in turn drives the network effect.
5 This gap may, as one review suggested, be closed by the works in the Keynesian–Kaleckian tradition; see e.g. Richardson and Courvisanos (2008).
6 If it had not been obvious before, it is clear from the recent and ongoing economic crisis, that the financial sector develops its own dynamics that may feedback into the other sectors. Promising models with respect to this are, as discussed above, for instance those by Minsky (1980) and Keen (1995).

References

Adkisson, R. V. (2004). Ceremonialism, intellectual property rights, and innovative activity. *Journal of Economic Issues*, 38(2):459–466.

Aghion, P. and Howitt, P. (1992). A model of growth through creative destruction. *Econometrica*, 60(2):323–351.

Aghion, P. and Howitt, P. (2009). *The Economics of Growth*, The MIT Press, Cambridge, MA.

Akerlof, G. A. (1970). The market for 'lemons' quality uncertainty and the market mechanism. *The Quarterly Journal of Economics*, 84(3):488–500.

Albin, P. S. (1982). The metalogic of economic predictions, calculations and propositions. *Mathematical Social Sciences*, 3(4):329–358.

Albin, P. S. (1987). Microeconomic foundations of cyclical irregularities or 'chaos'. *Mathematical Social Sciences*, 13(3):185–214.

Albin, P. S. and Foley, D. K. (1992). Decentralized, dispersed exchange without an auctioneer: a simulation study. *Journal of Economic Behavior & Organization*, 18(1):27–51.

Anderson, B. (1991 [1983]). *Imagined Communities: Reflections on the Origin and Spread of Nationalism*. Verso, London. Revised and extended edition.

Anderson, R. (2001). Why information security is hard – an economic perspective. Presented at the 17th Annual Computer Security Applications Conference, New Orleans, December 10–14, 2001. http://www.asac.org/2001/papers/110.pdf. accessed 07/07/2012.

Anderson, R. (2002). Security in open versus closed systems – the dance of Boltzmann, Coase and Moore. Presented at the Conference 'Open Source Software: Economics, Law, and Policy', Toulouse, June 20–21, 2002. http://www.idei.fr/display.php?a=2486, accessed 07/07/2012.

Andriani, P. and McKelvey, B. (2007). Beyond Gaussian averages: redirecting international business and management research toward extreme events and power laws. *Journal of International Business Studies*, 38:1212–1230.

Antonelli, C. (2005). Models of knowledge and systems of governance. *Journal of Institutional Economics*, 1(1):51–73.

Antonelli, C. (2009). The economics of innovation: from the classical legacies to the economics of complexity. *Economics of Innovation and New Technology*, 18(7): 611–646.

Antonelli, G. and de Liso, N. (1997). Introduction: An appraisal of the economic analysis of technological change: The path to the last decade. In Antonelli, G. and de Liso, N., editors, *Economics of Structural and Technological Change*, pages 1–47. Routledge, London and New York.

Arrow, K. J. (1962). The economic implications of learning by doing. *The Review of Economic Studies*, 29(3):155–173.

Arthur, W. B. (1983). Competing technologies and lock-in by historical small events: the dynamics of allocation under increasing returns. *International Institute for Applied Systems Analysis Paper Laxenburg, Austria. (Center for Economic Policy Research, Paper 43, Stanford)*, WP-83-92.

Arthur, W. B. (1988). Self-reinforcing mechanisms in economics. In Anderson, P., Arrow, K. J. and Pines, D., editors, *The Economy as an Evolving Complex System*, pages 9–31. Santa Fe Institute Studies in the Sciences of Complexity, Addison Wesley, Redwood City California.

Arthur, W. B. (1989). Competing technologies, increasing returns, and lock-in by historical events. *Economic Journal*, 99(394):116–31.

Arthur, W. B. (1990a). Positive feedbacks in the economy. *Scientific American*, pages 92–99.

Arthur, W. B. (1990b). 'Silicon Valley' locational clusters: when do increasing returns imply monopoly? *Mathematical Social Sciences*, 19(3):235–251.

Arthur, W. B. (1996). Increasing returns and the new world of business. *Harvard Business Review*, 74(4):100–109.

Arthur, W. B. (2007). The structure of invention. *Research Policy*, 36(2):274–287.

Arthur, W. B., Ermoliev, Y. M., and Kaniovski, Y. M. (1983 [1982]). A generalized urn problem and its applications. *Cybernetics*, 19:61–71. Translated from Russian (*Kibernetica* 1:49–56).

Arthur, W. B., Ermoliev, Y. M., and Kaniovski, Y. M. (1987). Path dependent processes and the emergence of macro-structure. *European Journal of Operational Research*, 30:294–303.

Arthur, W. B. and Lane, D. A. (1993). Information contagion. *Structural Change and Economic Dynamics*, 4(1).

Asosheh, A. and Ramezani, N. (2008). A comprehensive taxonomy of DDoS attacks and defense mechanism applying in a smart classification. *WSEAS Transactions on Communications*, 7(4).

Axelrod, R. (1984). *The Evolution of Cooperation*. Basic Books, New York.

Axelrod, R., Mitchell, W., Thomas, R. E., Bennett, D. S., and Bruderer, E. (1995). Coalition formation in standard-setting alliances. *Management Science*, 41(9):1493–1508.

Axtell, R. L. (2001). Zipf distribution of U.S. firm sizes. *Science*, 293(5536):1818–1820.

Axtell, R. L., Epstein, J. M., and Young, H. P. (2000). The emergence of classes in a multi-agent bargaining model. Center on Social and Economic Dynamics; Working Paper No 9.

Baldani, J. P., Bradfield, J., and Turner, R. W. (2005). *Mathematical Economics*. Thomson/South-Western, Mason, Ohio, 2nd edition.

Banerjee, A. V. (1992). A simple model of herd behaviour. *The Quarterly Journal of Economics*, 107(3):797–817.

Baptista, R. (1998). Clusters, innovation, and growth: A survey of the literature. In Swann, G. M. P., Prevezer, M., and Stout, D., editors, *The Dynamics of Industrial Clusters*, pages 13–51. Oxford University Press, Oxford and New York.

Bar-Yam, Y. (1997). *Dynamics of Complex Systems*. Addison-Wesley, Reading, MA.

Barro, R. J. and Sala-i-Martin, X. (2003). *Economic Growth*. MIT Press, Cambridge, MA, 2nd edition.

Baten, J. and van Zanden, J. (2008). Book production and the onset of modern economic growth. *Journal of Economic Growth*, 13:217–235.

Berger, S. (2009a). Circular cumulative causation à la Myrdal and Kapp. In Berger, S., editor, *The Foundations of Non-Equilibrium Economics: The Principle of Circular Cumulative Causation*. Routledge, London.

Berger, S. (2009b). Introduction. In Berger, S., editor, *The Foundations of Non-Equilibrium Economics: The Principle of Circular Cumulative Causation*. Routledge, London.

Berger, S. and Elsner, W. (2007). European contributions to evolutionary institutional economics: The cases of 'cumulative circular causation' (CCC) and 'open systems approach' (OSA). *Journal of Economic Issues*, 41(2):529–538.

Bitzer, J. and Geishecker, I. (2010). Who contributes voluntarily to OSS? An investigation among German IT employees. *Research Policy*, 39(1):165–172.

Bjørstad, T. E. (2008). An introduction to new stream cipher designs. Working Paper for 25th Chaos Communication Congress, Berlin, 27–30 December 2008.

Boccaletti, S., Latora, V., Moreno, Y., Chavez, M., and Hwang, D.-U. (2006). Complex networks: structure and dynamics. *Physics Reports*, 424(4–5):175–308.

Boldrin, M. and Levine, D. K. (2009). Does intellectual monopoly help innovation? *Review of Law & Economics*, 5(3).

Bormotov, M. (2009). Economic cycles: historical evidence, classification and explication. MPRA Paper 19616, University Library of Munich, Germany.

Bourdieu, P. (1986). The forms of capital. In Biggart, N. W., editor, *Readings in Economic Sociology*, pages 280–291. Blackwell Publishers Ltd, Malen, MA.

Bowles, S. and Choi, J.-k. (2002). The first property rights revolution. Working Paper 02-11-061, Santa Fe Institute.

Bowles, S. and Gintis, H. (2003). The evolution of cooperation in heterogeneous populations. Working Paper 03-05-031, Santa Fe Institute.

Bowles, S. and Sethi, R. (2006). Social segregation and the dynamics of group inequality. Working Paper 06-02-006, Santa Fe Institute.

Bresnahan, T. F. and Greenstein, S. (1999). Technological competition and the structure of the computer industry. *The Journal of Industrial Economics*, 47(1):1–40.

Bureau of Economic Analysis (2010). 2009 National GDP Accounts on the Website of the Bureau of Economic Analysis of the US Department of Commerce: http://www.bea.gov/national. published 2010; accessed 04/10/2010.

Bush, P. D. (1987). The theory of institutional change. *Journal of Economic Issues*, 21:1075–1116.

Butler, L. and Neuhoff, K. (2008). Comparison of feed-in tariff, quota and auction mechanisms to support wind power development. *Renewable Energy*, 33(8): 1854–1867.

Campbell-Kelly, M. and Garcia-Swartz, D. D. (2008). Pragmatism not ideology: IBM's love affair with open source software. http://ssrn.com/abstract=1081613. accessed 07/07/2012.

Carlsson, B. (2004). The digital economy: what is new and what is not? *Structural Change and Economic Dynamics*, 15(3):245–264.

Carlsson, B. and Stankiewicz, R. (1991). On the nature, function and composition of technological systems. *Journal of Evolutionary Economics*, 1:93–118.

Cass, D. (1965). Optimum growth in an aggregative model of capital accumulation. *The Review of Economic Studies*, 32(3):233–240.

Cass, D. (1966). Optimum growth in an aggregative model of capital accumulation: a turnpike theorem. *Econometrica*, 34(4):833–850.

Chen, J. (2010a). Switching costs in network industries. NET Institute Working Paper No. 09-25. http://www.netinst.org/Chen_09-25.pdf.

Chen, P. (2002). Microfoundations of macroeconomic fluctuations and the laws of probability theory: the principle of large numbers versus rational expectations arbitrage. *Journal of Economic Behavior & Organization*, 49(3):327–344.
Chen, P. (2005). Evolutionary economic dynamics: persistent cycles, disruptive technology and the trade-off between stability and complexity. In Dopfer, K., editor, *The Evolutionary Foundations of Economics*, pages 472–505. Cambridge University Press, Cambridge.
Chen, P. (2010b). *Economic Complexity and Equilibrium Illusion: Essays on Market Instability and Macro Vitality*. Routledge, London and New York. http://ssrn.org/abstract=1499862, accessed 07/07/2012.
Choi, J. P. (1994). Network externality, compatibility choice, and planned obsolescence. *The Journal of Industrial Economics*, 42(2):167–182.
Choi, J. P. (2004). Tying and innovation: a dynamic analysis of tying arrangements. *The Economic Journal*, 114(492):83–101.
Choi, J. P. and Kim, B.-C. (2010). Net neutrality and investment incentives. *The RAND Journal of Economics*, 41(3):446–471.
Clauset, A., Shalizi, C. R., and Newman, M. E. J. (2009). Power-law distributions in empirical data. *SIAM Review*, 51:661–703.
Coase, R. H. (1960). The problem of social cost. *Journal of Law and Economics*, 3:1–44.
Conlisk, J. (1989). An aggregate model of technical change. *The Quarterly Journal of Economics*, 104(4):787–821.
Cooper, R. and John, A. (1988). Coordinating coordination failures in Keynesian models. *The Quarterly Journal of Economics*, 103(3):441–463.
Cordes, C., Richerson, P., McElreath, R., and Stirmling, P. (2011). How does opportunistic behaviour influence firm size? An evolutionary approach to organizational behaviour. *Journal of Institutional Economics*, 7(1):1–21.
Cournot, A. (1897 [1838]). *Researches into the Mathematical Principles of the Theory of Wealth*. The Macmillan Company, New York. Translated from French by Nathaniel T. Bacon (Original Title: (*Recherches sur les principes mathématiques de la théorie des richesses*)).
Cowan, R., McKendall, A., Mili, A., Yang, L., Wang, L., Chen, D., Janardhana, V., and Spencer, T. (2002). Software engineering technology watch. *Information Sciences*, 140(3–4):195–215.
Coyne, C. J. and Leeson, P. T. (2006). The economics of computer hacking. *Journal of Law, Economics, and Policy*, 1(2).
Daly, H. E. (1968). On economics as a life science. *The Journal of Political Economy*, 76(3):392–406.
Darmon, E., Rufini, A., and Torre, D. (2009). Back to software 'profitable piracy': the role of information diffusion. *Economics Bulletin*, 29(2):543–553.
David, P. A. (1985). Clio and the economics of QWERTY. *The American Economic Review*, 75(2):332–337.
David, P. A. (1992). Heroes, herds and hysteresis in technological history: Thomas Edison and the battle of the systems reconsidered. *Industrial and Corporate Change*, 1(1): 129–180.
David, P. A. (2005). Path dependence in economic processes: implications for policy analysis in dynamical system contexts. In Dopfer, K., editor, *The Evolutionary Foundations of Economics*. Cambridge University Press, Cambridge.
David, P. A. and Bunn, J. A. (1988). The economics of gateway technologies and network evolution: lessons from electricity supply history. *Information Economics and Policy*, 3(2):165–202.

David, P. A. and Steinmueller, W. (1994). Economics of compatibility standards and competition in telecommunication networks. *Information Economics and Policy*, 6(3–4):217–241.

Dawkins, R. (1976). *The Selfish Gene*. Oxford University Press, Oxford.

Day, R. (1994). Do economies diverge? In Silverberg, G. and Soete, L., editors, *The Economics of Growth and Technical Change: Technologies, Nations, Agents*. Edward Elgar, Aldershot, Hants, Brookfield, VT.

de Groot, B. (2006). *Essays on Economic Cycles*. PhD thesis, Published by Rotterdam School of Management (RSM) Erasmus University, Erasmus Research Institute of Management (ERIM). ISBN 978-905892-123-9.

de Groot, B. and Franses, P. H. (2008). Stability through cycles. *Technological Forecasting and Social Change*, 75(3):301–311.

de Groot, B. and Franses, P. H. (2009). Cycles in basic innovations. *Technological Forecasting and Social Change*, 76(8):1021–1025.

De Langhe, R. and Greiff, M. (2010). Standards and the distribution of cognitive labour. *Logic Journal of IGPL*, 18(2):278–293.

de Liso, N. and Metcalfe, J. S. (1996). On technological systems and technological paradigms: some recent developments in the understanding of technological change. In Helmstädter, E. and Perlman, M., editors, *Behavioral Norms, Technological Progress, and Economic Dynamics, Studies in Schumpeterian Economics*, pages 331–371. The University of Michigan Press, Ann Arbor, MI.

Delre, S., Jager, W., and Janssen, M. (2007). Diffusion dynamics in small-world networks with heterogeneous consumers. *Computational Mathematical Organization Theory*, 13:185–202.

Demsetz, H. (1968). Why regulate utilities? *Journal of Law and Economics*, 11(1):55–65.

Dessalles, J.-L. and Phan, D. (2005). Emergence in multi-agent systems: cognitive hierarchy, detection, and complexity reduction. In Mathieu, P., Beaufils, B., and Brandouy, O., editors, *Artificial Economics*, pages 145–157. Springer, Berlin.

Devezas, T. C. and Corredine, J. T. (2002). The nonlinear dynamics of technoeconomic systems: an informational interpretation. *Technological Forecasting and Social Change*, 69(4):317–357.

Diebolt, C. and Doliger, C. (2006). Economic cycles under test: a spectral analysis. In Devezas, T., editor, *Kondratieff Waves, Warfare and World Security*, pages 39–47. IOS Press, Amsterdam.

Diker, V. G. and Scholl, H. J. (2001). The art of leveraging: how powerful nonlinear feedback processes can restructure rapidly growing technology and knowledge industries. Presented at the 34th Annual Hawaii International Conference on System Sciences (HICSS), Maui January 3–6, 2001. http://doi.ieeecomputersociety.org/10.1109/HICSS.2001.926301, accessed 07/07/2012.

Dixit, A. K. and Stiglitz, J. E. (1977). Monopolistic competition and optimum product diversity. *The American Economic Review*, 67(3):297–308.

Dixon, R. and Thirlwall, A. P. (1975). A model of regional growth-rate differences on Kaldorian lines. *Oxford Economic Papers*, 27(2):201–214.

Dolfsma, W. and Leydesdorff, L. (2009). Lock-in and break-out from technological trajectories: modeling and policy implications. *Technological Forecasting and Social Change*, 76(7):932–941.

Domar, E. D. (1946). Capital expansion, rate of growth, and employment. *Econometrica*, 14(2):137–147.

Dosi, G., Ermoliev, Y., and Kaniovski, Y. (1994). Generalized urn schemes and technological dynamics. *Journal of Mathematical Economics*, 23(1):1–19.

Dosi, G. and Kaniovski, Y. (1994). The method of generalised urn schemes in the analysis of technological and economic dynamics. In Silverberg, G. and Soete, L., editors, *The Economics of Growth and Technical Change: Technologies, Nations, Agents*. Edward Elgar, Aldershot, Hants, and Brookfield, VT.

Dunbar, R. I. M. (1993). Coevolution of neocortical size, group size and language in humans. *Behavioral and Brain Sciences*, 16(04):681–694.

Easley, D. and Kleinberg, J. (2010). *Networks, Crowds, and Markets: Reasoning About a Highly Connected World*. Cambridge University Press.

Economides, N. and White, L. J. (1994). Networks and compatibility: implications for antitrust. *European Economic Review*, 38(3–4):651–662.

Economides, N. S. and Tåg, J. (2011). Net neutrality on the internet: a two-sided market analysis. NET Institute Working Paper 07-45. http://ssrn.com/abstract=1019121.

Eggenberger, F. and Pólya, G. (1923). Über die Statistik verketteter Vorgänge. *ZAMM – Journal of Applied Mathematics and Mechanics/Zeitschrift für Angewandte Mathematik und Mechanik*, 3(4):279–289.

Egyedi, T. M. and Heijnen, P. (2005). Scale of standards dynamics: change in formal, international IT standards. In Bolin, S., editor, *The Standards Edge: Future Generation*. Bolin Group.

Elsner, W. (2012). The theory of institutional change revisited: the institutional dichotomy, its dynamic, and its policy implications in a more formal analysis. *Journal of Economic Issues*, 46:1–44.

Elsner, W. and Heinrich, T. (2009). A simple theory of 'meso'. on the co-evolution of institutions and platform size – with an application to varieties of capitalism and 'medium-sized' countries. *Journal of Socio-Economics*, 38(5):843–858.

Elsner, W. and Heinrich, T. (2011). Coordination on 'meso'-levels: on the co-evolution of institutions, networks, and platform size. In Mann, S., editor, *Sectors Matter. Exploring Mesoeconomics*, pages 115–163. Springer, Berlin, Heidelberg.

Elsner, W., Hocker, G., and Schwardt, H. (2010). Simplistic vs. complex organization: Markets, hierarchies, and networks in an organizational triangle – a simple heuristic to analyze real-world organizational forms. *Journal of Economic Issues*, 44(1):1–30.

EU KLEMS Project (2011). Website of the EU KLEMS growth and productivity accounts project's online database. http://www.euklems.net/ accessed 06/14/2011.

Evron, G. (2008). Battling botnets and online mobs: Estonia's defense efforts during the internet war. *Georgetown Journal of International Affairs*, 9(1).

Farrell, J. and Klemperer, P. (2007). Coordination and lock-in: competition with switching costs and network effects. In Armstrong, M. and Porter, R., editors, *Handbook of Industrial Organization*, volume 3 of *Handbooks in Economics*, pages 1967–2072. Elsevier, Amsterdam.

Farrell, J. and Saloner, G. (1985). Standardization, compatibility and innovation. *RAND Journal of Economics*, 16(1):70–83.

Farrell, J. and Saloner, G. (1986). Installed base and compatibility: innovation, product preannouncements, and predation. *The American Economic Review*, 76(5):940–955.

Farrell, J. and Saloner, G. (1992). Converters, compatibility, and the control of interfaces. *The Journal of Industrial Economics*, 40(1):9–35.

Farrell, J. and Shapiro, C. (2004). Intellectual property, competition, and information technology. Competition Policy Center, Working Paper Series 11353, Competition Policy Center, Institute for Business and Economic Research, UC Berkeley.

Fehr, E. and Fischbacher, U. (2004). Social norms and human cooperation. *Trends in Cognitive Sciences*, 8(4):185–190.

Fehr, E., Fischbacher, U., and Gächter, S. (2002). Strong reciprocity, human cooperation, and the enforcement of social norms. *Human Nature*, 13:1–25.

Fehr, E. and Gächter, S. (2000). Cooperation and punishment in public goods experiments. *The American Economic Review*, 90(4):980–994.

Fehr, E. and Gächter, S. (2002). Altruistic punishment in humans. *Nature*, 415:137–140.

Foley, D. K. (1998). Introduction to barriers and bounds to rationality. In Foley, D. K., editor, *Barriers and Bounds to Rationality: Essays on Economic Complexity and Dynamics in Interactive Systems, by Albin, P. S., with an Introduction by Foley, D. K.*, pages 3–72. Princeton University Press, Princeton, NJ.

Foley, D. K. (2010). What's wrong with the fundamental existence and welfare theorems? *Journal of Economic Behavior & Organization*, 75(2):115–131.

Föllmer, H. (1974). Random economies with many interacting systems. *Journal of Mathematical Economics*, 1:51–62.

Forrester, J. W. (1977). Growth cycles. *De Economist*, 125:525–543.

Foucault, M. (1976 [1978]). *The History of Sexuality. Volume 1: An Introduction*. Random House. (Original title: *Histoire de la sexualite I: La Volonte de savoir*).

Frankel, M. (1962). The production function in allocation and growth: a synthesis. *The American Economic Review*, 52(5):996–1022.

Freeman, C. and Perez, C. (1988). Structural crisis of adjustment, business cycles and investment behaviour. In Dosi, G., Freeman, C., Nelson, R., Silverberg, G., and Soete, L., editors, *Technological Change and Economic Theory*, pages 38–66. Pinter Publishers, London and New York.

Gallagher, S. and West, J. (2009). Reconceptualizing and expanding the positive feedback network effects model: a case study. *Journal of Engineering and Technology Management*, 26(3):131–147.

Gandolfo, G. (1997). *Economic Dynamics*. Springer, Berlin, 1st edition.

Ganley, P. and Allgrove, B. (2006). Net neutrality: a user's guide. *Computer Law & Security Report*, 22(6):454–463.

Georgescu-Roegen, N. (1975a). Dynamic models and economic growth. *World Development*, 3(11–12):765–783.

Georgescu-Roegen, N. (1975b). Energy and economic myths. *Southern Economic Journal*, 41(3):347–381.

German Federal Statistical Office (2010). 2009 Annual and Quarterly National GDP Accounts; German Federal Statistical Office. http://www.destatis.de. published February 2010; accessed 04/10/2010.

Gilbert, R. J. and Newbery, D. M. G. (1982). Preemptive patenting and the persistence of monopoly. *The American Economic Review*, 72(3):514–526.

Gintis, H. (2000). *Game Theory Evolving: A problem-centered Introduction to Modeling Strategic Behavior*. Princeton University Press, Princeton, NJ.

Glowniak, J. (1998). History, structure, and function of the internet. *Seminars in Nuclear Medicine*, 28(2):135–144.

Goodwin, R. M. (1967). A growth cycle. In Feinstein, C., editor, *Socialism, Capitalism and Economic Growth*, pages 54–58. Cambridge University Press, Cambridge.

Goyal, S. (2007). *Connections: An Introduction to the Economics of Networks*. Princeton University Press, Princeton, NJ.

Greiff, M. (2011). *Social Approval as an Incentive: the Voluntary Provision of Public Goods in Social Production Communities*. Peter Lang, Frankfurt am Main.

Grosche, G., Ziegler, V., Ziegler, D., and Zeidler, E. (1995). *Teubner-Encyclopedia of Mathematics. Pt. 2. (Teubner-Taschenbuch der Mathematik. T. 2.)*. Teubner, Leipzig.

Hall, J. and Günther, J. (2011). Claiming Paul David's path dependence as institutionalist theory. Research paper presented at the AFEE Annual Meeting, Denver, January 7–9, 2011.

Hands, D. W. (1993). More light on integrability, symmetry, and utility as potential energy in Mirowski's critical history. In De Marchi, N., editor, *Non-Natural Social Science: Reflecting on the Enterprise of More Heat Than Light*, pages 118–130. Duke University Press, Durham, NC.

Hands, D. W. (2006). Integrability, rationalizability, and path-dependency in the history of demand theory. *History of Political Economy*, 38:153–185.

Hargreaves Heap, S. P. and Varoufakis, Y. (1995). *Game Theory: A Critical Introduction*. Routledge, London.

Harrod, R. F. (1939). An essay in dynamic theory. *The Economic Journal*, 49(193):14–33.

He, Z. (1994). Diffusion of movable type in China and Europe: why were there two fates? *International Communication Gazette*, 53(3):153–173.

Helbing, D. and Johansson, A. (2010). Evolutionary dynamics of populations with conflicting interactions: classification and analytical treatment considering asymmetry and power. *Physical Review E*, 81(1):016112.

Hicks, J. R. (1935). Annual survey of economic theory: the theory of monopoly. *Econometrica*, 3(1):1–20.

Howitt, P. and Clower, R. (2000). The emergence of economic organization. *Journal of Economic Behaviour and Organization*, 41:55–84.

Hughes, T. P. (2000). Industrial revolutions: from canal systems to computer networks. University of Utah William R. and Erlyn J. Gould Distinguished Lecture on Technology and the Quality of Life at the University of Utah, September 13, 2000. http://content.lib.utah.edu/cdm4/item_viewer.php?CISOROOT=/uspace& CISOPTR=2763, accessed 07/07/2012.

ISC (2011). Internet Systems Consortium website. https://isc.org/solutions/survey/history, accessed 06/14/2011.

John, W., Tafvelin, S., and Olovsson, T. (2008). Trends and differences in connection-behaviour within classes of internet backbone traffic. In Claypool, M. and Uhlig, S., editors, *Passive and Active Network Measurement*, pages 192–201, Springer, Berlin, Heidelberg, New York.

Kahneman, D. and Tversky, A. (1973). Availability: a heuristic for judging frequency and probability. *Cognitive Psychology*, 5:207–232.

Kaldor, N. (1940). A model of the trade cycle. *The Economic Journal*, 50(197):78–92.

Kaldor, N. (1957). A model of economic growth. *The Economic Journal*, 67(268):591–624.

Kaldor, N. (1972). The irrelevance of equilibrium economics. *The Economic Journal*, 82(328):1237–1255.

Kaldor, N. (1975). What is wrong with economic theory. *The Quarterly Journal of Economics*, 89(3):347–357.

Kaldor, N. and Mirrlees, J. A. (1962). A new model of economic growth. *The Review of Economic Studies*, 29(3):174–192.

Kalecki, M. (1962). Observations on the theory of growth. *The Economic Journal*, 72(285):134–153.

Kaminsky, D. (2004). MD5 to be considered harmful someday. Cryptology ePrint Archive, Report 2004/357. http://eprint.iacr.org/2004/357, 6 December 2004, accessed 07/07/2012.

Kaminsky, D. (2006). Explorations in namespace: white-hat hacking across the domain name system. *Communications Of The ACM*, 49(6).

Kaminsky, D., Patterson, M., and Sassaman, L. (2010). PKI layer cake: new collision attacks against the global X.509 infrastructure. In Sion, R., editor, *Financial Cryptography and Data Security*, volume 6052 of *Lecture Notes in Computer Science*, pages 289–303. Springer, Berlin and Heidelberg.

Katz, M. L. and Shapiro, C. (1985). Network externalities, competition and compatibility. *The American Economic Review*, 75(3):424–440.

Katz, M. L. and Shapiro, C. (1986). Technology adoption in the presence of network externalities. *Journal of Political Economy*, 94(4):822–841.

Katz, M. L. and Shapiro, C. (1994). Systems competition and network effects. *Journal of Economic Perspectives*, 8(2):93–115.

Kauffman, S. (1993). *The Origins of Order: Self-organization and Selection in Evolution*. Oxford University Press, New York.

Kauffman, S., Lobo, J., and Macready, W. G. (2000). Optimal search on a technology landscape. *Journal of Economic Behavior & Organization*, 43(2):141–166.

Kauffman, S. and Macready, W. (1995). Technological evolution and adaptive organizations. Working Paper 95-02-008, Santa Fe Institute.

Keen, S. (1995). Finance and economic breakdown: modeling Minsky's 'financial instability hypothesis'. *Journal of Post Keynesian Economics*, 17(4):607–635.

Keen, S. (2001). *Debunking Economics: the Naked Emperor of the Social Sciences*. Pluto Press, Annandale, N.W.

Keen, S. (2009a). Mathematics for pluralist economists. In Reardon, J., editor, *Handbook for Pluralist Economics Education*, pages 150–168. Routledge, London and New York.

Keen, S. (2009b). A pluralist approach to microeconomics. In Reardon, J., editor, *Handbook for Pluralist Economics Education*, pages 120–149. Routledge, London and New York.

Keynes, J. M. (1936). *The General Theory of Employment Interest and Money*. Harcourt, Brace, World, New York. Reprint.

Kim, J. S., Goh, K.-I., Kahng, B., and Kim, D. (2007). Fractality and self-similarity in scale-free networks. *New Journal of Physics*, 9(6):177.

King, S. P. and Lampe, R. (2003). Network externalities, price discrimination and profitable piracy. *Information Economics and Policy*, 15(3):271–290.

Kirman, A. P. (1997). The economy as an interactive system. In Arthur, W. B., Durlauf, S., and Lane, D., editors, *The Economy as an Evolving Complex System II*, pages 491–531. Westview Press, Boulder, CO.

Klemperer, P. (1987). Markets with consumer switching costs. *The Quarterly Journal of Economics*, 102(2):375–394.

Knack, S. (1996). Institutions and the convergence hypothesis: the cross-national evidence. *Public Choice*, 87:207–228.

Kretschmer, T. (2004). Upgrading and niche usage of PC operating systems. *International Journal of Industrial Organization*, 22(8–9):1155–1182.

Kretschmer, T. (2008). Splintering and inertia in network industries. *The Journal of Industrial Economics*, 56(4):685–706.

Krugman, P. (1997). How the economy organizes itself in space: a survey of the new economic geography. In Arthur, W. B., Durlauf, S., and Lane, D., editors, *The Economy as an Evolving Complex System II*, pages 339–362. Westview Press, Boulder, CO.

Kuhn, T. S. (1996). *The Structure of Scientific Revolutions*. University of Chicago Press, Chicago, 3rd edition.

Kunert, U. and Kuhfeld, H. (2007). The diverse structures of passenger car taxation in Europe and the EU commission's proposal for reform. *Transport Policy*, 14(4): 306–316.

Kurz, H. D. and Salvadori, N. (2003). Theories of economic growth: old and new. In Salvadori, N., editor, *The Theory of Economic Growth: A 'Classical' Perspective*, pages 1–22. Edward Elgar, Cheltenham and Northhampton, MA.

Lakhani, K. R. and von Hippel, E. (2003). How open source software works: free user-to-user assistance. *Research Policy*, 32:923–943.

Le Texier, T. and Gordah, M. (2011). Peer-to-peer networks and complementary goods: the impact of openness and innovation on profitable piracy. *Economic Analysis*, 44(1–2):15–37.

Lee, E., Lee, J., and Lee, J. (2006). Reconsideration of the winner-take-all hypothesis: complex networks and local bias. *Management Science*, 52:1838–1848.

Lehmann-Waffenschmidt, M. (2007). *Economic Evolution and Equilibrium: Bridging the Gap*. Lecture notes in economics and mathematical systems, 591. Springer, Berlin, Heidelberg, New York.

Leibenstein, H. (1950). Bandwagon, snob, and Veblen effects in the theory of consumers' demand. *The Quarterly Journal of Economics*, 64(2):183–207.

Leiner, B. M., Cerf, V. G., Clark, D. D., Kahn, R. E., Kleinrock, L., Lynch, D. C., Postel, J., Roberts, L. G., and Wolff, S. (2004). A brief history of the internet. http://www.isoc.org/internet/history/brief.shtml.

León-Ledesma, M. A. (2000). Cumulative growth and the catching-up debate from a dis-equilibrium standpoint. Studies in Economics 0001, Department of Economics, University of Kent.

León-Ledesma, M. A. (2006). Cycles, aggregate demand and growth. In Arestis, P., McCombie, J., and Vickerman, R., editors, *Growth and Economic Development: Essays in Honour of A.P. Thirlwall*, pages 82–95. Edward Elgar, Cheltenham.

Leoncini, R., Rentocchini, F., and Vittucci Marzetti, G. (2008). You won the battle. what about the war? A model of competition between proprietary and open source software. Department of Economics Working Papers 0811, Department of Economics, University of Trento, Italia. Newer version forthcoming in the *Revue d'Economie Industrielle* (2011).

Lerner, J. and Tirole, J. (2002). Some simple economics of open source. *The Journal of Industrial Economics*, 50(2):197–234.

Liagouras, G. (2005). The political economy of post-industrial capitalism. *Thesis Eleven*, 81(1):20–35.

Liagouras, G. (2008). Almost everything you would like to know about technology and growth and you will never be told by neo-Schumpeterians. In Elsner, W. and Hanappi, G., editors, *Varieties of Capitalism and New Institutional Deals: Regulation, Welfare, and the New Economy*, pages 300–323. Edward Elgar, Cheltenham and Northampton, MA.

Liebowitz, S. (2005). Economiststs topsy-turvy view of piracy. Technical report, EconWPA. http://ssrn.com/abstract=738864, accessed 07/07/2012.

Liebowitz, S. J. and Margolis, S. E. (1990). The fable of the keys. *Journal of Law and Economics*, 33(1):1–25.

Liebowitz, S. J. and Margolis, S. E. (1994). Network externality: an uncommon tragedy. *The Journal of Economic Perspectives*, 8(2):133–150.

Liebowitz, S. J. and Margolis, S. E. (1995). Path dependence, lock-in, and history. *Journal of Law, Economics, & Organization*, 11(1):205–226.

Lindgren, K. (1997). Evolutionary dynamics in game-theoretic models. In Arthur, W. B., Durlauf, S., and Lane, D., editors, *The Economy as an Evolving Complex System II*, pages 337–367. Westview Press, Boulder, CO.

Lorenz, H.-W. (1987). Strange attractors in a multisector business cycle model. *Journal of Economic Behavior & Organization*, 8(3):397–411.

Lotka, A. J. (1925). *Elements of Physical Biology*. Williams & Wilkins Company, Baltimore, MD.

Lucas, R. E. (1972). Expectations and the neutrality of money. *Journal of Economic Theory*, 4(2):103–124.

Lucas, R. E. (1975). An equilibrium model of the business cycle. *The Journal of Political Economy*, 83(6):1113–1144.

Lucas, R. E. (1981). *Studies in Business Cycle Theory*. Basil Blackwell, Oxford.

Lucas, R. E. (1988). On the mechanics of economic development. *Journal of Monetary Economics*, 22(1):3–42.

Luksha, P. (2008). Niche construction: the process of opportunity creation in the environment. *Strategic Entrepreneurship Journal*, 2(4):269–283.

Lyapunov, M. A. (1907). Problème général de la stabilité du mouvment. *Annales de la Faculté des Sciences de Toulouse*, 9:203–474.

Mandelbrot, B. B. and Hudson, R. L. (2004). *The (Mis)behaviour of Markets*. Basic Books, New York.

Manenti, F. M. and Somma, E. (2002). One-way compatibility, two-way compatibility and entry in network industries. Royal Economic Society annual conference 2002, Royal Economic Society.

Markus, L. and Yamabe, H. (1960). Global stability criteria for differential systems. *Osaka Mathematical Journal*, 12(2):305–317.

Marshall, A. (1990 [1890]). *Principles of Economics: an Introductory Volume*. Porcupine Press, Philadelphia, PA, reprint, 8th edition.

Marx, K. (1963 [1885]). *Das Kapital: Kritik der politischen Ökonomie. Buch II: Der Zirkulationsprozeß des Kapitals*. Dietz Verlag, Berlin, GDR. Reprint in *Marx-Engels Collected Works* (*Marx-Engels Gesamtausgabe*), Volume 24.

Marx, K. (1964 [1867]). *Das Kapital: Kritik der politischen Ökonomie. Buch I: Der Produktionsprozeß des Kapitals*. Dietz Verlag, Berlin, GDR. Reprint in *Marx-Engels Collected Works* (*Marx-Engels Gesamtausgabe*), Volume 23.

Marx, K. (1972 [1894]). *Das Kapital: Kritik der politischen Ökonomie. Buch III: Der Gesamtprozeß der kapitalistischen Produktion*. Dietz Verlag, Berlin, GDR. Reprint in *Marx-Engels Collected Works* (*Marx-Engels Gesamtausgabe*), Volume 25.

Mas-Colell, A., Whinston, M. D., and Green, J. R. (1995). *Microeconomic Theory*. Oxford University Press, New York.

Matsuyama, K. (1991). Increasing returns, industrialization, and indeterminacy of equilibrium. *The Quarterly Journal of Economics*, 106(2):617–650.

May, R. M. (1976). Simple mathematical models with very complicated dynamics. *Nature*, 261:459–467.

Maynard Smith, J. and Price, G. R. (1973). The logic of animal conflict. *Nature*, 246: 15–18.

McGrattan, E. R. (1998). A defense of AK growth models. *Quarterly Review*, 22(4):13–27.

McKelvey, M. (2001). The economic dynamics of software: three competing business models exemplified through Microsoft, Netscape and Linux. *Economics of Innovation and New Technologies*, 10:199–236.

Minsky, H. P. (1980). Capitalist financial processes and the instability of capitalism. *Journal of Economic Issues*, 14(2):505–523.

Mira, C. (1997). Some historical aspects of nonlinear dynamics – possible trends for the future. *Journal of the Franklin Institute*, 334(5–6):1075–1113.

Mirowski, P. (1989). *More Heat than Light: Economics as Social Physics, Physics as Nature's Economics*. Cambridge University Press, Cambridge.

Mokyr, J. (1992). Technological inertia in economic history. *The Journal of Economic History*, 52(2):325–338.

Mokyr, J. (1993). Editor's introduction: the new economic history and the industrial revolution. In Mokyr, J., editor, *The British Industrial Revolution: An Economic Perspective*, pages 1–127. Westview Press, Boulder, CO.

Mulder, P., De Groot, H. L. F., and Hofkes, M. W. (2001). Economic growth and technological change: a comparison of insights from a neo-classical and an evolutionary perspective. *Technological Forecasting and Social Change*, 68(2):151–171.

Myerson, R. B. (1978). Refinements of the Nash equilibrium concept. *International Journal of Game Theory*, 7:73–80.

Nash, J. F. (1950). Equilibrium points in *n*-person games. *Proceedings of the National Academy of Sciences of the United States of America*, 36(1):48–49.

NATO (2009). NATO and cyber defence. Technical report. Committee Reports of the NATO Parliamentary Assembly, 2009 Annual Session, Report 173 DSCFC 09 E bis, Rapporteur: Sverre Myrli (Norway). http://www.nato-pa.int/default.asp?SHORTCUT=1782, accessed 07/07/2012.

Navon, A., Shy, O., and Thisse, J.-F. (1995). Product differentiation in the presence of positive and negative network effects. CEPR Discussion Papers 1306, London.

Nelson, R. R. (1968). A 'diffusion model' of international productivity differences in manufacturing industry. *The American Economic Review*, 58:1219–1248.

Nelson, R. R. (1995). Recent evolutionary theorizing about economic change. *Journal of Economic Literature*, 33(1):48–90.

Nelson, R. R. and Winter, S. G. (1974). Neoclassical versus evolutionary theories of economic growth: critique and prospectus. *Economic Journal*, 84(336):886–905.

Nelson, R. R. and Winter, S. G. (1982a). *An Evolutionary Theory of Economic Change*. Harvard University Press, Cambridge, MA.

Nelson, R. R. and Winter, S. G. (1982b). The Schumpeterian tradeoff revisited. *The American Economic Review*, 72(1):114–132.

Nelson, R. R., Winter, S. G., and Schuette, H. L. (1976). Technical change in an evolutionary model. *The Quarterly Journal of Economics*, 90(1):90–118.

Netcraft Ltd (2009). Netcraft SSL survey January 2009. http://news.netcraft.com/ssl-sample-report/, accessed 06/14/2011.

Newman, M. E. J. (2005). Power laws, Pareto distributions and Zipf's law. *Contemporary Physics*, 46(5):323–351.

Nowak, M. A. (2006). *Evolutionary Dynamics: Exploring the Equations of Life*. Belknap Press of Harvard University Press, Cambridge, MA.

Oancea, B., Rosca, I. G., Andrei, T., and Iacob, A. I. (2011). Evaluating Java performance for linear algebra numerical computations. *Procedia Computer Science*, 3:474–478.

Oberholzer-Gee, F. and Strumpf, K. (2010). P2Ps impact on recorded music sales. Working Paper. http://www.eecs.harvard.edu/p2pecon/confman/papers/s1p2.pdf, accessed 07/07/2012.

Odlyzko, A. (2003). The case against micropayments. In Wright, R., editor, *Financial Cryptography*, volume 2742 of *Lecture Notes in Computer Science*, pages 77–83. Springer, Berlin and Heidelberg.

O'Mahony, M. and Timmer, M. P. (2009). Output, input and productivity measures at the industry level: the EU KLEMS database. *The Economic Journal*, 119(538): F374–F403.

Ostrom, E. (2000). Collective action and the evolution of social norms. *Journal of Economic Perspectives*, 14(3):137–158.

Pacheco, J. M., Pinheiro, F. L., and Santos, F. C. (2009). Population structure induces a symmetry breaking favoring the emergence of cooperation. *PLoS Computational Biology*, 5(12):e1000596.

Pegoretti, G., Rentocchini, F., and Vittucci Marzetti, G. (2009). An agent-based model of competitive diffusion: network structure and coexistence. OPENLOC working paper series, No. 16/2009, Department of Economics, University of Trento, Italy.

Peitz, M. and Waelbroeck, P. (2006). Piracy of digital products: a critical review of the theoretical literature. *Information Economics and Policy*, 18(4):449–476.

Perdisci, R., Antonakakis, M., Luo, X., and Lee, W. (2009). WSEC DNS: protecting recursive DNS resolvers from poisoning attacks. In IEEE/IFIP International Conference on Dependable Systems Networks, 2009, pages 3 –12.

Perez, C. (1983). Structural change and assimilation of new technologies in the economic and social systems. *Futures*, 15(5):357–375.

Pitchford, J. D. (2002). Trevor Swan's 1956 economic growth 'seminar' and notes on growth. *Economic Record*, 78(243):381–387.

Prasad, A. and Mahajan, V. (2003). How many pirates should a software firm tolerate? An analysis of piracy protection on the diffusion of software. *International Journal of Research in Marketing*, 20(4):337–353.

Prechelt, L. (2000). An empirical comparison of seven programming languages. *Computer*, 33:23–29.

Prigogine, I. (1975). Dissipative structures, dynamics and entropy. *International Journal of Quantum Chemistry*, 9(S9):443–456.

Prigogine, I. and Stengers, I. (1984). *Order out of Chaos*. Bantam Books, New York.

Pritchett, L. (1997). Divergence, big time. *The Journal of Economic Perspectives*, 11(3): 3–17.

Pyka, A. and Fagiolo, G. (2005). Agent-based modelling: a methodology for neo-Schumpeterian economics. In Hanusch, H. and Pyka, A., editors, *The Elgar Companion to Neo-Schumpeterian Economics*. Edward Elgar, Cheltenham.

Quah, D. (1999a). Internet cluster emergence. CEP Discussion Papers 0441, London School of Economics.

Quah, D. T. (1999b). The weightless economy in growth. *The Business Economist*, 30: 40–53.

Raymond, E. S. (2000). The Cathedral and the Bazaar. http://www.catb.org/~esr/writings/cathedral-bazaar/cathedral-bazaar/, accessed 07/07/2012.

Reinganum, J. F. (1983). Uncertain innovation and the persistence of monopoly. *The American Economic Review*, 73(4):741–748.

Ricardo, D. (2001 [1817]). *On the Principles of Political Economy and Taxation*. Batoche Books, Kitchener, Ontario, 3rd edition. Reprint.

Richardson, C. and Courvisanos, J. (2008). Modeling Keynes with Kalecki. In Wray, L. and Forstater, M., editors, *Keynes and Macroeconomics after 70 Years*, pages 99–122. Edward Elgar, Cheltenham and Northamption, MA.

Romer, P. M. (1986). Increasing returns and long run growth. *Journal of Political Economy*, 94:1002–1037.

Romer, P. M. (1990). Endogenous technological change. *The Journal of Political Economy*, 98(5):S71–S102.

Romer, P. M. (1994). The origins of endogenous growth. *The Journal of Economic Perspectives*, 8(1):3–22.

Rubinstein, A. (1979). Equilibrium in supergames with the overtaking criterion. *Journal of Economic Theory*, 21(1):1–9.

Samuelson, P. A. (1939a). Interactions between the multiplier analysis and the principle of acceleration. *The Review of Economics and Statistics*, 21(2):75–78.

Samuelson, P. A. (1939b). A synthesis of the principle of acceleration and the multiplier. *The Journal of Political Economy*, 47(6):786–797.

Sato, R. (1981). *Theory of Technical Change and Economic Invariance*. Economic theory, econometrics, and mathematical economics. Academic Press, New York.

Saviotti, P. P. and Pyka, A. (2008). Micro and macro dynamics: industry life cycles, inter-sector coordination and aggregate growth. *Journal of Evolutionary Economics*, 18:167–182.

Schelling, T. C. (1971). Dynamic models of segregation. *Journal of Mathematical Sociology*, 1:143–186.

Schmookler, J. (1965). Technological change and economic theory. *The American Economic Review*, 55(1/2):333–341.

Schmookler, J. (1966). *Invention and Economic Growth*. Harvard University Press, Cambridge, MA.

Selten, R. (1975). Reexamination of the perfectness concept for equilibrium points in extensive games. *International Journal of Game Theory*, 4:25–55.

Shy, O. (1996). Technology revolutions in the presence of network externalities. *International Journal of Industrial Organization*, 14(6):785–800.

Shy, O. (2001). *The Economics of Network Industries*. Cambridge University Press, New York.

Shy, O. (2002). A quick-and-easy method for estimating switching costs. *International Journal of Industrial Organization*, 20(1):71–87.

Shy, O. and Thisse, J.-F. (1999). A strategic approach to software protection. *Journal of Economics & Management Strategy*, 8(2):163–190.

Silverberg, G., Dosi, G., and Orsenigo, L. (1988). Innovation, diversity and diffusion: a self-organisation model. *The Economic Journal*, 98(393):1032–1054.

Silverberg, G. and Lehnert, D. (1993). Long waves and 'evolutionary chaos' in a simple Schumpeterian model of embodied technical change. *Structural Change and Economic Dynamics*, 4(1):9–37.

Silverberg, G. and Lehnert, D. (1994). Growth fluctuations in an evolutionary model of creative destruction. In Silverberg, G. and Soete, L., editors, *The Economics of Growth and Technical Change: Technologies, Nations, Agents*. Edward Elgar, Aldershot, Hants and Brookfield, VT.

Silverberg, G. and Verspagen, B. (1996). From the artificial to the endogenous: modeling evolutionary adaptation and economic growth. In Helmstädter, E. and Perlman, M., editors, *Behavioral Norms, Technological Progress, and Economic Dynamics, Studies in Schumpeterian Economics*, pages 331–371. The University of Michigan Press, Ann Arbor, MI.

Silverberg, G. and Verspagen, B. (2005). A percolation model of innovation in complex technology spaces. *Journal of Economic Dynamics and Control*, 29(1–2): 225–244.

Silverberg, G. and Verspagen, B. (2007). Self-organization of R&D search in complex technology spaces. *Journal of Economic Interaction and Coordination*, 2:211–229.

Simcoe, T. (2006). Open standards and intellectual property rights. In Chesbrough, H., Vanhaverbeke, W., and West, J., editors, *Open Innovation: Researching a New Paradigm*, pages 161–183. Oxford University Press, Oxford.

Simon, H. A. (1955). A behavioural model of rational choice. *The Quarterly Journal of Economics*, 69(1):99–118.

Slutzky, E. (1937 [1927]). The summation of random causes as the source of cyclic processes. *Econometrica*, 5(2):105–146. Translated from Russian (*Problems of Economic Conditions*, 3:1, 1927, ed. by The Conjuncture Institute, Moscow).

Smith, A. (2005 [1776]). *An Inquiry into the Nature and Causes of the Wealth of Nations*. University of Oregon Libraries (Scholars' Bank). https://scholarsbank.uoregon.edu/xmlui/handle/1794/782, accessed 07/07/2012.

Smith, E. and Foley, D. K. (2008). Classical thermodynamics and economic general equilibrium theory. *Journal of Economic Dynamics and Control*, 32(1):7–65.

Smith, R. (2010). Beyond growth or beyond capitalism? *Real-world Economics Review*, (53):28–42.

Smith, V. L. (1989). Theory, experiment and economics. *The Journal of Economic Perspectives*, 3(1):151–169.

Solow, R. M. (1956). A contribution to the theory of economic growth. *The Quarterly Journal of Economics*, 70(1):65–94.

Solow, R. M. (1957). Technical change and the aggregate production function. *The Review of Economics and Statistics*, 39(3):312–320.

Solow, R. M. (1994). Perspectives on growth theory. *The Journal of Economic Perspectives*, 8(1):45–54.

Stango, V. (2004). The economics of standard wars. *Review of Network Economics*, 3(1):1–19.

Stango, V. and Greenstein, S. (2008). The economics and strategy of standards and standardization. In Shane, S., editor, *Handbook of Technology and Innovation Management*. Blackwell, Oxford.

Sundararajan, A. (2008). Local network effects and complex network structure. *The B.E. Journal of Theoretical Economics*, 7(1).

Swan, T. W. (2002 [1956]). Economic growth. *Economic Record*, 78(243):375–380. Lecture Notes by Trevor W. Swan, written before the publication of both his and Solow's articles on their growth models and published in 2002; for historical details, see Pitchford (2002).

Swann, G. M. P. (1998). Clusters in the US-computing industry. In Swann, G. M. P., Prevezer, M., and Stout, D., editors, *The Dynamics of Industrial Clusters*, pages 77–105. Oxford University Press, Oxford and New York.

Swann, G. P. (2000). The economics of standardization. Final Report for Standards and Technical Regulations Directorate, Department of Trade and Industry, London.

Swann, G. P. (2002). The functional form of network effects. *Information Economics and Policy*, 14(3):417–429.

Tanenbaum, A. S. (2001). *Modern Operating Systems*. Prentice-Hall, Upper Saddle River, NJ, 2nd edition.

Thirlwall, A. P. (1980). Rowthorn's interpretation of Verdoorn's law. *The Economic Journal*, 90(358):386–388.

Thirlwall, A. P. (1986). A general model of growth and development on Kaldorian lines. *Oxford Economic Papers*, 38(2):199–219.

Thirlwall, A. P. and Hussain, M. N. (1982). The balance of payments constraint, capital flows and growth rate differences between developing countries. *Oxford Economic Papers*, 34(3):498–510.

Tversky, A. and Kahneman, D. (1972). Subjective probability: a judgment of representativeness. *Cognitive Psychology*, 3:430–454.

Uchida, M. and Shirayama, S. (2008). Influence of a network structure on the network effect in the communication service market. *Physica A: Statistical Mechanics and its Applications*, 387(21):5303–5310.

US CERT (2009). US Computer Emergency Response Team website. http://www.cert.org/stats/, accessed 06/14/2011, not updated since February 2009.

US NIST (2011). US National Vulnerability Database of the US-National Institute of Standards and Technology. http://web.nvd.nist.gov/, accessed 07/07/2012.

Uzawa, H. (1965). Optimum technical change in an aggregative model of economic growth. *International Economic Review*, 6(1):18–31.

Varian, H. R. (1992). *Microeconomic Analysis*. W. W. Norton & Company, New York, 3rd edition.

Vaughn, R. and Evron, G. (2006). DNS amplification attacks. Preliminary Release. http://www.isotf.org/news/DNS-Amplification-Attacks.pdf, accessed 07/07/2012.

Veblen, T. (2008 [1899]). *Theory of the Leisure Class*. Project Gutenberg.

von Hippel, E. (2001). Innovation by user communities: learning from open-source software. *Sloan Management Review*, 42(4):82–86.

von Neumann, J. and Morgenstern, O. (1944). *Theory of Games and Economic Behavior*. Princeton University Press, Princeton, NJ.

Vrancken, J., Kaart, M., and Soares, M. (2008). Internet addressing standards: a case study in standards dynamics driven by bottom-up adoption. In Egyedi, T. M. and Blind, K., editors, *The Dynamics of Standards*. Edward Elgar, Cheltenham and Northampton, MA.

Waldeck, R. (2010). Segregation in social dilemmas. Research paper presented at the ECCS 2010 Sattelite Meeting on Emergent Properties in Natural and Artificial Complex Systems, Lisbon, September 15–16, 2010.

Waterman, A. M. C. (2002). Economics as theology: Adam Smith's wealth of nations. *Southern Economic Journal*, 68(4):907–921.

Watts, D. J. (1999). *Small Worlds: The Dynamics of Networks between Order and Randomness*. Princeton University Press.

Watts, D. J. and Strogatz, S. H. (1998). Collective dynamics of 'small-world' networks. *Nature*, 393(6684):440–442.

West, J. (2003). How open is open enough? Melding proprietary and open source platform strategies. *Research Policy*, 32(7):1259–1285.

West, J. (2005). The fall of a Silicon Valley icon: was Apple really Betamax redux? In Bettis, R. A., editor, *Strategy in Transition*, pages 274–301. Blackwell, Oxford.

West, J. (2007). Seeking open infrastructure: contrasting open standards, open source and open innovation. *First Monday*, 12(6).

West, J. and Dedrick, J. (2001). Open source standardization: the rise of Linux in the network era. *Knowledge, Technology, & Policy*, 14(2):88–112.

West, J. and Dedrick, J. (2004). Why firms adopt open source platforms: a grounded theory of innovation and standards adoption. MIS Q Workshop on Standard Making, Seattle, 2003. http://www.joelwest.org/misq-stds/proceedings/145_236-257.pdf, accessed 07/07/2012.

West, J. and Dedrick, J. (2006). Scope and timing of deployment: moderators of organizational adoption of the Linux server platform. *International Journal of IT Standards Research*, 4(2).

Wheat, I. D. (2009). Teaching economics as if time mattered. In Reardon, J., editor, *Handbook for Pluralist Economics Education*, pages 69–89. Routledge, London and New York.

Wikimedia Foundation (2011). Online statistics report on the Wikimedia wikis. http://stats.wikimedia.org, accessed 06/14/2011.

Williamson, O. E. (1973). Markets and hierarchies: some elementary considerations. *The American Economic Review*, 63(2):316–325.

Windrum, P. (2004). Leveraging technological externalities in complex technologies: Microsoft's exploitation of standards in the browser wars. *Research Policy*, 33(3): 385–394.

Winter, S., Kaniovski, Y., and Dosi, G. (2003). A baseline model of industry evolution. *Journal of Evolutionary Economics*, 13(4):355–383.

Winter, S. G. (2006). The logic of appropriability: from Schumpeter to Arrow to Teece. *Research Policy*, 35(8):1100–1106.

Yegorov, Y. (2003). Social dynamics and emergence of order: building theory of field in economics. Working Paper, published 01/25/2003, Santa Fe Institute.

Young, A. A. (1928). Increasing returns and economic progress. *The Economic Journal*, 38(152):527–542.

Ziewitz, M. and Brown, I. (2011). A prehistory of internet governance. In Brown, I., editor, *Research Handbook on Governance of the Internet*. Edward Elgar, Cheltenham. http://ssrn.com/abstract=1844720, accessed 07/07/2012.

Index